4

CLAY-BONDED FOUNDRY SAND

CLAY-BONDED FOUNDRY SAND

W. B. PARKES

APPLIED SCIENCE PUBLISHERS LTD
LONDON

APPLIED SCIENCE PUBLISHERS LTD
RIPPLE ROAD, BARKING, ESSEX, ENGLAND

ISBN: 0 85334 501 5

LIBRARY OF CONGRESS CATALOG CARD NUMBER 73-142298

WITH 128 ILLUSTRATIONS AND 20 TABLES

Printed in Great Britain by Galliard Limited, Great Yarmouth, Norfolk, England.

PREFACE

This book is based on the material which has been used in the courses on 'Control of Foundry Sand' which have been held at the British Cast Iron Research Association, Alvechurch, during the last twenty years and have been attended by some 1700 people, foundry managers, foundry technologists, metallurgists, foremen, sand technicians and occasionally apprentices.

The information used has been collected over a period reaching back to 1920, much of it at the British Cast Iron Research Association from 1936 onwards, during which time the author had the invaluable assistance of R. G. Godding and D. A. Taylor. Thanks are due to them, to the innumerable foundrymen who have listened so patiently, and to Dr J. G. Pearce, Dr H. Morrogh and Dr H. T. Angus for their encouragement. Where the results of other workers have been used, appropriate reference has been made.

Although the book deals mainly with clay-bonded sand, two chapters on oil–sand cores have been included. Oil–sand is no longer the only material suitable for the production of intricate cores, but it is still used on a large scale and over a wider field than is available for those methods which become viable when very large numbers of identical castings are required.

July, 1971 W. B. PARKES

CONTENTS

CHAPTER 1

Sand Testing and its Application

CLAY-BONDED SAND

Systematic sand testing began when the American Foundrymen's Association first published methods recommended for measuring moisture content, permeability, strength and 'fineness'. The recommendations were prepared by a Committee which began work in 1921, and were published in 1924. The original methods still form the basis of testing everywhere.

In the intervening period, slight changes in procedure have been made when experience has shown these to be desirable, and additional tests have been devised to give information which could not be obtained from the tests in the original series. However, the 1924 publication would still serve most of the needs of a foundry testing clay-bonded sand. Several countries prepare their own recommendations, but these are based on those of the American Committee.

In America, the American Foundrymen's Society publishes a *Foundry Sand Handbook* which includes not only tests which have been accepted as standard, but also 'tentative' standards which are going through a probationary period before being accepted as full standards, together with some 'non-standard' tests. There is also information on the application of test results in the foundry. The handbook reached its seventh edition in 1963.

British recommendations are drawn up by the Joint Committee on Sand Testing, the first report being published in 1945, the second in 1954 and the third in 1966.

In both countries tests are kept under constant review and the appropriate publications should be consulted to obtain precise details of the procedure to be adopted, since changes and additions may be made whenever experience shows these to be desirable.

TESTING OF PREPARED CLAY-BONDED SAND

Up to the present, all recommendations of the Joint Committee have dealt with prepared sands only, and the following have been described:

(1) preparation of sand for testing;
(2) mechanical analysis;
(3) moisture content;
(4) preparation of test piece;
(5) measurement of permeability;
(6) measurement of strength;
(7) shatter test;
(8) drying of test piece;
(9) interpretation of test results.

SECTION 1 — PREPARATION OF SAND FOR TESTING

The degree of aeration or compaction of a sand has a significant influence on the results of the tests. If, for example, a sample is taken from a mill not fitted with an aerator, a mixture of loose sand and coherent aggregates is obtained and test results differ appreciably from those which would have been obtained if the sand aggregates had been broken down by aerating or sieving. Aerating a partly compacted sand results in an increase in permeability and a decrease in strength.

If the sand has been aerated at the mill some re-aggregation may occur before it reaches the moulding machines, particularly if it is stored in a deep hopper in which the lower layers are compacted by pressure when the hopper is full or by impact when it is almost empty. This effect is much less pronounced than that due to compaction in the mill, but even so it is sometimes necessary to sieve sand on to the pattern when castings with a very good surface finish are required. In all cases, the sand to be tested should be treated in the same way as the sand in the foundry, *i.e.* if the foundry sand is sieved on to the pattern, the sample should be put through a sieve with the same aperture size.

It does not follow that samples should be taken from the exit to the hoppers over the machines. Should the sand be out of control, the earliest possible information is required, and, indeed, rapid methods for the estimation of moisture content have been devised just to avoid the delay of an hour or so which is involved when the standard method is used. This early warning can only be obtained by sampling at the mill and, if this sample differs appreciably from that at the machines, tests should be made to determine the difference. Apart from the above difference in compaction, the most likely change between mill and machines is in moisture content.

A similar factor may make it difficult to obtain concordant results if samples are exchanged by different laboratories. To show the extent of differences due to transport of sand, two samples were taken from a mill with an efficient aerator and placed in air-tight tins. One was allowed to stand on a laboratory bench and the other placed in the boot of a car and driven for fifty miles. The two samples were then tested on the same machine by the same operator and it was found that the ride had reduced permeability and increased strength by 15–25%. The difference was eliminated if both samples were sieved through a 10-mesh sieve. Sieving or any other form of aeration always results in some loss of water and the operation should therefore be kept to a minimum.

SECTION 2 — MECHANICAL ANALYSIS

The test is carried out to determine (1) the proportion of clay present, and (2) the size distribution of the sand grains.

The separation into 'clay' and 'sand' is purely arbitrary and is mainly a separation into two fractions, one finer and one coarser than 20 μ. Non-clay particles of appropriate size are included with the clay, and, on this account, the fine fraction is not described as 'clay' but as 'AFS clay' in America and 'clay grade' in Britain.

Separation depends on the rate of settling in water. When a solid particle is suspended in water it settles, or rises, at a rate which depends on —

(1) the difference in specific gravity between solid and water;
(2) the viscosity of the water;
(3) the size of the particle.

The main constituents of foundry sands — silica, clay, and perhaps felspar — have specific gravity very near to 2·65 and (1) is therefore constant.

The viscosity of water decreases as temperature rises and for very accurate work the test should be carried out at 15°C, but for control purposes, water at laboratory temperature can be used.

Since the first two factors are constant, the rate of settling depends only on size, and particles of sand with a diameter of 20 μ settle in water at a rate of 1 in min^{-1}.

Particles finer than this are included in the clay grade. Exceptionally, if coal or other material with a specific gravity lower than that of quartz is present, particles somewhat finer than 20 μ will be included in the sand grade and, conversely, if heavy minerals such as zircon are present, some coarser than 20 μ will be included in the clay grade.

The method of separation adopted is to suspend the sand in water, to remove all particles which settle at a rate of less than 1 in min^{-1}, and to divide the residue into a number of fractions by means of standard sieves.

Method of testing

Preparation of sample

The sand is dried at a temperature of 105–110°C and a quantity of about 50 g taken for testing.

When the sample contains only a very small amount of clay, or none at all, it is extremely difficult to obtain from a larger quantity of dried sand a smaller quantity which is truly representative. In this case the sample should be moistened and mixed, and a quantity which contains about 50 g of dry sand taken. This is dried and weighed, and the whole of it taken for testing.

Separation of clay from sand

The sample is added to about 750 cm^3 of boiling water to which 5 cm^3 ammonia (specific gravity 0·880) has been added. A stainless steel or porcelain dish 10 in diameter is a suitable vessel. Boiling is then continued for 10 min. This frees the clay from the sand grains and breaks it down into individual particles.

The sand and water are cooled and transferred to a straight-sided jar about 4 in diameter and 8 in deep. This jar has two marks on its vertical sides, one at 1 in, and the other at 6 in above the bottom. Water is added to the upper mark and the whole mixed and allowed to stand for 10 min. Some very hard waters cause the clay to flocculate, forming large aggregates with a high settling rate and, if this occurs, distilled or demineralised water should be used.

The upper 5 in of the suspension are removed by means of a syphon, the end of which is turned over horizontally to avoid sucking up coarse particles from the bottom of the jar. By this means, much of the material with a settling rate of less than 0·5 in min^{-1} is removed. The arrangement is shown in Fig. 1.

Water is again added to the upper mark and, after mixing and allowing to settle for 5 min, the upper 5 in are again removed. This procedure, with a settling time of 5 min, is repeated until the upper 5 in are clear after standing for 5 min. The remaining sand and water are returned to the dish, as much of the water poured off as is possible without removing sand grains, and the residue dried and weighed. The loss in weight is reported as 'clay grade' and is expressed as a percentage of the weight of dry sand taken.

The preliminary separation with a settling time of 10 min is necessary when an appreciable amount of clay is present. Clay raises the viscosity of water and reduces the rate of settling of particles suspended in it, and if the first settling time is restricted to 5 min, some particles which should have settled to below the 1 in mark will remain above it and be lost. When the clay content exceeds 5% it is advisable to have two preliminary settling periods of 10 min.

In addition to particles with the appropriate settling rate, the procedure described also removes anything soluble in boiling water. If the sand contains, for example, cereal binder, this will be included in the clay grade, but as the amount is usually small and must be controlled within narrow limits, the error is not serious and a suitable correction can be made.

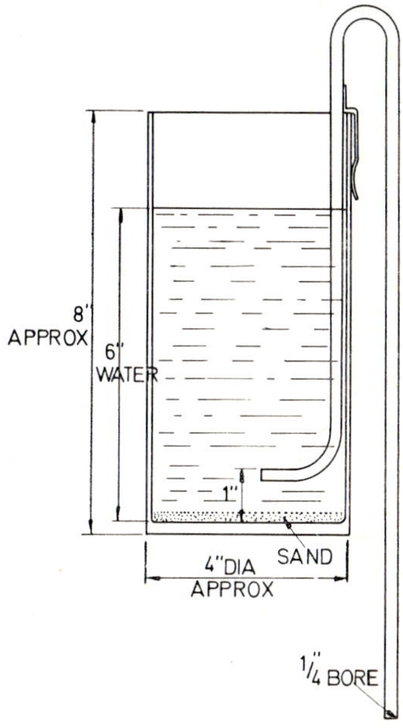

FIG. 1. Separation of clay grade.

The method originally devised and still recommended by AFS differs in one detail. Instead of being boiled with ammonia, the sand is shaken or stirred with a solution of caustic soda prepared by adding 25 cm^3 of a standard solution (30 gl^{-1}) to 475 cm^3 water. Subsequent procedure is the same. The two methods give similar results and the sole reason for preferring the ammonia method is one of convenience. It is not necessary to make up a standard solution, and if a large excess of ammonia is added in error, such excess is removed during the first 2 min boiling and the accuracy of the result is not impaired.

Sieving

The sieves used are selected from British Standard sieves and are given in Table 1.

The finest sieve is placed on the pan and the remainder added in aperture size order. The sand freed from clay is placed on the uppermost sieve and the cover placed in position. The sieves are placed on a shaking machine and shaken for a time which depends on the nature of the sand and must be determined by trial (see below).

When sieving is complete, each sieve in turn is inverted over a sheet of glazed paper and brushed with a fairly stiff bristle brush. No attempt must be made to force embedded grains through the sieve in the direction of sieving or permanent damage will result. The individual fractions are weighed and the results calculated as a percentage of the original sample.

Table 1. *British Standard sieves used for sand testing*

Aperture size (μ)	Former sieve number
1676	10
1003	16
699	22
500	30
353	44
251	60
211	72
152	100
104	150
76	200
53	300

Hand sieving of foundry sands is usually unsatisfactory. For determination of the oversize or undersize using a single sieve, hand sieving is the most accurate and often the quickest method, but many foundry sands have significant proportions on eight sieves and it is difficult and laborious to handle the whole nest, while sieving on each sieve in turn is lengthy.

Reporting the results

The results may be reported in either of two ways.

(1) The sieve aperture prefixed by a plus sign and followed by the actual proportion of sand on the sieve, *e.g.*

$$+500\ \mu\text{–}30{\cdot}2\%$$

indicates that 30·2% passes a 699 μ sieve and is caught on a 500 μ sieve.

(2) The sieve aperture prefixed by a plus sign and followed by the cumulative proportion of sand which will not pass the sieve, *e.g.*

$$+500\,\mu\text{–}45{\cdot}0\,\%$$

indicates that 45 % of the sand is caught on all sieves with apertures greater than 500 μ combined. To avoid confusion, the method of reporting should be clearly stated.

Determination of the end point of sieving

When a sieve containing sand is shaken, particles much smaller than the sieve aperture pass through readily. There remain coarser particles together with those, known as near mesh particles, which will only just pass through the sieve and these are separated slowly.

Since it is not possible to manufacture a sieve with completely consistent apertures and moreover, since many sand particles are irregular in shape and will only pass a sieve when presented to the appropriate aperture with the appropriate orientation, sieving is never complete and an arbitrary end point must be chosen. The method to be adopted and the end point selected are the subject of British Standard 1796 (1952), *Methods for the Use of BS Fine Mesh Test Sieves* which should be consulted on all matters concerning sieving. As a guide, sieving on any one sieve is considered to be complete if after alternate shaking and tapping for 2 min, the weight of sand passing the sieve is less than 0·2 % of the original sample.

Two factors increase the time required for sieving:

(1) an increase in the amount of sand on the sieve;
(2) a decrease in aperture size.

For routine tests on a number of sands, sieving time can be fixed by taking the finest of the sands to be tested and selecting the finest sieve on which a substantial proportion remains. The sand is sieved for, say, 15 min and the selected sieve subjected to alternate tapping and shaking for 2 min. If more than 0·2 % of the original sample passes in this time, machine sieving must be continued until the 'rate' test is satisfied.

This time can be adopted for all sands which are not finer than that used for the test. Sieving time will be unnecessarily long when coarser sands are tested, but the results will not be seriously in error.

Fineness number

The fineness number of sand, an American Standard, is obtained by multiplying the percentage of sand on each sieve and in the pan, by the sieve number of the sieve through which the sand has passed. The figures so obtained are added, and the sum divided by the percentage of granular material in the sand.

There will be only minor differences if a sand is tested using British and American Standard sieves in turn, these arising mainly because the mesh numbers of sieves with the same aperture size are different in the two series.

The fineness number of a sand should be used with caution. From the method of calculation, it can be seen that the fineness number of a coarse sand is lower than that of a fine sand, and the fineness number is sometimes used as an index of permeability or some other property depending on grain size. This can lead to serious errors as is shown by the following extreme cases.

Sand A contains 72% on 50 mesh (through 40 mesh)
Sand A contains 28% on 270 mesh (through 200 mesh)

Fineness no.	72 × 40	2880
	28 × 200	5600
		8480

$$\frac{8480}{100} = 84{\cdot}8$$

Sand B contains 50% on 100 mesh (through 70 mesh)
Sand B contains 50% on 140 mesh (through 100 mesh)

Fineness no.	50 × 70	3500
	50 × 100	5000
		8500

$$\frac{8500}{100} = 85{\cdot}0$$

These sands have the same fineness number, but that is the only resemblance.

Where the sands have the same kind of distribution then the lower the fineness number, the coarser the sand, and the number can therefore be used to compare sands from the same deposit or sands produced from one deposit, by a particular classifying plant. However, it would be simpler to use permeability as a basis for comparison.

International Standard sieves

British and American Standard sieves are very similar. In the full series, the aperture sizes form a geometrical progression, each differing from its immediate neighbours by the factor of $4\sqrt{2}$. The aperture sizes are identical except for a little 'rounding' of individual figures which has been carried out in order to simplify manufacture. In America, Tylor sieves, which also differ slightly from the US Standard, are frequently used.

Differences in aperture size are not large enough to impair the usefulness of comparisons between analyses made using the different sieves. There may be a difference in mesh number owing to the use of wire of different diameter when constructing the sieves. In neither country is the full series of standard sieves used, as the sieving procedure would become too cumbersome. For some sands it would be necessary to use up to twenty sieves and the additional information so obtained would not justify the additional time required for carrying out a test. In consequence only half of the sieves are used.

Table 2. *American, British and International Standard sieves*

American μ		British μ		International μ
2362	(8)			2000
1651	(12)	1676	(10)	1400
1167	(16)	1003	(16)	1000
833	(20)	699	(22)	710
589	(36)	500	(30)	500
414	(40)	353	(44)	355
295	(50)	251	(60)	250
208	(70)	211	(72)	180
147	(100)	152	(100)	125
104	(142)	104	(150)	90
74	(200)	76	(200)	63
53	(270)	53	(300)	45

In America alternate sieves are used. Selection of a British series was more complicated in that early work on the sizing of sand was carried out, either by elutriation, using the size limits drawn up by Boswell,[1] or by sieving using the now obsolete IMM sieves which were entirely different from the present British and American Standards. When the IMM sieves were replaced by the new Standard sieves, a series was selected which corresponded as nearly as possible with the grades previously used. As a result, the British and American series used in sand testing correspond almost exactly for 211 μ and finer, but differ in the coarser range.

On the Continent a completely different series, the Preferred Number series, is used. However, every alternate sieve in the British and American series corresponds very closely with a sieve in the Continental series, and it has been agreed that these sieves shall be adopted as an International Standard. Both the British Standards Institution and the American Society for Testing Materials recommend that the new International Standard shall be everywhere adopted. Aperture sizes, in μ, of the three series are given in Table 2.

Eventually the sieve number will disappear, since the number of meshes per inch is meaningless in a metric series. For the present it is retained as a subsidiary title, but BSI has made it quite clear that this is purely a temporary measure and that within a limited period, sieves will be described by the aperture size only.

Unfortunately it has not yet been possible to obtain international agreement on the adoption of the new series of sieves for testing foundry sands and at the present time three are in use, British and American which agree at the fine end of the range but are otherwise different, and Continental which corresponds with neither. Eventually all three will be abandoned in favour of the International series.

SECTION 3 — MOISTURE CONTENT

The test for moisture content is the most important of the tests on foundry sand, for two reasons.

(1) The moisture content can vary rapidly. As a rule, other changes are comparatively slow because the factors which control them do not change rapidly. For example, clay loses its bonding power and coal dust is decomposed rather slowly at temperatures as high as 400°C, but the temperature of the sand cannot exceed 100°C before all the water is removed.

(2) Moisture affects all the physical properties of a sand and most defects which are due to failure of the mould materials could be due to failure to control moisture.

It is important to know the moisture content as early as possible and in order to do this, three rapid methods of measurement have been devised to supplement the rather lengthy standard method. The available methods are —

(1) the standard method;
(2) drying at a higher temperature;
(3) a self-contained instrument independent of laboratory facilities;
(4) instruments giving instantaneous measurement of moisture content.

Standard method

A convenient weight of sand, usually 20 g, is heated at a temperature between 105 and 110°C until its weight is constant. The sample must be cooled in a desiccator before each weighing. The loss in weight, expressed as a percentage of the weight of sand taken, is the moisture content.

This method is common to all standard methods of testing sand. It should always be used when the results of other methods have to be checked.

Drying at a high temperature

The rate of drying is increased if the sand temperature is raised. However, prolonged heating at the higher temperature may result in decomposition of organic matter or clay, and both temperature and time of drying must be controlled.

The test can be carried out without any special apparatus, for example, by using an oven heated to 250°C or a hot plate, but it is more convenient to use one of the pieces of equipment designed for the purpose. There are two of these.

In the Ridsdale–Dietert moisture tester (Fig. 2) a 50 g sample of sand

FIG. 2. Ridsdale–Dietert moisture tester. (Courtesy of Ridsdale and Co. Ltd.)

is placed on a fine mesh sieve which is attached to the body of the instrument by means of a clip. When the instrument is switched on, a fan blows heated air through the sand sample which, when dry, is allowed to cool and then re-weighed.

In the Fischer tester (Fig. 3) a 50 g sample of sand is placed in a pan and attached to the body of the instrument where it is heated by means of an infra-red heater. When dry, the sand is cooled and re-weighed.

Both instruments are provided with a time switch so as to avoid the risk that the sand may be left in the drier for too long and become overheated.

FIG. 3. Fischer moisture tester. (Courtesy of G. Fischer Ltd.)

The sample of 50 g is an advantage since some sands are not entirely homogeneous and consequent errors are minimised by taking a large sample. On the other hand the large sample requires a longer cooling time.

Self-contained instrument

The 'Speedy' moisture tester requires no laboratory facilities and is used by foundries which have no laboratory. It may also be used, even where a laboratory is available, to obtain results in the foundry more quickly than would otherwise be possible.

The instrument (Fig. 4) consists of a stout metal container, A, and a cap, B, with a gasket to permit an air-tight joint between the two.

The container carries a hinged stirrup, C, and a screw, D. When the cap is placed on the container, the stirrup is rotated to a position such that the screw can be used to press cap and container firmly together.

The container is placed on its side and enough powdered calcium carbide to fill the measuring spoon provided, is added. The instrument includes a balance and a counterpoise which permits 6 g of sand to be weighed out. This sand is placed in the cap.

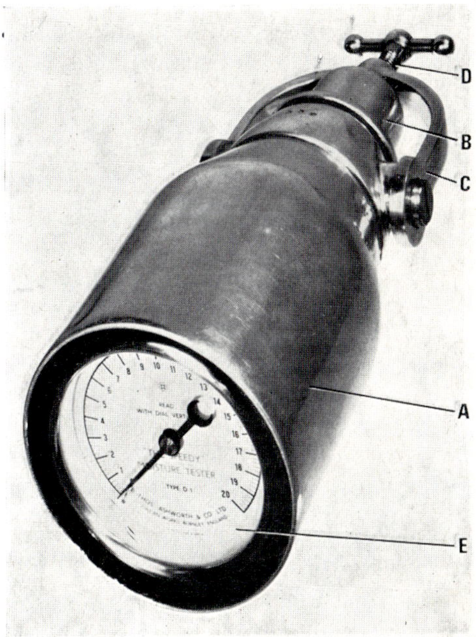

FIG. 4. Speedy moisture tester. (Courtesy of Thos. Asworth and Co. Ltd.)

Cap and container, still on its side, are brought together and secured and the apparatus shaken to obtain intimate mixing of sand and carbide. Since there is a large excess of carbide, acetylene is produced in proportion to the amount of water present. This exerts a pressure which is recorded on a pressure gauge, E, in the base of the container, this gauge being calibrated to give a direct reading of moisture content.

Since excess of carbide is used, some remains at the conclusion of a test. The instrument must therefore be carefully cleaned before each test, and since it is easier to ensure complete absence of residual carbide in the cap, the sand must always be placed here and the carbide in the container.

A great advantage of this instrument is that it is easily transported and can be set up wherever required. A disadvantage is the small weight of sand taken. Unless weighing and transferring to the cap are carried out quickly, there may be a significant loss of water.

Instantaneous measurement of moisture content

Many properties of sand are influenced by the moisture content and, as some of these can be measured in a second or so, the use of one of these properties as a basis for determination of moisture content has been tempting.

As early as the 1920's an instrument, the Grubb moisture meter, which measured conductivity and could be calibrated to read moisture content direct, was available. It consisted of two electrodes which were pushed into the sand, a battery and a resistance indicator. The conductivity of sand increases as moisture content increases, but is also profoundly influenced by packing density and, since pushing the electrodes into sand changed the packing density, the instrument was not a success. Nevertheless, it was revived under various names at intervals for some years.

A method of overcoming the effect of packing density has been described by Godding.[2] A sample of sand is drawn from a belt every few seconds, is compacted, and the moisture content of the compact is obtained from measurement of microwave power absorption. The whole action is automatic and enables an almost continuous record of moisture content to be kept. An alternative form,[3] suitable for giving a single reading on the sand in a mill, has also been developed.

SECTION 4 — PREPARATION OF TEST PIECE

The test piece used for measuring the physical properties of clay-bonded sand is a cylinder 2 in diameter and 2 in long. It is compacted by means of three blows from a 14 lb weight, falling from a height of 2 in on to the sand in a container.

This is the test piece designed by the American Committee 1921–1924. The rammer used is substantially that designed by Dietert at the same time. Where the metric system is used the dimensions are 5 cm instead of 2 in, but otherwise the same test piece is in universal use.

Rammer

Two forms of rammer are shown in Figs. 5 and 6.

They differ only in detail. Each consists of a stand A with two guides, B_1, B_2. A steel rammer rod C which slides in these guides, carries a stop D which rests on the supporting arm for guide B_1. The head of the rammer rod is a sliding fit into the container used for holding the sand. A 14 lb weight H rests on the stop D. The cam F, fitted with a handle, is so arranged

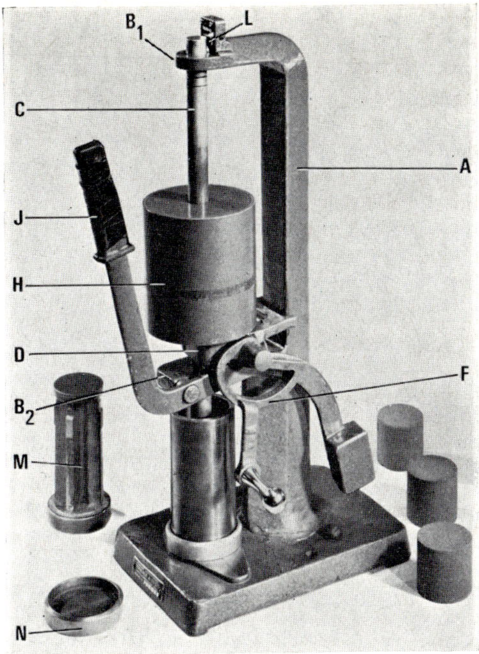

FIG. 5. Ridsdale–Dietert rammer. (Courtesy of Ridsdale and Co. Ltd.)

that by turning the handle the weight is raised to a height of 2 in and allowed to fall freely on to the stop. The total weight of the ramming assembly — rod, head, stop and weight — is 17 lb. A second handle J is provided for raising the assembly clear of the container. A recess in the base assists accurate location of the container beneath the rammer head.

In Fig. 5, a scale L, attached to the upper arm of the stand, carries three marks such that when the upper end of the rod is directly opposite the central mark, the test piece is exactly 2 in long. The other marks are $\frac{1}{32}$ in above and below the central mark. For routine testing it is sufficient for the upper end of the rammer rod to be between the outside marks. In Fig. 6 a plate L with a notch is fixed beneath the stop and the test piece is exactly 2 in long when the red line on the rammer rod is opposite the centre of the notch.

Support for rammer

Standard methods of testing recommend that the rammer should be fastened to a brick or concrete pillar and this is the most satisfactory method of mounting. If it cannot be adopted, other means of obtaining a

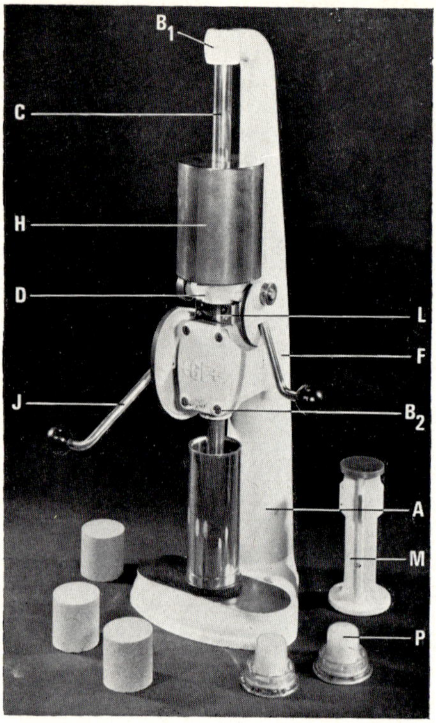

FIG. 6. Fischer rammer. (Courtesy of G. Fischer Ltd.)

solid support must be found; for example, if the rammer is placed on a bench it should be directly over one of the bench supports and it is useful if a solid iron base, 2 in thickness, is placed beneath the rammer. If the rammer is placed in the centre of a light bench, high permeabilities and low strengths are found and with tough sands the results can be in error by a factor of two or more.

Metal container for forming test piece

This is a tube 5 in high and 2 in internal diameter with a minimum thickness of $\frac{1}{16}$ in, and made from hard and corrosion-resisting material. Sands with low moisture content are difficult to strip and for these a container split into two halves longitudinally and provided with locating pins and a locking ring should be used.

In the Ridsdale–Dietert apparatus, the container fits into a cup N which retains the sand during ramming. This cup carries a boss which fits into the locating recess in the base plate. In the Fischer rammer the container is closed by means of a plug P.

Stripping post
The stripping post, used to remove the test piece from the container, is a suitably supported metal disc M which is a sliding fit into the container.

Procedure
The appropriate quantity of sand is weighed out and placed in the container, using a funnel to obtain uniform distribution of the sand. If a container closed by means of a cup is used, it can be inverted two or three times with a cup on each end, thus obtaining still more uniform distribution.

The container is placed beneath the rammer head and the ramming assembly lowered gently until supported by the sand. The handle operating the cam is turned three times with a 1 s interval between the turns. After the third ram, the top of the rammer rod, or the red mark on the rod, should come between the tolerance marks. If it does not, the test piece should be discarded and another rammed up with a suitable adjustment to the weight of sand taken.

The discarded test piece should not be roughly broken down and the sand re-used, or low permeability and high strength results will be obtained. If, for any particular reason, it is necessary to re-use the sand, the error can be minimised by rubbing the test piece through a 10-mesh sieve. However, such a procedure invariably results in loss of moisture and when obtaining a sample it is necessary to take a quantity sufficient for all tests.

Alternative procedure for forming test piece
For routine testing the foregoing procedure should always be used. The test piece is not, however, uniformly rammed and it is unsuitable for some research purposes. Indeed, it is recommended by the American Foundrymen's Society that, when measuring green strength, the test piece should always be placed in the machine with the end which was uppermost during ramming uppermost during testing, a procedure difficult to follow when some testing machines are used.

A more uniform test piece can be obtained if, when the container is closed by a plug, it rests on a split ring 2 in internal diameter and $\frac{1}{8}$ in deep, fitted with handles to facilitate withdrawal. The sand is introduced and given one blow with the rammer. The ring is then withdrawn and the remaining two blows given with the container unsupported.

This procedure cannot be used when the container is closed by a cap.

SECTION 5 — MEASUREMENT OF PERMEABILITY

The permeability number of a sand is defined as the volume of air in cm^3, at a pressure of 1 cm water gauge, which would pass through a test piece in 1 min if that test piece were 1 cm long and 1 cm^2 in cross-sectional area.

In the actual test, the standard test piece 5·08 cm long and 20·268 cm^2 in cross-sectional area is used, and the time measured is that required for 2000 cm^3 to pass. The pressure is a nominal 10 cm water gauge, but slight variations may occur when the apparatus has been in use for some time.

The only variables are the permeability number, the time and the pressure. The constants can be substituted in the formula

$$\text{permeability number} = \frac{\text{volume of air} \times \text{height of specimen}}{\text{cross section} \times \text{time in min} \times \text{pressure}}$$

$$= \frac{2000 \times 5{\cdot}08}{20{\cdot}268 \times \text{time in min} \times \text{pressure}}$$

$$= \frac{501{\cdot}2}{\text{time in min} \times \text{pressure}}$$

or, as time is always measured in s

$$= \frac{30\,072}{\text{time in s} \times \text{pressure}}$$

The accuracy of measurement is such that the fourth significant figure is irrelevant. If the pressure is 10 cm water gauge, the permeability number is obtained by dividing 3000 by the time in s.

Apparatus used for permeability test

The apparatus is shown in outline in Fig. 7.

It consists of four parts:

(1) An air reservoir or bell A, which floats on water in tank B;
(2) a three-way valve and indicator C;
(3) a seal D which ensures that air from the bell passes through the test piece in its container;
(4) a manometer E.

The tank, 7 in diameter and 13 in deep, is fitted with a central vertical outlet tube connected with the three-way valve, which is marked 'on, off, vent', to show whether the tube is connected with the test piece container, is closed or is set to allow air to enter the by-pass. The tank is filled to an internal indicator mark, when it is about half full.

The bell has a vertical tube which slides smoothly over the air outlet to the tank. Near the top of this tube are four holes with a combined area of not less than $\frac{1}{4}$ in^2, which allow air to be forced from the bell to the valve. The bell is graduated with four marks on its external surface, the lowest being marked 'X' and the others '0, 1000 and 2000'. Starting with the '0' mark just visible above the edge of the tank, disappearance of the third or fourth mark shows that 1000 cm^3 or 2000 cm^3 of air has passed through the valve. The 'X' mark, which is about $\frac{3}{4}$ in below the '0' mark, serves as a convenient indicator when commencing a test. The weight of the bell is such that a pressure of 10 cm water gauge is recorded on the manometer when the exit from the seal is closed.

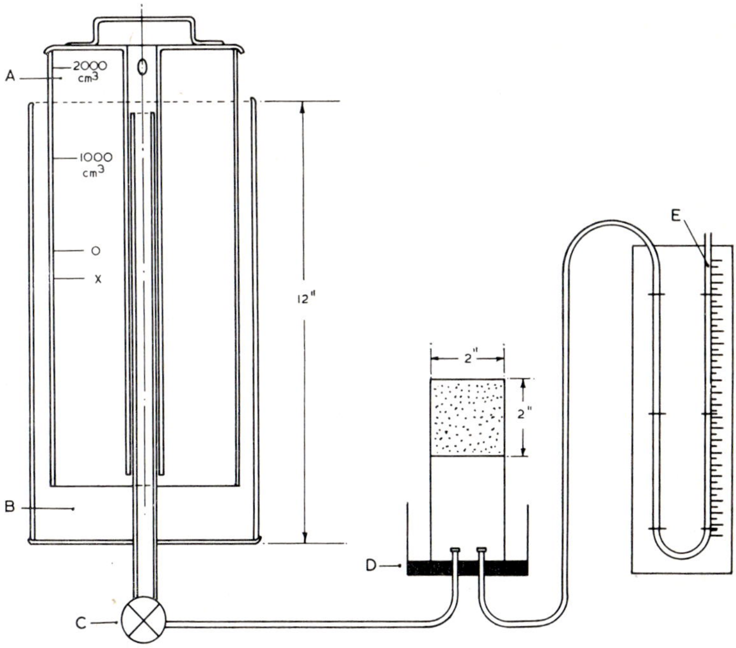

FIG. 7. Arrangement of apparatus for permeability test.

The bell and tank must be made of a material which does not corrode after prolonged contact with water.

Procedure for the determination of permeability

The three-way valve is turned to the 'vent' position and the bell is raised slowly. When the 'X' mark is visible, the valve is closed and the test piece in its container is attached to the sealing device. The valve is turned to the 'on' position. When the '0' mark disappears, a timer is started and the pressure on the manometer noted. The timer is stopped when the '2000' mark disappears and the time for the passage of 2000 cm^3 of air through the test piece obtained. the permeability is calculated using the formula given on p. 18.

Rapid method of measuring permeability by means of a standard orifice

If a small orifice is placed in the circuit between valve and sealing device, the pressure at the valve remains at 10 cm water gauge, but the pressure at the sealing device falls. If the resistance to the passage of air through the orifice is much greater than the corresponding resistance through the test piece, the pressure recorded by the manometer falls to a low figure; if the resistance of the orifice is small in relation to that of the test piece, pressure will fall only slightly. If a standard orifice is used, then the higher the permeability of the test piece, the lower the pressure at the sealing device, and if this pressure is known, the permeability can be derived.

The sensitivity of the test falls off as the pressure moves away from 5 cm water gauge and when it is above 8 cm or below 2 cm water gauge the method is worthless. In order to widen the range of sands which can be tested, two orifices are used, one for sands with high permeability and one for sands with low permeability. By this means, all the sands used in the foundry can be covered with an error not exceeding 10%.

With the large orifice in position, the time required to pass 2000 cm^3 of air through the sealing device should be 30 s, with the small orifice $4\frac{1}{2}$ min. This should be checked occasionally.

Types of apparatus available

(1) *Ridsdale–Dietert apparatus*

The apparatus is shown in Fig. 8.

The whole of the apparatus is mounted on a single stand with the tank above the sealing device. A mercury seal is used and the space beneath the test piece is connected by metal tubes with the manometer. This manometer, which is placed in front of the tank, consists of a vertical tube and a reservoir, the relative areas of cross-section being such that when the level in the tube falls from 10 cm to 0, there is no appreciable change in the level in the reservoir and pressure can be read directly from the tube. A transparent scale is centred just above the tube and when pressure is to be measured, the scale is rotated until a straight section is vertical. A screw is provided to adjust the 0 setting. The open ends of the connections to the valve and manometer are protected by a dust cover.

When an orifice is to be used, it is screwed into position beneath the dust cover. To obtain a direct reading, the test piece container is placed in position on the sealing device and the valve turned to the 'on' position. The scale is rotated until the red line on the curved section coincides with the water level in the manometer and the permeability number read off.

There are two scales, one reading from 0–90 for use with the small orifice, and one from 50–800 for use with the large orifice.

When a number of sands differing widely in permeability have to be tested in sequence, frequent changes of orifice may be necessary. In this case it may be quicker, as well as more accurate, to remove the orifice and use the standard method.

FIG. 8. Ridsdale–Dietert permeability apparatus. (Courtesy of Ridsdale and Co. Ltd.)

(2) *Fischer permeability apparatus*

This is shown in Fig. 9.

The apparatus is mounted on a single stand with the sealing device, three-way valve and gauge on the base beside the tank. The sealing device consists of a rubber pedestal through which pass connections to the valve and manometer.

Pressure is recorded on a gauge with three scales, one for recording pressure, one for recording low permeability using the small orifice and the third for recording high permeability using the large orifice.

The orifices are secured in position beneath the sealing device and a change from one to another made by means of a switch.

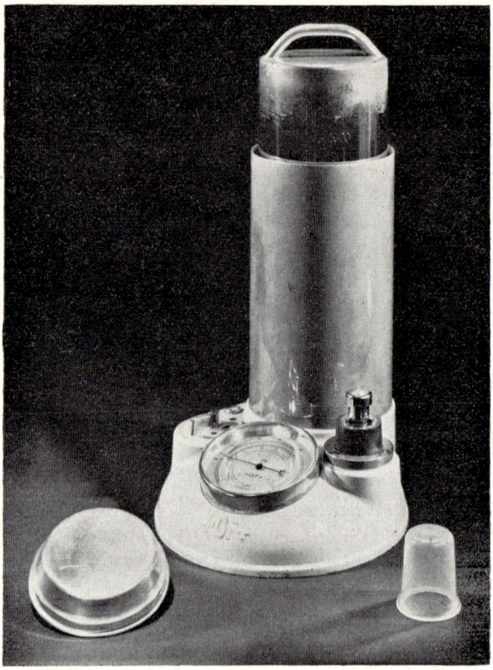

FIG. 9. Fischer permeability apparatus. (Courtesy of G. Fischer Ltd.)

The precautions given by the manufacturer must be observed when the apparatus is being used or tested, otherwise the gauge may be permanently damaged.

Testing the permeability apparatus

The most frequent causes of inaccuracy are leaks, partial blockage of the tube leading from the valve to the sealing device, and damage to the orifice.

Leaks

The bell should be raised until either the '0' or '1000' mark is just visible above the edge of the tank and the valve closed. The exit to the sealing device is closed by any convenient means, for example, a rubber bung in the test piece container and the valve turned to the 'on' position. No appreciable movement of the bell should take place in 15 min.

Partial blockage

This is usually due to water or dirt in the tube leading from the valve to the sealing device. If it is clear, the time required for 2000 cm^3 of air to pass through the open sealing device should not greatly exceed 2 s.

If there is a partial blockage, the tube should be blown out by means of compressed air. (When blowing out the Fischer apparatus, the gauge must be isolated or it will be permanently damaged.)

Should the time for the passage of air greatly exceed 2 s after cleaning the tubes, it may be that the air holes in the tube in the bell are too small. This may lead to an error in measuring very high permeability.

Orifices

The orifices should be checked by measuring the time required to pass 2000 cm^3 air. If this is too high, the orifice may be partly blocked and should be washed out with water and dried by means of compressed air. It must not be cleaned with a hard object.

If the time is low, the orifice has been damaged and should be replaced.

Measurement of dry permeability

Dry permeability is only rarely measured for purposes of routine control and so far as the sands used for green sand moulding are concerned, there is little difference between green and dry permeability. When, however, the sand is used for moulds which are to be dried, a much higher moisture content is usual and green permeability is appreciably lower than dry permeability.

Nevertheless, it is usual to use the test on the green sand for control purposes, while recognising the effect of drying in subsequently raising permeability.

To measure dry permeability, it is necessary to dry the test piece and this must first be removed from the container. Methods of drying are described in Section 8.

When the test piece is dry, it cannot be replaced in the container, or, if it can, then there is not an air-tight joint between container and test piece. It is therefore necessary to use a holder for the test piece, which will permit passage of air along the axis and prevent its egress through the curved sides. Two such pieces of apparatus are available for use with the appropriate permeability apparatus.

Ridsdale–Dietert dry permeability tube

The apparatus is shown in Fig. 10.

It consists of a tube K 5 in long with an internal diameter of $2\frac{1}{4}$ in in the upper 3 in. At a depth of 3 in from the top the tube narrows to $1\frac{3}{4}$ in, forming a flat shelf at A, which carries a soft rubber ring B. The dry test piece is placed on this ring. A crosshead, hinged at C, can be secured at D, in a position above the test piece and carries a screw E, to which is attached a small perforated pressure plate F. Four ribs L locate the test piece in the centre of the tube.

To carry out a test, the test piece is placed in the tube, the crosshead fixed and the screw turned so as to force the test piece down on to the rubber ring. To seal the curved surface, mercury is poured into the space H between the test piece and tube. No mercury must rest on the surface of the test piece.

The tube is then placed in the mercury seal of the permeability apparatus and the test completed in the usual way.

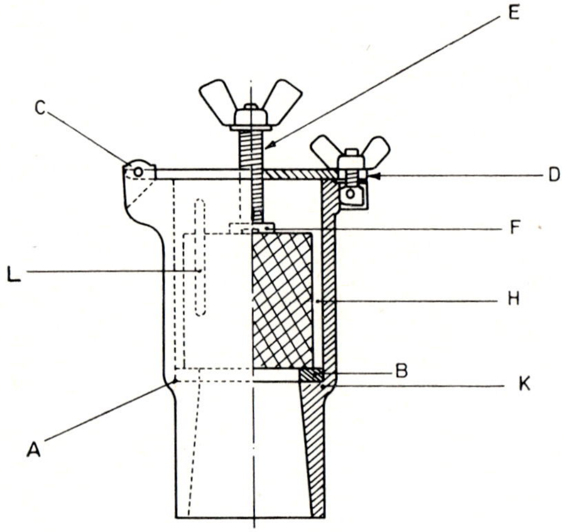

FIG. 10. Ridsdale–Dietert dry permeability tube.

Procedure with sands with low dry strength

If the dry strength of the sand is low, the edge of the test piece in contact with the rubber ring may crumble before the applied pressure is sufficient to produce a mercury tight joint. In this case the procedure for preparing the test piece should be modified as follows.

A steel ring, $\frac{1}{8}$ in square in section and a sliding fit into the test piece container, is placed in the latter so that it rests on the base of the cup. The sand sample is placed in the container and the test piece rammed in the usual way. The weight of sand taken is adjusted so that the overall length of the test piece including the ring is 2 in. The test piece is stripped, dried and allowed to cool. It is then placed in the dry permeability tube with the steel ring in contact with the rubber ring and the joint made good by means of the screw.

Fischer dry permeability tube

The apparatus is shown in Fig. 11.

It consists of a cylindrical holder $2\frac{1}{16}$ in diameter and $5\frac{1}{2}$ in long, which contains a rubber sleeve $2\frac{1}{8}$ in deep. A small tube fitted with a valve passes through the outer wall and allows the sleeve to be inflated. A hand pump is used to supply air.

To carry out a test, the dry test piece is placed in the tube so that the rubber sleeve overlaps equally top and bottom. The sleeve is inflated, sealing the curved surface of the test piece, and the valve closed.

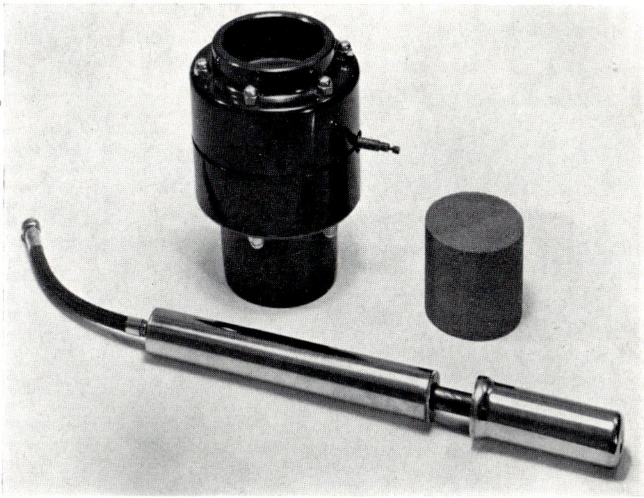

FIG. 11. Fischer dry permeability tube. (Courtesy of G. Fischer Ltd.)

The holder is then placed on the rubber pedestal of the permeability tester and the test completed as for green permeability.

SECTION 6 — MEASUREMENT OF STRENGTH

Green strength

For routine control purposes, green strength is always measured in compression. Dry strength may be measured in compression or shear.

Green strength is ascertained by applying a steadily increasing force to the plane ends of the test piece and along its axis until failure occurs. The maximum load recorded is the green strength. It is permissible to use the test piece previously used for the permeability test, provided that it is stripped and tested immediately.

Removal of the test piece from the container

The test piece container is placed over the stripping post and pushed downward with a steady pressure. This leaves the test piece on the stripping post.

If the test piece is difficult to strip, the split container described in Section 4 should be used.

Testing procedure
The test piece is tested immediately it has been stripped, otherwise it may air dry and an inaccurate result will be obtained. The suggested rate of loading is 30 lb/in^2 min but quite appreciable changes can be made without affecting the test result to a significant degree. On the other hand, increasing the rate of loading by a factor of three or more can lead to inaccurate results.

The American Foundrymen's Society recommends that the end of the test piece which is uppermost in ramming should be uppermost in testing, but this recommendation is difficult to follow. In the AFS handbook two machines are described, a spring balance machine in which the test piece is pressed from above, and a pendulum machine in which the test piece begins in a horizontal position and moves through an arc during testing. In Britain, in addition to the pendulum machine, a hydraulic machine in which the test piece is always horizontal and a spring balance machine in which the test piece is pulled up against a stop are also used.

The effect of the orientation of the test piece is small unless the sand has a high shatter index, but nevertheless, when carrying out routine tests, the test piece should always be placed in the particular machine being used, with the same orientation.

Types of apparatus for measuring green strength

Ridsdale–Dietert universal tester
The machine is shown in Fig. 12.

It consists of a frame on which are mounted a pendulum weight and a pusher arm, each carrying a compression head. When a test piece is placed in these compression heads, raising the pusher arm forces the test piece to raise the weight, the load on the test piece increasing as the top of the arc is approached. When the test piece breaks, a magnetised rider remains on the scale of the frame and gives the compression strength in lb in^{-2}.

The machine is usually supplied with a motor for raising the weight, but a handwheel can be substituted if required.

The machine is not very satisfactory for measuring strengths below 3 lb in^{-2}. It may also be difficult to maintain the recommended rate of loading with very strong sands when using the handwheel.

With the compression heads in the position shown in Fig. 12 the maximum load which can be applied is 18·5 lb in^{-2}. If the strength of the sand exceeds this, the test must be carried out with the compression heads in an alternative position nearer the fulcrum. This increases the maximum load by a factor of 5, and, if a motor drive is used, increases the rate of loading to the same extent. A change of this order influences the result obtained, and when testing very strong sands two precautions must therefore be observed.

(1) If the test piece does not break when the maximum load of 18·5 lb in^{-2} has been reached, it must be discarded and a new one rammed up for testing with the compression heads in the alternative position.

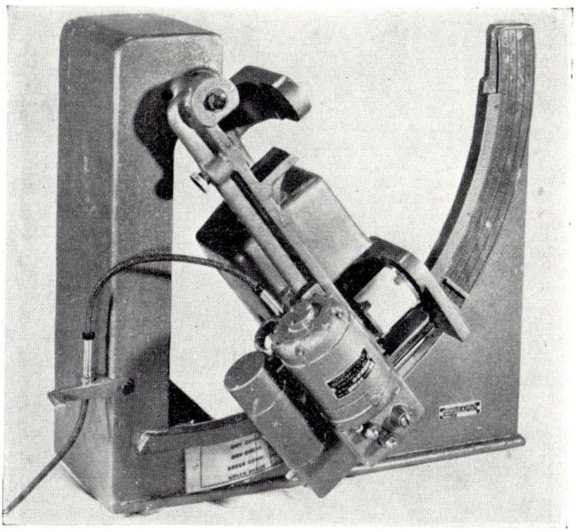

FIG. 12. Ridsdale–Dietert universal strength machine. (Courtesy of Ridsdale and Co. Ltd.)

(2) If the sand to be tested has a strength sometimes above and sometimes below 18·5 lb in^{-2}, all tests must be carried out with the compression heads nearer the fulcrum. Otherwise the difference in the rate of loading is so great that the results are not comparable.

Fischer universal strength machine

The machine is shown in Fig. 13.

It consists of a fixed arm to which one compression head is attached and a hydraulic press with two pistons. One piston carries the second compression head and the other is attached to a handwheel. When the wheel is turned, pressure is developed on the forward movement and is transmitted to the compression head and test piece.

The oil reservoir can be connected by means of a three-way valve with (a) a high pressure gauge, (b) both the high and a low pressure gauge, or (c) a release. The maximum load which can be recorded on the low pressure gauge is 22·5 lb in^{-2}; if the strength of the sand exceeds this, the gauge must be isolated by means of the three-way valve or the indicator will be permanently damaged.

It is difficult to obtain a uniform rate of loading. When the load is first applied, a single turn of the handwheel may apply a load of 1 lb in^{-2}; when the load has reached 20 lb in^{-2}, a single turn of the wheel may

FIG. 13. Fischer universal strength machine. (Courtesy of G. Fischer Ltd.)

double it. When carrying out a test, the movement of the pointer should be observed and the handwheel turned at a decreasing rate so that the rate of application of the load is as uniform as possible.

Spring balance machine

The machine is shown in Fig. 14.

It consists of a tension spring balance from which is suspended a frame carrying the lower compression head. By turning a handwheel, the frame can be raised or lowered. The upper compression head is fastened to a crosshead supported by two uprights rigidly attached to the base of the machine.

The handwheel is turned to its lowest position and the test piece, preferably in a retaining cup to hold the broken test piece, is placed on the lower compression head. The handwheel is turned so as to raise the frame and when the test piece reaches the upper compression head, an increasing load is applied to the test piece until failure occurs. Strength is indicated by a pointer.

FIG. 14. Spring balance machine. (Courtesy of Ridsdale and Co. Ltd.)

For carrying out tests on a wide variety of sands, a balance of 100 lb capacity, equivalent to 32 lb in^{-2}, is the most convenient. For very weak sands, a balance with a capacity of 14 lb is more satisfactory.

Selection of machine for green strength test

Provided that a reasonable attempt is made to maintain the recommended rate of loading of 30 lb in^{-2} min^{-1}, the three types of machine give concordant results on sands with strength in the range 3–17 lb in^{-2}.

It is, however, difficult to maintain this rate of loading with very low or very high strengths on the hydraulic machine, while the pendulum machine is unsatisfactory for strengths below 3 lb in^{-2} and has the objectionable step in the rate of loading at 18·5 lb in^{-2}. The most useful machine is that with the spring balance.

Usually, however, the choice of type of machine is dependent on the range of tests which are to be made. The spring balance is useful only for measuring green strength, whereas the other machines can be used, for example, for measuring the much higher dry strength or the tensile strength of sand with organic binders, provided the requisite attachments are fitted.

Dry strength

Test pieces must be dried and cooled in the manner described in Section 8. The standard cylindrical test piece is used and may be tested either in compression or shear.

Compression test

The test is made by applying a steadily increasing compressive load to the plane faces of the test piece until it fails. The maximum load is the compression strength, which is reported in lb in^{-2} on the cross-sectional area of the test piece.

The range of strength is wide — from less than 30 lb in^{-2} for sands used for green sand moulding and bonded with a kaolinite, to more than 300 lb in^{-2} for some of the sands used for making large moulds. This leads to some difficulty in maintaining a uniform rate of loading, which in the American specification is 140 $\pm$ 20 lb in^{-2} min^{-1}.

Either the pendulum or hydraulic machine may be used. The maximum load which can be obtained on the pendulum machine is less than 100 lb in^{-2}, which is insufficient for many sands, but a high dry strength attachment is available which increases this maximum by a factor of 3. This is adequate for most sands, but if the machine is motor driven, fitting the attachment also increases the rate of loading to the same degree. The high pressure gauge on the hydraulic machine has a maximum load of 160 lb in^{-2}, but an attachment is available which raises this to 1000 lb in^{-2}. Maintaining a constant rate of loading is very difficult.

The effect of the rate of loading is not usually serious in routine testing. The rate of loading does affect the result and, with either of the machines, it is almost impossible to obtain a constant rate of loading over the whole of the range of strength to be met with in foundry sands. However, it is rarely important to know whether a sand which, when tested on the machine available, has a strength of 300 lb in^{-2} is exactly ten times as strong as one which has a strength of 30 lb in^{-2}. Usually, the test is carried out to show whether there has been a change in the strength of the sand in use, or whether some other sand has similar properties, and in this case, so long as a similar rate of loading is maintained for the comparative tests, the absolute rate is less important.

Shear test

Sometimes shear, instead of compression strength is determined. There is little to choose between the two tests so far as their usefulness is concerned, and adoption of the shear test is usually a matter of convenience.

For example, if the pendulum machine is used and compression strength exceeds 90 lb in^{-2}, the high dry strength attachment has to be used and this puts the machine out of action for green strength tests. This is a nuisance. Shear strength is much lower than compression strength and is within the compass of the machine for almost all clay-bonded sand, without using the high strength attachment, while replacing compression heads by shear heads requires only a few seconds.

Test piece and testing apparatus required for the shear test are the same as for the compression test except for test heads. These have semi-circular faces in place of the circular faces of the compression heads and the test piece fails by shearing along its longitudinal axis.

The result is reported in lb in^{-2} on the longitudinal cross-section and both the pendulum and hydraulic machines have the appropriate scales.

SECTION 7 — SHATTER TEST

The shatter test is included in the series of standard tests only in Britain. In many ways it is an unsatisfactory test and was adopted as a stop-gap in 1950 after lengthy trials in the laboratory and foundries, awaiting the time when it could be replaced by a more satisfactory test. It is an index, but only an index, of the toughness of the sand under impact, a property which is not easily measured.

The shatter index is defined as the proportion by weight of the undried standard compression test piece which is retained by a 0·5 in aperture British Standard sieve when the test piece has been shattered by falling from a height of 6 ft on to a solid anvil.

Apparatus for carrying out the shatter test

The components of a suitable apparatus are shown in Fig. 15.

The anvil E, 2 in diameter, is rigidly attached to the base D of the machine. It carries a removable head F, 3 in diameter, which prevents sand from the broken test piece from piling up round the anvil. The head is removable to enable the sieve to be lifted easily from the machine.

The inverted stripping post A is above the centre of the anvil at a height of 6 ft 2 in. When a test piece is removed by pushing the container upwards over the stripping post, the lower end of the test piece is 6 ft above the anvil.

The stripping post is supported by an arm B attached to a column C. The top of this column *must* be attached to a wall or other rigid object. Alternatively, the arm can be attached directly to the wall.

The anvil projects through the centre of a 0·5 in British Standard sieve G with its pan H.

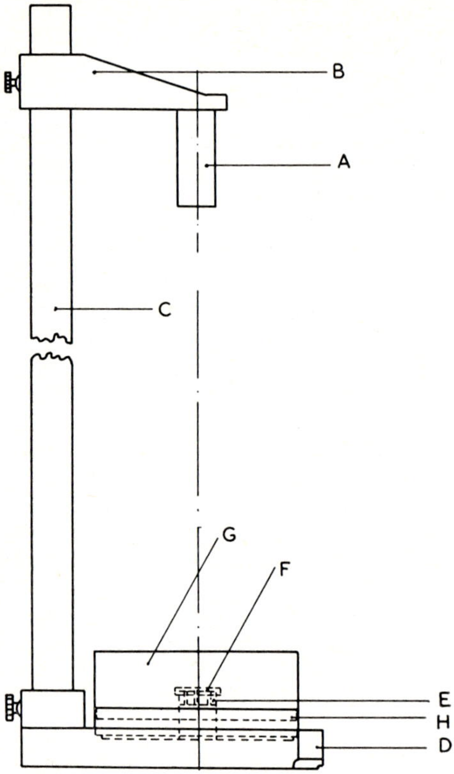

FIG. 15. Apparatus for shatter test.

Procedure for determination of shatter index

The test piece container, with the test piece, is pushed steadily upwards over the stripping post. The test piece falls and shatters when it strikes the anvil.

The anvil head is removed to permit removal of the sieve and the sand retained is weighed. This quantity is to include the cone of sand remaining on the anvil and this should be transferred to the balance and not to the sieve.

Alternatively, the sand passing through the sieve can be weighed and the oversize determined by difference.

Apparatus available

The Ridsdale shatter test apparatus is shown in Fig. 16.

The apparatus differs from that shown in Fig. 15 in that the container is supported on a platform having a circular hole through which the test piece is ejected. Stripping is carried out by means of a lever-operated plunger.

Precautions to be observed

A reliable result can only be obtained if the specified conditions are adhered to rigidly and the following precautions observed.

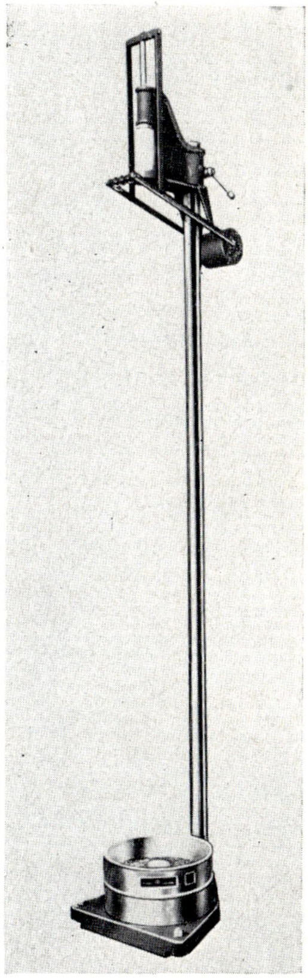

FIG. 16. Ridsdale shatter test apparatus.

(1) The test piece must fall in such a way that, at the moment of impact, the whole of its lower plane surface makes contact with the anvil simultaneously. In this case, the cone of sand left on the anvil is symmetrical. If this condition is not fulfilled the result of that test should be rejected.

In order to fulfil this condition the following details must be maintained.

(a) The stripping post or plunger must be mounted with its axis vertical and its lower plane surface horizontal and directly over the centre of the anvil.

(b) The stripping device must be quite rigid since otherwise the test piece may begin to turn over when stripped. When the arm is supported on a column the top of the column must be attached to a rigid object.

(c) The pressure to eject the test piece must be applied steadily, all jerkiness being avoided.

It may be difficult to fulfil this condition when the moisture content of the sand is low, particularly when stripping is done by hand. In this case, it is permissible to push the test piece partly out of the container using a stripping post on the bench and complete stripping on the apparatus.

(2) The anvil must be rigidly attached to the base of the machine, not simply rest on it.

(3) When the test piece is stripped by means of a plunger, the operation must be carried out gently.

(4) The sieve must not be shaken since otherwise oversize pieces may be broken down and pass through the sieve. In order that this requirement shall be fulfilled, it is necessary that the sieve and pan be kept in good condition so that they can be separated easily.

(5) The lower face of the stripping post and the face of the anvil must be clean and dry.

If these precautions are observed there is no difficulty in obtaining test results within the specified limit of $\pm 5\%$.

SECTION 8 — DRYING OF TEST PIECES

Drying

The drying of clay-bonded sand is simply the removal of water and this can conveniently be done at a temperature a little above 100°C. This procedure is adopted in the American Standard where drying at 105–110°C is specified.

In the case of sands containing sand and clay only, drying temperatures up to 300°C have little influence on the result of the test, but some foundry mixtures contain organic binders which decompose slowly at temperatures exceeding 200°C. Most foundry drying stoves are maintained at temperatures well above 200°C and if the test is required to give an index of stove performance, this factor must be taken into account. In the British recommendations it is laid down that test pieces should be dried at various temperatures between 150 and 250°C and that others should be dried in the foundry stove alongside moulds.

Some natural clay-bonded sands contain yellow hydrated iron oxide, which gives a very low green strength but a very high dry strength when dried at temperatures below 200°C. At higher temperatures, the oxide loses its combined water and turns red, its dry strength falling to 0. Only a very few sands contain sufficient of the yellow oxide for this effect to be important, and they are generally unsuitable for foundry use. It is advisable to dry at temperatures above and below 200°C when examining a bright yellow sand from an hitherto unknown deposit.

Cooling

The recommendation in both British and American specifications is that after drying, test pieces should be allowed to cool to room temperature in a desiccator.

If a dry clay-bonded sand is allowed to stand in air at room temperature, it absorbs water and loses strength; the amount of water absorbed and the consequent loss in strength, as well as the rate of change, depend on atmospheric humidity. However, the variations with temperature in the strength of clay-bonded sand, so long as it is quite dry, are very small indeed and it matters little whether the sand is at 30, 50 or 100°C. The absence of a desiccator is therefore no bar to the measurement of dry strength, and all that is necessary is to make the test before the test piece is cold. This is permitted in the British specification which states that the test piece can be cooled in air provided it is tested while still warm to the touch.

SECTION 9 — INTERPRETATION OF TEST RESULTS

Mechanical analysis

The sieve test affords the only means available for showing why there has been a change in those properties of a sand which are mainly dependent on grain size distribution. Of these the most important are permeability and surface finish.

A pronounced fall in permeability is frequently due to an increase in the proportion of very fine material present and an increase of a few per cent of such material has a marked effect.

Finish on flat or rounded surface becomes smoother as permeability decreases, but when any fine detail is present, the presence of a few large grains may have a serious effect. In both cases, interest is mainly concentrated on the relatively small amounts at the extreme ends of the range. The change in permeability which results from the addition of a sand with high permeability to one with low permeability is also indicated by the size distribution of the two sands concerned. If addition of the more open sand results in a decrease in the proportion of fine grains without at the same time increasing the width of size distribution, an increase in permeability follows. If, however, the number of sieves at the coarse end of the range on which an appreciable amount of sand is caught is increased by two, the increase in permeability is small until the more open sand forms half of the mixture.

When dealing with sands of uniform grain size, too much importance should not be attributed to the bulk of the grains being found on one sieve rather than two, since this may merely be the result of the arbitrary selection of sieve sizes. For example, a sand might consist entirely of grains which pass 211 μ and remain on 152 μ using the British sieves hitherto recommended, but might be divided into two fractions by the 180 μ sieve which will be used when the ISO recommendations are adopted, although in both cases the aperture sizes of neighbouring sieves differ by the factor of $\sqrt{2}$. Similarly, a sand, which consists entirely of grains appearing on one sieve in the ISO series might be found on two using the American or British series.

Moisture content

The moisture content is the actual percentage of free moisture in the sand, but by itself does not indicate whether the sand has a higher or lower moisture content than is desirable.

For example, a mixture of a silica sand and a clay, using all new materials, may be suitable for green sand moulding with a moisture content of 2 or 3%, depending on the amount and kind of clay present, but if that sand, with about the same proportion of active clay, has been in use in the foundry for some time, the minimum moisture content needed to produce a mouldable mixture may exceed 4%. This is because the dead clay present has no bonding power, but still absorbs almost as much water as active clay. When a sand has been in use under uniform conditions for some time, the amount of dead clay becomes reasonably constant and it is then possible to fix a definite level for moisture content.

Different clays require different proportions of water to develop plasticity. In particular, the clays in some natural sands require a much higher moisture content than any of the clays available for use in synthetic sands.

In green sand moulding, moisture content is usually maintained at a level only a little above that needed to produce a sand suitable for moulding, so as to avoid those casting defects, which are due to high moisture content. When the moulds are to be dried, this limitation does not apply and as dry strength increases with increase in moisture content, it is usual to increase the amount considerably.

Permeability

The permeability number is the actual volume of gas which will pass through rammed sand of definite dimensions, rammed in a specified manner, and twice as much gas will pass when the permeability number is 100 than when it is 50.

However, moulds made from the same sand do not necessarily have the same permeability. From the way in which the permeability number is calculated, it is clear that permeability is inversely proportional to sand thickness, but a factor which is frequently of even greater importance is mould hardness, and a soft mould may have a permeability four times that of an otherwise identical hard mould made with the same sand. This is discussed in Chapter 8.

Green strength and shatter index

The green strength of a sand is simply the load it can withstand without fracture and this is all that the test, by itself, reveals. Even so, the test has a use — for example, to decide whether or not the top of a mould will break under upward pressure when it is full or collapse under its own weight when the support of the pattern is withdrawn.

The shatter index was originally used by Graham[4] who described it as an index of flowability. This is a term which has many definitions, but Graham was seeking a test which would tell him whether with constant flowability, hardness distribution in a particular mould would remain constant, provided the airline pressure and the number of jolts on a certain machine remained constant. If the controlling property is defined as flowability, then the shatter index is an index of flowability, although the relation is an inverse one. The shatter index is an index of toughness and the higher the toughness of sand, the more difficult it is to ram on a jolter, and the greater the hardness differences in different parts of a mould. It is doubtful whether the test would have been so widely accepted if this had been its only value; its greatest use is, in combination with the green strength, as an index of moulding properties.

Stripping a mould from a pattern

The kind of sand which gives a good strip from a complicated pattern was, and still sometimes is, known as a strong sand, and the green strength test was adopted as a measure of the property or properties concerned. The result was not successful. If the amount of clay in a sand is increased and there is a corresponding increase in water content, green strength is raised and stripping properties improve, but it is just as easy to find conditions where stripping properties are better when strength is low.

Two examples should suffice.

(1) If the normal amount of water in a particular foundry sand is, say 5 %, and the moisture content is reduced to 2·5 %, green strength increases, but it becomes impossible to obtain a good strip from a pattern more complicated than a flat plate.

(2) A not uncommon practice where difficulty is experienced in stripping moulds from an occasional pattern is to place a little oil–sand from the core shop in the place where a bad lift is being obtained, fill up with clay-bonded sand and ram. This practice usually enables a good lift to be obtained. In the language of the moulder, the clay-bonded sand is too weak but the oil–sand is not, and yet the 'weak' clay-bonded sand is likely to have a strength of 5–10 lb in^{-2} and the 'stronger' oil–sand a strength of 5–10 oz in^{-2}. Clearly, some property other than green strength is involved.

In order to illustrate the controlling factor, two sands were prepared, each with a green strength of 8 lb in^{-2}, one of which gave a good lift from a certain pattern while the other failed. Test pieces were made from the two sands and broken on a spring balance machine. Films were taken

FIG. 17. Breaking of sand with poor lifting properties — on application of load.

to show how the sand behaved when the load was applied, and stills from these films are shown in Figs. 17–25.

Figures 17–20 are those obtained when the sand did not give a good lift. Figure 17 shows application of the load; Fig. 18 shows the appearance of the first crack; Fig. 19 is $\frac{1}{50}$ s later; and Fig. 20 shows the residue after the end of the test. Figures 21–25 are those obtained with the sand which gave a good lift. Figure 21 shows application of the load and should be compared with Fig. 17; Fig. 22 shows the appearance of the first crack and should be

FIG. 18. Breaking of sand with poor lifting properties — first crack appears.

FIG. 19. Breaking of sand with poor lifting properties — 1/50 s after appearance of crack.

FIG. 20. Breaking of sand with poor lifting properties — residue after fracture.

FIG. 21. Breaking of sand with poor lifting properties — on application of load.

FIG. 22. Breaking of sand with poor lifting properties — first crack appears.

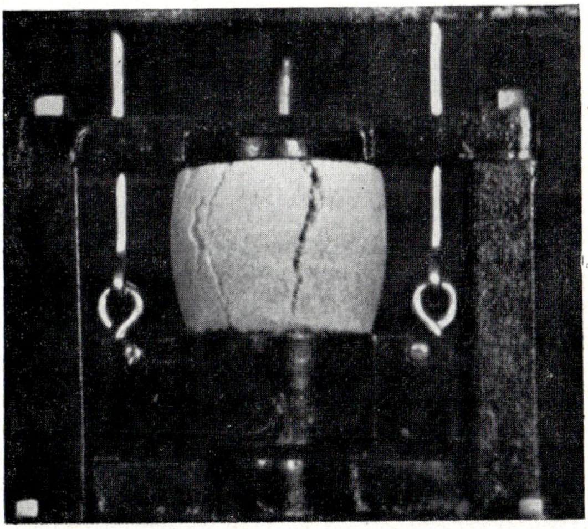

FIG. 23. Breaking of sand with poor lifting properties — 1/10 s after appearance of crack.

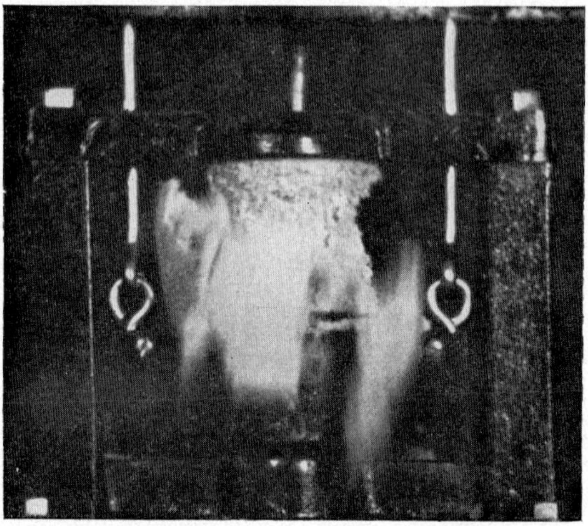

FIG. 24. Breaking of sand with poor lifting properties — after further interval of time.

FIG. 25. Breaking of sand with poor lifting properties — residue after fracture.

compared with Fig. 18; Fig. 23 is $\frac{1}{10}$ s later and has no counterpart in the first series; Fig. 24 is after a further interval and should be compared with Fig. 19; and Fig. 25 shows the residue after the test and should be compared with Fig. 20.

The difference between the two sands is clear. The unsatisfactory sand cracked before it could be deformed appreciably and shattered almost immediately after the first crack appeared. The satisfactory sand deformed before it cracked and then continued to deform with a gradual widening of the crack before breaking up into a small number of large pieces. One factor which may be measured is the extent to which the sand can be deformed before it cracks.

Consideration of the series of operations performed after a mould is rammed leads to the same conclusion. Friction between sand and pattern is high and if a vertical, or near vertical, pattern face is to be separated from a sand face, the force required to overcome friction is high, and the upward load on the layer of sand on the mould surface near the pattern exceeds its transverse strength. This layer of sand therefore breaks away with the pattern. A hand moulder therefore strikes the pattern — rapping plates should be provided to enable this to be done without damage — in such a way as to enlarge the mould cavity and reduce friction so as to permit withdrawal. On a machine a vibrator serves the same purpose. In this operation the thickness of the layer of sand between pattern and box must be reduced in length without cracking and can be compared with the test piece in Figs. 17–25 turned on its side.

The importance of deformation was recognised by Dietert[5] and he produced an instrument to be attached to the pendulum machine to measure this property. The test has not been widely adopted. The deformation is small and not easily measured on a sand test piece and the end-point is suspect, but the most serious limitation of this method of testing is that the extent of deformation depends on the rate of deformation. What is required is a measurement with the very rapid rate deformation obtained when a hammer or vibrator is used, not with the relatively slow loading of a testing machine.

No method of measuring deformation under impact and suitable for use with foundry sand has been devised, but the shatter index can be used in conjunction with the green strength to obtain a comparison between the deformation of different sands.

The shatter index is an index of toughness, which is proportional to the area under the stress/strain curve or as a first approximation, to the product of strength and deformation. If, therefore, shatter index and strength are kept constant, then deformation is kept constant.

If one parameter is constant and the other varies, the change in deformation can be seen — for example if two sands have the same shatter index and different green strengths.

Change in deformation of sands A and B

	A	B
Shatter index	70	70
Toughness	x	x
Strength ($lb\ in^{-2}$)	5	10
Deformation	$x/5$	$x/5$

A has double the deformation of B and is less likely to give trouble in stripping. It is not necessary to know either the value of x or the units in which it is measured.

This method of comparing sands has three limitations.

(1) Although the shatter index is obtained under impact conditions, green strength is obtained with a slow and steady rate of loading. It has been seen that green strength changes as the rate of loading is increased and there is no method for measuring it under conditions of impact loading.

This is a serious criticism, and lengthy trials were made, first in the laboratory and then in foundries before the test was adopted.

(2) The test does not distinguish between very tough sands. For example, many of the sands used in high pressure squeeze machines have a shatter index exceeding 95, frequently 100, and the test is worthless.

(3) It is not possible to compare the deformation of two sands when both shatter index and green strength are different — for example if, for sands A and B above, the test results had been 80 and 10 lb in^{-2} and 60 and 5 lb in^{-2}.

An alternative is therefore desirable. Nevertheless, for routine control tests, where the object is to maintain a sand of constant properties, the test is adequate unless the toughness of the sand is outside the range covered by the instrument.

It would be possible to make the shatter test quantitative and extend its range by discarding the 6 ft height of fall and instead using the height required to give a suitable proportion of sand, 50 or 60%, on the sieve. However, many foundry sands have a shatter index exceeding 60 and a few exceeding 90, and the change would therefore require a large increase in the height of fall. Apart from the resulting inconvenience, an objection to this practice is that the test piece is likely to turn over a little in its fall.

Limiting effect of sand adhesion

Deformation of a clay-bonded sand increases with increasing moisture content, and stripping becomes correspondingly easier, but only up to a certain point. The adhesion of sand to a pattern also increases and if a certain moisture content, which depends on the nature of the sand, is exceeded, a clean strip is not obtained, some sand adhering to the pattern.

At somewhat lower moisture contents, the sand does not adhere to the pattern in this way, but nevertheless the adhesive force is sufficient to

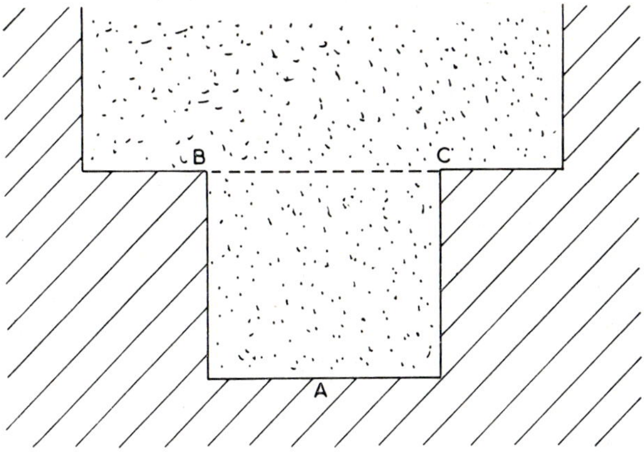

FIG. 26. Lifting a cod of sand.

cause difficulty. Suppose a heavy cod such as that shown in Fig. 26 is to be lifted. There are three factors which hinder removal:

(1) friction between sand and the vertical walls of the pattern;
(2) the weight of the cod;
(3) adhesion of sand to pattern on the face at A.

If the load imposed on the sand section BC by these three factors combined, exceeds the tensile strength at that section, the cod will remain in the pattern when the mould is stripped.

In order to reduce (1), the pattern must be vibrated so as to reduce the size of the cod itself and give a clearance between sand and pattern. This becomes progressively easier as the moisture content is raised. On the other hand, increasing moisture content, even at a level lower than that at which small patches of sand begin to adhere to the pattern, leads to an increase in sand adhesion and to a quite rapid fall in tensile strength.

One factor which assists stripping can only be improved at the expense of one or both of the others and a compromise must be adopted. In the case of a cod such as that shown in Fig. 26, moisture content should be slightly lower than usual, the gain resulting from higher tensile strength and lower adhesion being more than sufficient to compensate for lower deformation. Under very difficult conditions it may be necessary to add a small amount of dextrin which increases deformation without increasing adhesion or reducing tensile strength.

Dry strength

Drying a rammed sand results in a large increase in strength, but at the same time there is an increase in brittleness. The surface of a mould may dry slowly, usually only when the sand is warm, before closing and dries rapidly during pouring.

In spite of the high strength, erosion of sand during pouring or damage to the mould during coring up may occur. One factor which is of some importance is that when a mould dries slowly and a layer of dry sand only a few grains thick is present on the surface, that surface is more friable than it would be if the mould were dried to a considerable depth.

This condition is frequently found when the prepared sand is warm, and may develop between the time when the mould is first stripped and when the cores are placed in position. Some sand grains may be dislodged when the cores are placed and appear later as an inclusion in the casting.

Erosion by the metal stream and damage during core placing increase as dry strength decreases. On the other hand, a large increase in dry strength may give trouble at the shake-out and in the sand preparing plant.

OTHER TESTS ON CLAY-BONDED SANDS

In addition to those which are the subject of a standard, there are several other tests which are used with some frequency, although for some reason no standard method has been prepared. Usually, they are tests which do not form part of the routine control programme, but may nevertheless be of considerable importance.

Green mould hardness

Although, strictly speaking, green mould hardness is not a property of the sand itself, its influence on the properties of a sand mould is so important that no system of sand control is complete without it. The main controlling factor is the method of ramming, but sand properties have a secondary influence, in that sand toughness has an influence on the way in which sand packs when rammed, whilst the maximum hardness obtainable depends on sand strength.

The appropriate test method is included only in the American Standards, but the test itself is very widely used and the testing equipment employed is the same almost everywhere. The test is a modified Brinell test in which a spring-loaded hemispherical indentor is pressed against the mould surface and the depth of penetration measured.

In the original form of the instrument, the indentor has a radius of 0·100 in, and protrudes 0·100 in beyond the base of the instrument. This indentor is held in position by a spring such that a pressure of 237 g is just sufficient to press the tip flush with the base, *i.e.* a distance of 0·100 in. A dial gauge indicates the displacement in 100 units of 0·001 in.

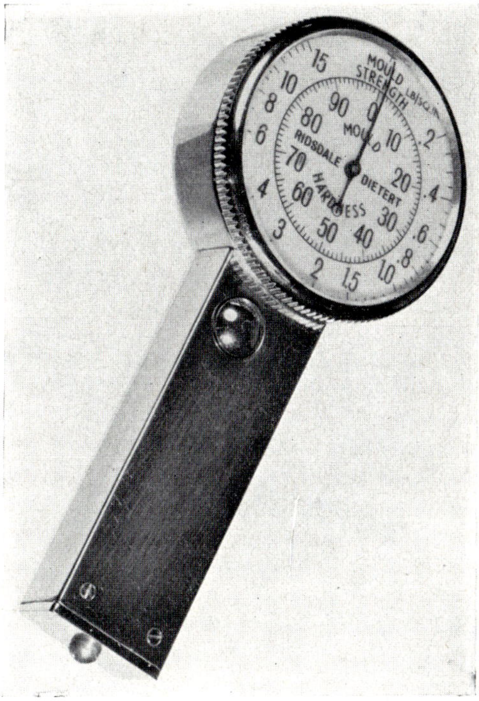

FIG. 27(a). Hardness tester — 0·100 in indentor. (Courtesy of Ridsdale and Co. Ltd. (Ridsdale/Dietert Design.))

In operation, the base of the instrument is pressed firmly against the mould surface. When the mould is so soft that the indentor penetrates to the full depth of 0·100 in, the hardness number is 0; when no penetration occurs and the indentor is forced back flush with the base, the hardness number is 100.

An alternative form of the instrument has an indentor with a tip of 0·500 in radius, which again protrudes 0·100 in beyond the base. In this case a load of 980 g is needed to force the indentor back through 0·100 in.

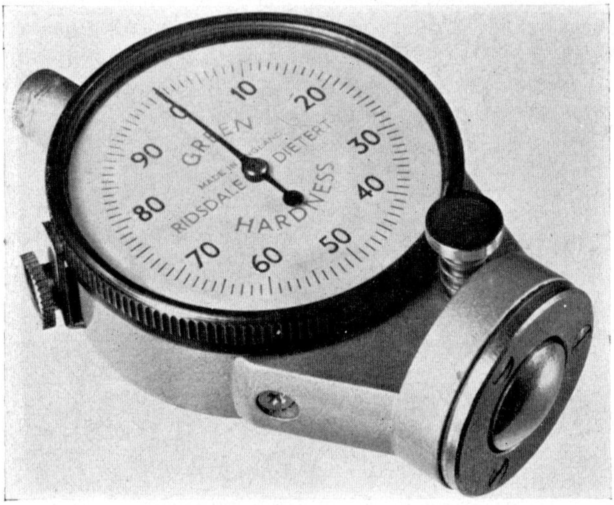

FIG. 27(b). Hardness tester — 0·500 in indentor. (Courtesy of Ridsdale and Co. Ltd. (Ridsdale/Dietert Design.))

FIG. 27(c). Hardness tester — conical indentor. (Courtesy of Ridsdale and Co. Ltd. (Ridsdale/Dietert Design.))

The two instruments give similar results. That with the larger indentor gives more uniform results because it is less affected by the small soft areas which are frequent when the sand has not been aerated.

Both instruments are useful for the general run of moulds made by hand or machine, but are of less value when the moulds are made by high pressure squeezing, in which case hardness approaches 100. For these hard moulds a third type of instrument with a conical indentor is available. The three instruments are shown in Fig. 27 (a, b and c).

Mould hardness, mould density and degree of compaction

Mould hardness can be used to show differences in the extent to which sand is compacted in different parts of a mould or by different methods of ramming. However, in a general investigation into the behaviour of sand during ramming, using a variety of sands, mould hardness has two limitations.

(1) Mould hardness depends not only on the degree of compaction, but also on the green strength of the sands;

(2) when mould hardness varies over a wide range, two different instruments have to be used and there is no direct relation between the results given by them.

To overcome these limitations, some observers have used the apparent density of the mould as an index of ramming. This method removes the awkward step which is present when one hardness tester has to be replaced by the second, but is useful only provided there is no change in the kind of sand being used. Mould density depends not only on the degree of compaction, but also on the grading of the sand and amount of clay and coal dust present. For example, a natural sand spread on eight sieves in the standard series and containing 15% of clay — a not uncommon condition with some rotten rock sands — can be rammed to an apparent density of 1·8 by a moderate squeeze, but a synthetic sand, with grains on only two neighbouring sieves and containing 5% bentonite, could not be rammed to this density without breaking the sand grains.

This difficulty could be avoided by ramming each of the sands under consideration by means of 10 blows on the standard rammer, measuring the apparent density of the test piece so obtained, and using the factor actual density/test piece density as an index of ramming.

Refractoriness

There are no agreed tests for refractoriness. The AFS handbook for some years included tests for obtaining the A and B sintering points by pressing a heated platinum strip against a sand test piece and noting adhesion of sand to the strip, but the test was never widely used.

No single test for refractoriness can cover all requirements, since the behaviour of a mixture of materials differing in fusion point depends on the rate of heating and on the holding time at temperature. At one extreme, with very rapid heating, each of the constituents melts in turn as the appropriate temperature is reached; at the other, with very slow heating, reactions between the various constituents occur and the whole may be liquid at a comparatively low temperature.

For the range of castings produced in green sand moulds, the most important factor is the presence or absence of an appreciable proportion of grains which fuse at a temperature below that at which the metal solidifies. A suitable test is to heat the clay-free sands to the appropriate temperature and note whether an appreciable proportion of the grains have fused.

No British sands contain grains with a fusion point low enough to cause concern in the iron foundry and testing is unnecessary. On the other hand, many contain grains fusible at the higher temperatures necessary in the steel foundry.

Estimation of coal dust

No standard method for estimating the amount of coal dust present in a sand has been described, and methods which are available are of doubtful accuracy, but under particular conditions can be helpful.

(1) *Loss on ignition*

This is the simplest but least reliable test. The loss on ignition includes both coal and coke, less ash, and combined water in the clay. If organic binders are used these are also included. The amount of coke present can vary widely and the loss on ignition cannot be used as a basis for calculating the amount of coal dust.

However, in a mechanised foundry producing only castings of a similar kind and making constant additions of new sand, the amount of coke tends to become almost constant and loss on ignition can be used as a control.

Elsewhere, as much as 6% coke has been found in the sand in a foundry making fairly heavy castings and less then 1% in a light castings foundry with down-draught dust control on the shake-out.

(2) *Carbon in volatile matter*

The volatile matter in the coal dust used in foundries contains a reasonably constant proportion of carbon. If, therefore, the amount of carbon present in the volatile matter can be determined, the volatile matter and the amount of coal dust can be calculated.

This method has been described with various modifications.[6,7] Other materials containing volatile carbonaceous matter, for example, cereal binders or calcium carbonate, must be first removed by boiling with water or hydrochloric acid.

Great care is necessary in estimating carbon, since the calculation is based on a small difference between two much larger quantities, and an error of 0·1% in the carbon in the volatile matter leads to an error of nearly 1% in the calculated amount of coal dust.

(3) *Volatile matter*

A result sufficiently accurate for control purposes can be obtained more simply from an estimation of the amount of volatile matter in the sand. The test is made by heating a weighed quantity of the dried sand to a convenient temperature between 700 and 1000°C in an atmosphere of nitrogen and then re-weighing. The loss in weight is due to the volatile matter from the coal dust and the combined water in the clay.

An approximate correction can be made for the combined water — 10% of the clay grade if a natural sand is used and 5% if the clay is a montmorillonite — and the amount of coal dust can be calculated as in the following example in which the coal dust itself contained 30% volatile matter.

Clay	6% montmorillonite
Combined water	$\frac{6 \times 5}{100} = 0{\cdot}3\%$
Total volatile matter	$= 2{\cdot}7\%$
Volatile hydrocarbon	$= 2{\cdot}4\%$
Coal dust	$= 2{\cdot}4 \times \frac{100}{30} = 8\%$

The correction for combined water may be slightly in error, since there is no means of estimating how much of the clay is active and how much is 'dead'. In practice the error is not serious, since an error of 5% in the amount of montmorillonite leads only to an error of 1% in the amount of coal dust, and the amount of active clay must be controlled to far closer limits than 5% in order to maintain moulding properties.

For control purposes it is not necessary to calculate the actual amount of coal dust present; working to a fixed volatile matter is all that is required. An increase of 0·3% in volatile matter indicates that the amount of coal dust has risen by 1% and the addition can be suitably modified.

If the sand contains calcium carbonate, the test should be made at a temperature of 700°C so as to avoid decomposition and evolution of CO_2. Calcium carbonate is usually present either in the form of shell in shore sands or small amounts of calcite in some clays.

Cereal binders also interfere, and if these are present, the sand should be boiled with water and filtered through a medium free from volatile matter before carrying out the test. If the sand also contains calcium carbonate, dilute hydrochloric acid can be used in place of water in order to remove the carbonate and avoid any restriction on the testing temperature.

Appearance of castings

In a foundry where there are no facilities for estimating coal dust in the above manner, the only guide is the appearance of the castings. This is discussed in Chapter 6.

Estimation of pitch

When pitch is used in place of coal dust, the amount required is frequently about 1%. Methods of control based on carbon content, volatile matter, or loss on ignition are not entirely satisfactory, since uncertainties due to the combined water in the clay, although of minor importance in the estimation of 6% coal dust, may be serious when dealing with 1% pitch.

The pitch used as an addition to foundry sands consists of a number of organic compounds together with free carbon. The organic compounds are soluble in trichlorethylene, and the amount can be determined by extraction in a Soxhlet extractor followed by evaporation of the solvent.[8] If the soluble matter in both sand and pitch are determined, the amount of pitch in the sand can be calculated.

This procedure is not suitable for testing used sands. When pitch is heated, some of the organic compounds volatilise and some decompose, but others polymerise forming new compounds with higher molecular weight. These are not soluble in trichloroethylene, but are as useful as the original pitch in reducing certain defects in castings. In consequence, some foundry sands which are quite satisfactory in service and give off the typical heavy yellow fume when heated, are free from compounds soluble in trichlorethylene.

These compounds are soluble in pyridine and control can be established by using this in place of trichloroethylene. Great care is necessary when using pyridine since it is exceedingly poisonous with a maximum allowable concentration of 5 ppm (15 mg m^{-3}), the only safeguard being that even in such low concentrations the smell is most objectionable. The operation should be carried out in a fume cupboard with a good draught and the effluent should be discharged well above ground level and away from open windows. Under no circumstances must an inexperienced operator be allowed to carry out the test.

Estimation of wood flour

When wood flour and its decomposition products are the only insoluble organic materials present in a sand, the amount present can be derived from the volatile matter as with coal dust. The proportion of volatile matter in wood varies with different woods, but is usually more than 70% and a reasonably accurate result can be obtained, even when the amount present is as little as 0·5%.

Frequently, however, both coal dust and wood flour are present and in this case it is not possible to apportion the volatile matter in the correct proportion between the two.

Several chemical methods have been devised. These depend on the presence in wood of organic compounds, which can be converted quantitatively to other compounds which can be estimated.

In one such method, the sand is treated with sulphuric acid which converts certain compounds to sugar, and this is then estimated. In a second, the sand is treated with hydrochloric acid and furfural estimated. Whichever method is used, the test must be standardised using the same wood.

These methods are not really satisfactory. Few foundries have the necessary equipment and even when this is available, it is exceedingly difficult to obtain reliable results. It must be accepted that once the sand has been mixed, accurate estimation is improbable and the foundry must take steps to see that the minimum amount necessary is not exceeded. This is discussed in Chapter 6.

pH value

Some clays are decidedly alkaline, notably the sodium montmorillonites, and others are acid. When an alkaline clay is strongly heated, the dead clay is less alkaline than the active clay and may be acid, and the alkalinity or acidity of a used foundry sand is usually different from that of a new sand. Coal dust and cereal binders and their decomposition products also influence alkalinity and acidity.

Alkalinity and acidity are measured on the pH scale. The pH of a solution is the logarithm of the reciprocal of the concentration of hydrogen ions in g ions l^{-1}. If this concentration is $1/10^7$, the pH is 7; if $1/10^8$, the pH is 8, and so on. A change of 1 in the pH shows an increase or decrease in the ion concentration by a factor of 10.

Pure water has a pH of 7 and is neutral; if the pH is greater than 7, the solution is alkaline, if the pH is less than 7, the solution is acid.

pH is usually determined on an aqueous extract of the sand, *e.g.* 50 g sand with 100 cm^3 water, and the concentration should be recorded with the results. Roberts[9] made measurements on rammed sand compacts, as well as on aqueous extracts, and found that the former had a pH of about 2 units lower. However, the result obtained on compacts varies with the degree of compaction, and the difference might have been less if it had been possible to ram the sand so that no voids were present. In view of the uncertainty of the results of tests on compacts, it is more informative to use aqueous extracts.

When the extract has been obtained, a very rough estimate of the pH can be made by means of test papers which take on a colour depending on pH. The accuracy of this method is not high and the results obtained cannot safely be compared with those obtained elsewhere. For accurate work use should be made of one of the direct reading pH meters available.

REFERENCES

1. BOSWELL, P. G. H. (1918). *British Resources of Refractory Sands*. Ch. 2, Taylor and Francis, London.
2. GODDING, R. G. (1963). *BCIRA J.*, **11,** Sept., 641–661.
3. GODDING, R. G. Unpublished work at BCIRA.
4. GRAHAM, E. L. (1948). *Foundry Trade J.*, **84,** Apr. 15, 365–367.
5. DIETERT, H. W. and WOODLIEF, E. E. (1939). *Foundry*, **67,** Sept., 28–29.
6. APTEKAR, R. E. (1934). *Sea Coal Analysis as applied to Black Sand. Trans. A.F.A.*, **42,** 543–561.
7. MCCONNELL, A. McD. and MCPHENT, J. *Estimation of Coal Dust in Foundry Sand. Foundry Trade J.*
8. MAYER, A. (1949–1951). *BCIRA J. Res. Devel.*, **3,** 671–673.
9. ROBERTS, W. R. (1961), *BCIRA J.*, **9,** 327–346.

CHAPTER 2

Formation and Occurrence of Foundry Sands

The sands used in foundries in this country have mainly been produced by the breakdown of crystalline rocks, usually granite. Frost and rain were the most effective agents and by their action the parent rock was broken down into smaller and smaller aggregates and to individual grains.

The size of these grains controls the size of the sand grains used in the foundry. It is not easy to find a natural agency, glaciers excepted, which will reduce the size of single grains when these are present in large numbers and are already as fine as sands. They may, however, be decomposed by chemical action or undergo rounding of edges and corners if they are disturbed for a long period.

If the sand is not removed from its place of origin, it consists of grains of all the sizes present in the original rock crystals except for those which have been decomposed by chemical action. However, the sands useful in the foundry have been removed from the original site, usually by water, although Boswell[1] considered that the red colour of the Bunter sands may be the result of a long period under desert conditions.

Moving a sand by means of a current of water tends to place grains of one size in one place, although the separation is never perfect: at the same time rubbing of one grain against another results in a certain amount of rounding.

Separation of sand grains by water

When obtaining the size distribution of a sand, the fraction finer than 0·02 mm diameter was removed because it settled more slowly in water than did the coarser grains, as described in Chapter 1. By making the separation using faster rates of settling, it is possible to obtain other fractions coarser than the finest grade without using sieves. In his examination of British sand deposits, Boswell[1] used a wet method of separation for all sizes up to 0·25 mm diameter.

If, instead of being suspended in still water, the sand grains had been suspended in a current, the coarser grains would reach the bottom before travelling very far, and the fine grains, remaining in suspension for a longer time, would travel a longer distance before settling out. The distance travelled in water of a particular depth depends on the size of the grain and the speed of the water. The rates of settling in water of grains of various sizes, together with the distance they would travel before settling, are given in Table 1.

Table 1. *Movement of sand by water*

Diameter (mm)	Settling rate (ft min^{-1})	Distance (ft) moved while sinking 1 ft in a current of		
		30 mph	5 mph	1 mph
0·4 (28 mesh)	9	300	50	10
0·25 (60 mesh)	5	500	100	20
0·1 (150 mesh)	$1\frac{1}{3}$	2000	300	60
0·05 (300 mesh)	$\frac{1}{3}$	8000	1300	30

It is clear that movement by water must tend to separate grains of different size from each other, but in practice the above table could not be used to calculate where grains of a particular size would be deposited. The velocity of a stream of water is by no means uniform even across a single section, whilst changes in velocity from season to season are likely to lead to deposition in one place when the water is low and sluggish, followed by removal elsewhere when the stream is faster. In addition, eddies may bring up fine grains which have nearly reached the bottom and it is doubtful whether a particle fine enough to travel 500 ft while settling 1 ft will reach the bottom until the velocity of the current falls.

It is not necessary for the grains to be suspended in water before movement begins. They may be rolled along the bottom or may move by 'saltation'. In the latter case, a grain is carried some distance and when it reaches the bottom disturbs one or more other grains which in their turn are carried forward. It is in this way also that sands are moved by wind and Bagnold[2] worked out the initial velocity of wind required to initiate movement in settled sand of definite size and the lower velocity necessary to keep the disturbed sand moving.

The most uniform sands in this country are produced by a combination of the action of water and wind. The shore sands found on the west coast from the Clyde estuary southwards have been washed ashore by the tide and blown up into dunes by the wind.

Chemical decomposition of minerals

In the period which has elapsed since the parent rocks were broken down into sand grains, the less resistant minerals have been decomposed, mainly by water and usually with the assistance of the organic acids formed when plants decay. From the point of view of the foundryman, the most important minerals were quartz, felspar and mica, which were the main constituents of the granite from which many of our sands were formed. Numerous other minerals were present, but in such small quantity that they are of interest to the geologist rather than the foundryman.

Quartz is unaffected and forms the bulk of almost all foundry sands. Of the various felspars and micas which exist, potash felspar and mica were the most abundant. These have the composition:

Felspar $K_2O.Al_2O_3.6SiO_2$

Mica $K_2O.3Al_2O_3.6SiO_2$

When these decompose the end products are the potassium salts of any organic acid present in the water, free silica and a clay. The potassium salts are soluble in water and have been leached out, but the silica and clay remain along with the undecomposed quartz crystals. The clays are compounds of alumina, silica and water, those formed from felspar belonging to the kaolinite group and having the basic formula $Al_2O_3.2SiO_2.2H_2O$ and those formed from mica to the illite or secondary mica group with the basic formula $Al_2O_3.4SiO_2.H_2O$.

The extent to which decomposition of felspar has taken place varies widely from sand to sand. Many sands are entirely free from felspar, whereas others may contain up to 20%. Mica exists as thin plates which have a relatively large surface area, favouring chemical attack, whilst such plates offer little resistance to fracture when a sand is being transported. It is therefore unusual to find more than an occasional flake of mica in a foundry sand and its presence in the parent rock is usually deduced from the type of clay formed when it decomposes.

The composition of foundry sands

The sand first formed by the breakdown of rock, and consisting of quartz, felspar and mica in a wide range of grain size, is unsuitable for foundry use: decomposition *in situ* would remove felspar and mica, but would not change the size distribution of the quartz. Removal over a short distance, by means of a mountain torrent emptying into a lake, would also have little effect on this size distribution and Boswell observed the existence of such unsorted sands near the granite in Cornwall and round Dartmoor. For example, sand from St Keverne contained roughly equal proportions of the grades 0·5–0·25, 0·25–0·1, 0·1–0·05 and 0·05–0·01 mm diameter. On the other hand, transport by water under less extreme conditions usually resulted in some degree of sorting, although the degree of uniformity obtained varies widely in different localities.

Conditions controlling chemical decomposition also varied and the residual felspar content of sands ranges from 0 to more than 20%. Clay formed by such decomposition may remain with the sand, but may be separated during subsequent transport by water.

As a result, it is possible to have a whole series of sands differing in three important factors:

(1) the presence or absence of clay;

(2) the uniformity of grain size;

(3) the amount of felspar remaining in the sand grains.

In the foundry, the presence or absence of clay decides the use to which the sand can be put and it is usual to divide sand into two classes:

1. Clay-free sands, usually known as 'silica' sands, but occasionally as 'sea', 'river', or 'core' sands.
2. Clay-bearing sands, usually known as 'natural' sands because in their natural state they contain sufficient clay to enable the grains to cohere when they have been pressed together.

There is no sharp distinction between the two. Some silica sands, such as those from Leighton Buzzard and Reigate and the shore sands, are free from clay, but others such as an exceptionally fine sand from the Leighton Buzzard deposits, contain small amounts of clay. On the other hand, the red Bunter sands are our most important 'natural' sands, but some of them contain less than 2% of clay, or less than some of the 'silica' sands.

Treatment of sand at the quarry

The price which foundries are prepared to pay for sand is not high enough to permit expensive treatment by the producer and many sands are sold in the 'as-mined' condition. However, in other cases a fairly simple and cheap process does enhance the value of the product sufficiently to justify the additional cost.

Natural sands

Some natural sands are extremely hard and would not easily be broken down in the kind of sand preparation plant used in a foundry. These are crushed at the quarry.

Scottish rotten rocks are so compacted that this treatment is essential. The most frequent method of preparing the sands for foundry use was to mill them in a mill with very large and heavy rollers and a pan with slotted holes. The operation was slow and its cost was an appreciable part of the total cost of the sand.

The red Bunter sands in the Midlands, although not as hard as the Scottish rock sands are also hard enough for blasting to have sometimes been necessary when quarrying, although they are now usually extracted more cheaply by mechanical diggers. The sand so obtained consists of aggregates, many of considerable size, which are not broken down completely under normal conditions in a foundry mill and if the sand is used in this condition, the prepared sand contains numerous pellets with a diameter of $\frac{1}{8}$ in or more. These may impair the surface finish of castings and make sieving onto the pattern desirable. This was a nuisance when a hand-moulder made forty castings in a day; when forty moulds are made in an hour on a machine it is impossible.

At one time some of the sand was milled at the quarry, but the only type of mill which could be relied upon to break down all the aggregates was that with a revolving pan and heavy rollers rotating on their own axes, in which case the milled sand had to be removed by means of a shovel and the cost was high. Nowadays the operation is carried out quite as effectively by means of disintegrators. These have a high rate of output and can be fed automatically with the result that the cost of treatment is only about 20 or 25p per ton.

Red sand should always be bought in the treated condition. The whole of the clay in the sand is then effective and consumption can be reduced sufficiently to offset the additional cost, but the greatest advantage is that the treatment enables a more uniform product to be obtained. Most Bunter sand deposits vary vertically and laterally in the quarry, but a considerable amount of mixing takes place in a preparation plant and variations can be smoothed out. Some suppliers can guarantee sand composition to an extent not formerly possible — a welcome change from the time when the only distinguishing feature was the depth of colour.

Treatment of this kind is neither necessary nor desirable when dealing with uncompacted sands such as those from Erith or Southampton.

Silica sands

Washing with water is a cheap process used for removing small quantities of clay and other very fine material from some silica sands. The finer the sand, the greater the time required to separate the unwanted material and usually only sands coarser than 100 μ (150 mesh) are so treated.

Under some conditions, for example, in an area remote from cheaper sources of suitable sands and therefore subject to high transport costs, more elaborate treatment may be justifiable.

A demand for a grade of sand which can be found only in admixture with other grades may also justify more elaborate treatment. It can be deduced from Table 1 that removing the fraction finer than 0·02 mm diameter from a sand with a diameter less than 0·07 mm is a slow and difficult operation and, in this country at least, the natural conditions which would produce such a fine sand free from clay have not existed in nature. The very fine sands, therefore, contain appreciable quantities of clay which can only be separated at great cost. However, some silica sands with a very low clay content contain considerable proportions of these finer grains and separation is less difficult. Usually the fine grains form only a minor fraction of the whole and a comparatively large amount of sand has to be treated. The coarser grains have to be sold in competition with grains of similar size obtained from quarries where no costly treatment is needed, and so the whole of the cost of separation has to be borne by the small amount of very fine sand.

SAND DEPOSITS (1): THE NATURAL SANDS

The number of individual quarries which supply clay-bonded sand to foundries is so large that it is impracticable even to mention more than a small proportion. In some cases a very large deposit is worked by a number of quarries spread over a large area and some characteristics are common to all; at the other extreme, a small quarry in a deposit which has no counterpart elsewhere has a purely local importance.

The red Bunter sands

The deposits of this sand cover a very large area. To the east of the Pennines it stretches from Nottingham in the south to York in the north, and to the west from central Worcestershire to Lancashire, with further areas at Carlisle and Belfast.

The areas over which the sand outcrops, with only 1–2 ft of overburden, are so large, particularly in north Worcestershire and south Staffordshire, that the site selected for quarrying depended on the means of transport used. Until the early 1920's, much of the sand used in foundries in the Midlands was carried by canal and quarries were situated on the canal banks, so that the sand could be loaded directly onto barges. As canal transport was replaced by motor transport the old quarries were abandoned and others opened up adjacent to roads, sometimes only a few hundred yards from the original site.

The depth of sand is usually much greater in the south than in the north. In the area between Bromsgrove and Wolverhampton some quarry faces are more than 150 ft deep. The main reasons for this limit of working are that it is a little more costly to dig out a hole and that trouble with water is possible during periods of heavy rainfall. In Yorkshire and Lancashire the depth of quarries rarely exceeds 20 ft.

Variable nature of red sands

The composition of the red sand varies very widely and not all of it is suitable for use in foundries. Some is coarse, with grains up to $\frac{1}{8}$ in diameter, and with a low clay content. Sand of this kind was used for pig beds and by the building industry. That used in the foundry is finer and has a higher clay content. The two types of sand may occur almost side by side and occasionally one may be above the other. In one quarry in the Mansfield area, a considerable thickness of the coarse material has to be removed before the foundry sand is exposed. With such large differences, there is no difficulty in segregating the different kinds of sand.

Smaller differences are encountered in quarries where all the sand is suitable for use in foundries. Indeed, it is not to be expected that conditions would have remained constant over the whole of the time needed to lay down a deposit 200 ft in thickness. Usually it is possible to find a layer of very fine material separating sand of two different grades; frequently the layer is very thin, but occasionally a clay layer 2 ft thick exists. The change from one grade to another is quite sharp and there is no universal order. The most open sand may be at the top, at the bottom, or in between two layers of closer sand, whilst the plane of separation may be horizontal or tilted, sometimes at a steep angle. Occasionally a fault might have occurred right across the sand face and in one such quarry it was possible to obtain sands with permeabilities of 60, 40 and 20 within a few feet of each other. Working at this quarry has been abandoned.

Permeability and clay content

As the permeability of the sand rises, clay content falls. The relation is not an exact one — it is not possible to say that if permeability is 25, clay content is 12·6 — but is well marked. In general, if permeability is 60 or above, as is the case with many of the Yorkshire sands, the clay content is rather too low for the sand to be used alone in the foundry, and it is usual to add either a clay or a stronger sand. If the permeability is 20–25, clay content is usually about 15%.

Quarrying

When, as is frequently the case, there are several different grades of sand in a quarry, two different methods of working are possible.

In the first, which was at one time widespread, different grades of sand, *e.g.* 'close', 'medium' and 'open', may be produced. This is satisfactory unless the demand for the grades at the bottom is higher than for the upper layers. The price at which sand is sold is so low that it would be uneconomic to remove the upper layers to reach the lower, unless all the sand extracted could be sold.

In the second, the different layers are worked simultaneously, loads from each layer being tipped in sequence at the preparation plant where they become mixed and a uniform sand is obtained.

Mineral composition of the sand grains

A typical chemical composition of the sand fraction of a red sand is given in Table 2.

Clay was removed by washing after boiling with ammonia as in the estimation of the amount of clay, described in Chapter 1. The red sands are coated with a very firmly adherent coating of ferric oxide in which some clay is included. This raises the alumina and loss on ignition of the granular fraction slightly, but has a negligible effect on the alkalies.

Table 2. *Chemical analysis of sand grains — Bunter sand*

	% Composition
Silica	89·97
Ferric oxide	0·60
Alumina	5·21
Titania	0·18
Lime	0·12
Magnesia	0·27
Alkalies	3·08
Loss on ignition	0·47

The major impurity is potash felspar which has the composition $K_2O.Al_2O_3.6SiO_2$ and contains 16–17% potash with a little soda felspar containing 11–12% soda. The total amount of felspar is about 20%. The proportion of felspar is reasonably constant in the coarse and fine sand grains and in sand from different sources. Since the fusion point of felspar is about 1340°C, the red Bunter sands are not suitable for use in the steel foundry.

Mechanical composition

The mechanical composition of a number of red sands is given in Table 3.

Sand number 1 is a very fine-grained sand, much sought after by brass founders, but with a permeability rather too low for machine-made iron foundry moulds.

Sand number 2 is a sand of the kind most widely used in iron foundries; sand number 3 is a sand with a high permeability and a low clay content and is free from the very coarse grains usually found in the more open red sands. Its permeability is not very dissimilar from that of Southport sand to which 10% of a sand such as number 2 has been added — a mixture occasionally used for cores which need a little green strength. This red sand was used in place of the Southport red sand mixture during the war when supplies of Southport sand were uncertain. It is now used on a small scale as a base for a synthetic sand. There are several such deposits in the Midlands and in Yorkshire. The very low clay content is unusual in red sands.

Sand number 4 is typical of the coarser red sands. It is used on a small scale as a constituent of semi-synthetic sands for moulds for very large iron castings.

Table 3. *Mechanical composition of red sands*

	Sand number			
Aperture (μ)	1	2	3	4
+699 (22 mesh)	—	—	1·6	6·4
+500 (30 mesh)	—	—	4·4	4·0
+353 (44 mesh)	—	1·3	10·8	21·1
+251 (60 mesh)	2·0	9·2	27·4	17·4
+211 (72 mesh)	4·2	8·3	10·4	13·2
+152 (100 mesh)	8·7	17·0	21·0	12·8
+104 (150 mesh)	32·4	37·6	17·2	9·4
+ 76 (200 mesh)	18·3	10·5	4·0	6·2
+ 53 (300 mesh)	12·3	4·2	2·0	4·0
− 53	7·0	2·0		
Clay	15·0	10·0	1·2	5·5
Permeability number	20	40	140	180

Physical properties of red sands

Strength and, to a smaller degree, permeability of all clay-bonded sands depends on the moisture content and on the milling treatment. The properties over a range of moisture content of sand number 1 in Table 3, milled for 10 min in a laboratory size Simpson mixer are given in Table 4.

Table 4. *Physical properties of a red sand*

Moisture (%)	Permeability number	Green compression (lb in^{-2})
2·4	20	20·0
2·6	21	19·0
3·2	22	17·0
3·9	24	15·8
6·0	21	11·0
9·0	19	9·0

With the lower moisture contents, the sand is too brittle for mould making. The most suitable moisture content, using all new sand is 4–5% if the moulds are to be cast green, and 6–9% if they are to be dried.

The moisture range of red sands

Sand number 1 above could be used for making moulds with a moisture content ranging from 4 to nearly 15%. Sand number 2, with its lower clay content, could be used with a moisture content varying over a range of more than 8%.

Other natural sands have a somewhat narrower moisture range, whilst synthetic sands have a much narrower range still. This is of considerable practical importance in several kinds of mould making (see Chapter 5).

Estuarine sands

There are several deposits of sand which exist by river estuaries, particularly at the mouth of the Thames. The most important of these is Erith 'loam', a term used because the sand is fine-grained.

These sands contain far less felspar than the red sands, but the granular fraction contains appreciable quantities of glauconite,[3] which has a low melting point and makes the sand unsuitable for steel founding.

Erith loam

The deposit occurs near the Thames, below London. At one time it was widely used on the north east coast and in Scotland, particularly the Falkirk area, by makers of light castings. It was relatively cheap as it was used as ballast in boats which were returning after bringing down coal. It is widely used in the south east and along the south coast.

There is a gradual change in properties as the sand is taken from increasing depths and five grades are supplied — yellow, mild, medium, strong and extra strong. The grains become finer and the clay content higher as the depth below the surface increases.

The composition of the various grades is given in Table 5.

Table 5. *Composition of Erith loam*

Aperture (μ)	Yellow	Mild	Medium	Strong	Extra strong
+152 (100 mesh)	0·4	0·6	0·9	2·0	1·8
+104 (150 mesh)	61·4	55·1	23·7	14·0	12·0
+ 76 (200 mesh)	21·0	23·0	40·1	34·8	35·0
+ 53 (300 mesh)	5·0	6·7	13·3	21·1	28·1
− 53	3·2	4·0	8·0	11·0	12·0
Clay	9·0	10·7	14·0	19·0	21·1
Permeability number	44	35	19	15	12

There are other similar, although usually still finer, deposits in the same area, but these are not widely used. Another sand of the same type, but with coarser sand grains, is found near Southampton.

Rock sands

The rock sands quarried in Scotland are found in the area between Falkirk and Glasgow.

Their characteristics are completely different from those of the red and estuarine sands. Frequently they occur in bands a few feet in thickness, separated by other bands of no commercial value, and the cost of extraction is high. They usually exist as very hard sandstone which would not be broken down by a foundry mill, and milling at the quarry is essential. The composition of the prepared sand depends to some extent on the efficacy of the treatment, since some samples contain an appreciable proportion of aggregates.

The size distribution of the sand grains is different from that of most other types of natural sand in that some sand may be found on every sieve

Table 6. *Composition of rock sands*

Aperture (μ)	Scottish sand		Durham sand
+699 (22 mesh)	6·7	8·0	3·6
+500 (30 mesh)	3·3	7·2	4·0
+353 (44 mesh)	4·5	14·6	9·3
+251 (60 mesh)	9·9	22·2	23·9
+211 (72 mesh)	13·4	8·0	15·0
+152 (100 mesh)	20·0	8·6	12·0
+104 (150 mesh)	11·2	7·8	7·8
+ 76 (200 mesh)	5·2	2·8	2·4
+ 53 (300 mesh)	3·5	2·4	1·8
− 53	3·0	2·0	3·6
Clay	19·3	16·4	16·6
Permeability number	44	47	42

from 2·0 μ (8 mesh) down, with many compound grains. Although much of the sand is coarse, permeability is not particularly high. Felspar is present, but in a much smaller amount than in the red sands, and the rotten rocks were frequently included in sand mixtures in the steel foundry.

The clay content is high — up to 20%. The combination of a high clay content and coarse grains makes it difficult to produce moulds with a very smooth surface finish, and the light castings industry in Falkirk preferred to obtain other sands from as far away as the Midlands and the Thames estuary. The rock sand is widely used for somewhat heavier castings made in green sand and for large dry sand work.

A sand of similar origin and composition is found in Durham. The composition of typical Scottish and Durham sands is given in Table 6.

Other natural sands

There are numerous sand quarries in the country which are not included in the above three main groups, and most geological ages have left sand deposits behind. Indeed, few foundries in England are situated as much as 40 miles from the nearest deposit, but many are of purely local interest. Their importance tends to decline as foundries are more selective in their demands and are also more insistent on obtaining a uniform material.

SAND DEPOSITS (2): THE SILICA SANDS

The greater part of the silica sand used by iron foundries is derived from one of three deposits:

1. A sand which outcrops at widely separated places from Surrey and Kent, in Bedfordshire and in Norfolk. The sands from the different areas are very similar. The largest quarries are in the Reigate area, in Leighton Buzzard and near King's Lynn. A whole series of sands differing widely in size distribution is obtained, and most suppliers provide several different grades.

2. Sand in north Cheshire and south Lancashire, the most important quarries being at Congleton and Chelford.

3. Shore sands. The largest source of supply is near Southport, and smaller quantities are obtained from Ayrshire and both sides of the Severn estuary on the west coast and Durham on the east.

Sands from Reigate, Leighton Buzzard and King's Lynn

From each area several sands differing in grain size are produced. All have a high quartz content, sometimes exceeding 99%, and some are used for glass-making, where close control of composition and grading is essential.

Foundries are less interested in chemical composition — for example, 0·01% of iron oxide is of no importance to a core maker, but may have serious consequences for a maker of high quality glass — but do require constant grading. In this respect they benefit from the demand of the glass maker.

Care in selection on the part of the supplier is particularly necessary, since from each area several grades, which may differ widely in grain size, are produced.

Typical mechanical analyses of grades used by foundries are given in Table 7.

There are several suppliers and a large number of grades is available. However, Table 7 covers the range of sands useful in the foundry, but does not include all, or even a majority, of the individual grades available.

Table 7. *Composition of foundry sands*

Aperture (μ)	Reigate	Lynn	Reigate	Lynn	Reigate	Reigate	Leighton Buzzard	
+699 (20 mesh)	—	—	1·3	0·7	0·6	6·6	2·4	13·4
+500 (30 mesh)	—	—	4·2	1·0	2·9	12·6	19·2	23·2
+353 (44 mesh)	—	1·6	11·6	6·4	18·6	27·3	58·9	41·8
+251 (60 mesh)	1·0	5·2	22·4	28·4	44·7	33·8	15·6	26·3
+152 (100 mesh)	22·2	28·2	41·6	54·4	24·0	14·6	2·4	4·0
+104 (150 mesh)	30·2	55·4	11·2	7·8	7·2	3·5	1·4	1·3
+ 76 (200 mesh)	31·8	8·6	7·5	1·0	1·1	0·8	1·1	—
− 76	14·8	1·0	0·8	0·2	0·9	0·6	—	—
Clay	—	—	—	—	—	—	—	—

Many of the above sands are washed and the clay content is very low. The two finest sands are used mainly for shell moulding and the others with organic binders, for conventional, hot box and air-set cores, and in the CO_2 process. The coarser sands are available from any district where the sand outcrops, but sources of the finest sands are more limited.

The suppliers provide details of all the sands they produce and any grade a foundry is likely to need can be obtained.

Sands from Lancashire and Cheshire

Two different types of sand are obtained from this area:

1. A sand from the Congleton area, in which most of the grains occur on the 251, 150 and 104 μ sieves. The sand from some quarries has a clay content between 0·5 and 1 %, but some is almost clay-free.
2. A sand from Chelford which has an appreciably wider range of grain size. This can be obtained as-quarried, but a considerable amount is washed and divided into two grades, coarse and fine.

Typical compositions are shown in Table 8.

Table 8. *Lancashire and Cheshire sands*

Aperture (μ)	Congleton	Chelford coarse	Chelford fine
+500 (30 mesh)	0·8	3·0	—
+353 (44 mesh)	3·5	12·2	—
+257 (60 mesh)	22·6	46·1	4·0
+152 (100 mesh)	50·3	34·6	46·5
+104 (150 mesh)	19·1	4·1	35·2
+ 76 (200 mesh)	2·7	—	11·1
− 76	0·3	—	3·2
Clay	0·7	—	—

Shore sands

All the sands from the west coast are composed of grains of uniform size. A typical analysis of Southport sand is given in Table 9.

Samples taken over a period of more than 40 years showed negligible changes — no greater than would be found if the same sand had been tested using two different sets of acceptable standard sieves.

All the shore sands contain shell. That from the Southport area usually contains rather less than 2·5%, and again has been very uniform over a long period. The sand from Ayrshire has a similar amount of shell, but is sometimes contaminated with fine-grained materials brought down the Clyde. The sand from the Severn estuary contains much more shell — 6% has been found in samples from south Wales, and 10% in some from Somerset. This seriously reduces their value to foundries. The 2·5% present in Southport sand does not appear to cause any difficulty.

Table 9. *Typical mechanical analysis of Southport sand*

Aperture (μ)	% Composition
353 (+44 mesh)	0·2
251 (+60 mesh)	7·1
152 (+100 mesh)	81·9
104 (+150 mesh)	10·2
76 (+200 mesh)	0·2
−76 (−200 mesh)	0·3
Clay	0·1

There is a possibility that parts of a dune may contain vegetable matter — usually grass of the type used for preventing or reducing removal of the sand by wind, rain or tidal action.

Although the clay content of the sand is negligible, the quartz content is not high. This is only partly due to the presence in the sand of fusible minerals, and much of the non-quartz fraction consists of particles sufficiently fine for inclusion in the clay grade which adhere to the sand grains so strongly that they cannot be removed by boiling with ammonia or shaking with soda. If, however, the sand is treated with hydrochloric acid, a small proportion of the adherent coating dissolves, and the remainder is detached from the sand grains. Many silica sands have a thin, and frequently patchy, adherent coating, but its weight is usually less than 0·1%, whereas that on Southport sand is more than 2%. The effect of this coating is very small, even when the sand is used in the production of conventional oil–sand cores.

Other silica sands

There are many other smaller deposits of silica sand which are worked — in Kent, in the east and north Midlands, and in Yorkshire. The majority of these provide a sand suitable for conventional cores which is used in place of similar sands from the other deposits whenever it is more convenient or less costly to do so. A few are very fine sands; for example, the sands from Ryarsh in Kent and West Wittering in the east Midlands, but most of the very fine sands contain a proportion of clay, which may be quite low, but is usually variable. This limits their usefulness.

Where no deposit of a suitable silica sand exists locally, it may be profitable to take a sand which is contaminated with clay or has too wide a range of grain size and wash and grade it to obtain the kind of sand needed.

Table 10. *Levenseat sands — mechanical analysis*

Aperture (μ)		Sand number		
		1	2	3
+500	(30 mesh)	—	23·0	54·7
+353	(44 mesh)	—	18·2	29·1
+251	(60 mesh)	0·5	31·1	13·2
+152	(100 mesh)	32·6	24·1	3·0
+104	(150 mesh)	46·1	3·6	—
+ 76	(200 mesh)	12·0	—	—
− 76	(−200 mesh)	8·8	—	—
Clay		—	—	—

Such conditions occur in Scotland. There are some, usually small, excellent deposits of silica sand in places remote from likely customers and badly served with transport facilities. Shore sand is available, but is sometimes contaminated and is frequently found to be unsatisfactory. However, in view of the high cost of transport it is found possible to take a sand from Levenseat and wash and grade it to produce three grades of sand of suitable quality. The sand itself does not at first sight appear promising. It is extremely hard and rather difficult to extract, has an appreciable clay content and the grains are spread in more or less random fashion over all the sieves used in sand testing. This sand is roughly crushed in a large mill and passed through a bar disintegrator. At this stage, washing readily removes the clay leaving clean grains which are wet-screened to provide the three grades shown in Table 10.

SAND RESOURCES

Natural sands

The reserves of the Bunter sands suitable for foundry use are enormous. The deposit covers several hundred square miles and when the depth of quarries reaches 150 ft, as it does in the southern part of the deposit, each square yard supplies 50 tons. There is a tendency for the amount used to decline, but the change is in no way connected with a lack of suitable material, the chief reasons for the change being an increasing use of synthetic in place of natural sands and the adoption of new processes for which natural sands are not suitable. For example, many moulds and cores for large castings which were at one time made almost exclusively from clay-bonded sand and dried before use, are now made from silica sand bonded with sodium silicate or air-setting organic binders. Some small castings formerly made in green sand moulds, using a fine-grained natural sand to obtain a good surface finish, are now shell moulded, the process requiring silica sand in place of clay-bonded sand.

Occasionally complaints are received to the effect that supplies of a particular grade of a Bunter sand are worked out, but usually what has happened is that a quarry in which that particular grade of sand occurs under only a few feet of overburden, and which is situated near a good road within easy reach of the foundry, is worked out. Frequently the quarrying has to cease because the surrounding land is covered by buildings. In his survey of sources of sand Boswell noted that up to 1914 a quarry near the centre of Birmingham supplied not only local foundries, but exported to many countries abroad; by 1918 most of the sand not covered by buildings had been worked out and the last remaining open area in the city — situated about a mile from the first — was nearing exhaustion. Closure of these quarries was overlapped by the opening of others some ten miles distant in open country in north Worcestershire where many millions of tons are available.

Other kinds of natural sands are less abundant than the Bunter sands, but supplies of the more important are sufficient for the foreseeable future.

Silica sands

The demand for silica sands tends to increase, since the newer methods of producing castings — silicate bonded sands, air-setting sands and shell — use silica sands where natural sands were used earlier. Occasionally deposits of one particular grade from one source may be worked out — one coarse grade from Leighton Buzzard is no longer available and for a short time there was difficulty in obtaining one grade of Congleton sand — but it is usually possible to obtain an equally satisfactory grade from the same area.

There has been a tendency for the quarrying of silica sands to be concentrated in fewer hands, without necessarily reducing the number of deposits being worked, and certainly without reducing the number of grades available. Many silica sands are improved by treatment at the quarry and some of the newer processes, particularly shell moulding, require very close control which can be achieved only with a suitable preparation plant and adequate laboratory facilities, both of which become more readily available with larger units. Many silica sands can now be obtained with a guaranteed grading and the larger producers issue data sheets giving the appropriate details.

It is not possible for a user to draw up his own specification unless it happens to coincide with what is actually present in a quarry or can be obtained by a cheap and simple treatment. In order to satisfy all grading specifications, it would be necessary to separate each of a number of sands into fractions and recombine these fractions in suitable proportions, a simple procedure if sand were sold by the pound, but impossibly expensive when dealing with large quantities.

There is a tendency, as yet small, but nevertheless growing, to recover silica sand for re-use in the foundry. This is not due to any difficulty in obtaining supplies, which are at present ample, but solely for consideration of cost. Most silica sands, except the few mainly very fine sands which are only obtained by a costly treatment, are cheap at the quarry, but transport costs may be heavy. Indeed, taking the country as a whole, it costs more to carry the sand to the user than to buy it at the quarry. Moreover, dumping used materials becomes continually more expensive, and recovery becomes more attractive.

REFERENCES

1. Boswell, P. G. H. (1918). *British Resources of Refractory Sands*. Ch. 8, Taylor and Francis, London.
2. Bagnold, R. A. (1941). *The Physics of Blown Sands and Desert Dunes*. Ch. 3, Methuen and Co., London.
3. Skerl, J. G. A. (1932–1934). Unpublished work at BCIRA.

CHAPTER 3

The Influence of the Sand Grains

A clay-bonded mould consists of two fractions, sand grains and clay, the grains forming the bulk of the mould and the clay holding them in position. The two constituents are of equal importance, but have different functions, certain properties of the mould being controlled by the grains and others by the clay, and although a property controlled by one may be influenced to a minor degree by the other, it is convenient to consider the two separately.

The sand grains include all the particles, however fine, except those in the clay grade, and not only those which would be classed by a geologist as sand. Occasionally, grains finer than 100 μ are described as 'silt',[1] an undesirable distinction since there is no sudden transition in behaviour as the grain size changes from above to below 100 μ. This nomenclature was borrowed from the geologists.

The properties which are almost completely dependent on the granular fraction are:

refractoriness;
permeability;
surface finish;
expansion.

In addition, the following properties are influenced to a minor degree:

strength;
moisture requirement;
flowability.

REFRACTORINESS

The major constituent of foundry sands is quartz, which has a fusion point of about 1700°C. Some sands also contain appreciable quantities of more fusible minerals such as felspar, which fuses at about 1350°C. Few sands contain minerals which are not sufficiently refractory for the iron foundry, but many, in particular certain natural sands, are unsuitable for use in the steel foundry. In consequence, at one time most steel foundry green sand was based on imported Belgian or French sand. It can be shown that the refractoriness of the two sands depends only on the sand grains.

A red Bunter sand, similar in grading to a Belgian sand, but containing only 2% of clay, was milled with sufficient of a refractory fireclay to raise its clay content to the level of the Belgian sand and the two were used for preparing moulds for a cubic casting weighing about ½ ton. The two moulds were poured with mild steel at a temperature exceeding 1600°C. A reasonable casting was obtained from the Belgian sand, but that from the Bunter sand was covered with a layer, more than 1 in thick, of a mixture of sand and metal.

The clay was then washed from the two sands and replaced by several different clays, including that from the Belgian sand, and the sands thus obtained were used to face moulds for the ½ ton casting. In all cases the results were the same — a reasonable casting with the Belgian sand grains and gross penetration with the Bunter sand. The kind of clay had no obvious influence.

The refractoriness of sand grains

The granular fraction of the Belgian sand contained nearly 99% silica and consisted almost entirely of quartz with a fusion point of 1700°C. The total silica content of the Bunter sand was 90%, the chief constituents other than silica being alumina (over 4%) and potash (3·5%). The sand contained 20% felspar with a size distribution similar to that of the quartz. One grain in five had a fusion point of 1350°C.

With random distribution of the different kinds of grain there are, among the many millions of grains which form the surface layers of a mould, many groups of several felspar grains close to each other. When the castings were poured, the mould was heated to a temperature exceeding 1350°C before solidification of the casting commenced, and melting of the felspar grains permitted easy and deep penetration by molten metal.

The presence of a small proportion of fusible grains does not result in penetration of this kind and at one time many steel foundry moulding mixtures contained up to 15% of a red sand. In this case felspar grains formed only 3% of the sand and the connected groups which fused and permitted deep penetration were absent.

When the castings are small, the mould surface may not reach the temperature at which felspar fuses and steel castings weighing a few pounds have been produced in Bunter sands containing a small amount of coal dust. The surface finish was superior to that usual on steel castings, but the practice is not to be recommended for general use.

The refractoriness of the clay

The clay in a properly prepared moulding sand is present as a thin layer on the sand grains and perhaps as wedges near grain junctions. Fusion of the clay may lead to a slight increase in the very small passages between the sand grains, but does not produce large cavities of the kind formed when groups of grains fuse. The use of a fusible clay does not lead to metal penetration and when preparing a synthetic sand to be used for green sand moulding, steel founders almost invariably prefer a montmorillonite clay, although it has a lower fusion point.

When a natural sand is used, part of the clay may occur as aggregates and if these are not dispersed in the mill, they may be found on the mould surface, where they will almost certainly fuse when the casting is poured. This may lead to surface imperfections and sometimes to slag-like inclusions elsewhere in the casting, but not to metal penetration to an extent which makes sand removal difficult.

Changes in refractoriness with use

A sand may contain substances, not necessarily easily fusible themselves, which react with each other or with quartz or clay, to form readily fusible compounds. For example, calcium carbonate may occur in a fine state of division in some clays or as shell fragments in some sands, and should this be sufficiently near to the casting surface to reach a temperature exceeding 800°C it decomposes, forming the oxide. This in turn can react at high temperatures with sand and clay to form fusible silicates, which tend to increase in quantity as the sand is used until equilibrium is reached between the amount produced in a casting cycle and that removed with the sand discarded.

The fusible material so formed is finely divided and uniformly distributed throughout the sand. It does not therefore cause metal penetration but may fuse at the temperature reached by the mass of sand between, say, two heavy bosses. On cooling down it solidifies, forming a glass which bonds the mass of sand so firmly that removal is difficult.

PERMEABILITY

When gas passes through a mould or core, it must pass through the pores and permeability to gas therefore depends on the number and size of the pores, or on their number and total volume.

The volume of the pores in a granular material depends on the size distribution of the grains and on the way in which they are packed. If it is assumed that: (1) all grains are spherical; (2) the grains are so packed that pore space is minimal, certain important relationships between size distribution and permeability can be calculated. The validity of any conclusions reached can be checked by means of tests on mixtures of sand fractions.

The packing of spherical grains

The relationship between the size distribution and pore space of close-packed spheres has been described by several observers[2,3] and the important factors have been well established.

1. When spheres of identical size are packed in such a way that they occupy the smallest possible bulk volume, the total pore space is independent of the size of the spheres and is 26% of the total volume. If the spheres are large there is a relatively small number of large pores; if small, a larger number of smaller pores.

2. When smaller spheres are added to the packed uniform spheres, the effect on pore space depends on the relationship between the size of the spheres in the two fractions. If the smaller spheres are of such size that they can fit into the existing pores without disturbance, then pore space is reduced. If, however, an attempt is made to introduce spheres which are just too large, the whole structure is disturbed and pore space is little changed.

When all the largest pores have been filled, smaller spheres can be introduced into smaller pores, and the process continued until all the pores are filled, when permeability is 0, whatever the size of the original spheres may have been.

In view of the influence of pore space on permeability, it is important to know the size of the largest sphere which can be accommodated within the pores of close-packed uniform grains. This can be calculated and if the diameter of the original spheres is 'd', the largest inclusive sphere has a diameter of $(\sqrt{(2)} - 1)d$ or roughly $0{\cdot}4d$. The next largest sphere which can fit into the remaining pores has a diameter of $0{\cdot}2d$.

It would not be difficult, although tedious, to work out a series of sizes which would in theory give 0 permeability, commencing with spheres of any selected size. However, to obtain minimal permeability it would be necessary to place each sphere in the appropriate place. If all were simply mixed, a sphere with a diameter of $0{\cdot}2d$ might be found in a pore large enough to hold one of $0{\cdot}4d$, an arrangement which would be quite stable, but would prevent the closest form of packing.

Nevertheless, although the maximum theoretical effect is not obtained, mixing two sets of uniform spheres differing from each other in diameter, does produce the kind of change calculated. If the smaller spheres have a diameter less than $\frac{2}{5}$ that of the larger, there is a reduction in pore space; if the diameter is more than $\frac{2}{5}$, pore space is unchanged.

The permeability of close-packed uniform spheres

A gas passing through a granular material passes through a number of small tubes. These are not of uniform diameter and none of them passes through the mass in a straight line, with the result that the gas has to change direction when it has travelled a short distance.

The resistance to the passage of a gas along a tube increases rapidly as the size of the tube decreases, so much so that the resistance of a number of narrow tubes arranged in parallel is higher than that of a single tube with an area of cross section equal to that of the smaller tubes combined.

Resistance also increases when there are (a) changes in the diameter of the tube, or (b) bends in the tube. Formulae are available for calculating the effects on resistance of changes in diameter and of various kinds of bends, T-junctions and the like in piping. Gas passing through packed spheres of the size of sand grains has to pass through a number of tubes which vary rapidly in diameter and contain numerous bends and junctions, and resistance is high or, in other words, permeability is low.

The combined effect of the small diameter of the pores in sand, of the various constrictions and of the numerous bends, is enormous. For example, the permeability number of Southport sand, which, for a sand, is of very uniform grain size, is 150 using the standard method of ramming. Under these conditions the grains are by no means close-packed and the total pore space is about 50%. If all the pores were joined to form a straight tube through the centre of the test piece, the diameter of the tube would be 1·4 in. A tube having a diameter of $\frac{1}{8}$ in, and therefore having a cross section less than one-hundredth that of the combined pores bored through the test piece would produce a permeability too high to be measured on the standard apparatus, but exceeding 1000. The permeability of a sand is therefore lower by several orders of magnitude than it would be if the pores were joined to form a single straight tube. This point is of considerable importance when considering the venting of cores.

As the size of the spheres decreases, the pores become smaller, the diameter of the air passages becomes less and the number of changes in direction increases. Permeability therefore decreases as size decreases.

The permeability of non-uniform spheres

When two sets of uniform spheres are mixed, pore space is unchanged if the ratio of the diameters is less than 5:2 and reduced if it is greater than 5:2.

Suppose, now, that the larger spheres are added in gradually increasing quantity to the smaller.

1. If the size ratio is less than 5:2, there is no change in total pore space, but replacing small grains by larger, leads to a reduction in the number of

grains and therefore in the number of pores. The pores must become larger, resistance must decrease and permeability increase.

2. If the size ratio exceeds 5:2 there is a reduction in pore space which may more than offset the above increase in permeability, especially since the small spheres preferentially occupy the largest pores.

This leads to the apparently anomalous result that the addition of spheres with a diameter of 152 μ (100 mesh) to uniform spheres of 76 μ diameter (200 mesh) must raise permeability, but the addition of spheres with a diameter of 250 μ (60 mesh) may reduce it.

The packing of sand grains

When considering the packing of spheres, it was assumed that:

(1) perfect spheres were used;
(2) each fraction consisted of spheres of identical size;
(3) the spheres were assembled in close packing formation.

None of these assumptions can be made when dealing with foundry sands assembled by any form of ramming.

1. Spherical grains are rare. The grains of most foundry sands are sub-angular or, at best, 'rounded'. In many cases the three major diameters are not equal.

2. The most uniform sands available contain grains differing in diameter by more than 2:1. Even a sand obtained by sieving, using the two adjacent sieves in the series used for sand testing has a spread of diameter of 1·4:1.

3. Foundry sands are never close-packed. A sand obtained by sieving between two adjacent sieves cannot be rammed to produce an assembly with a pore space of less than 40%.

It is therefore necessary to see how far the conclusions drawn from a study of close-packed spheres can be applied to rammed sands.

To do this, four samples of sand were obtained by sieving washed silica sands and taking fractions between neighbouring sieves in the IMM series. These sieves are no longer used, but were particularly convenient for the purpose, since the wire diameter was equal to aperture size, and aperture size was therefore inversely proportional to the mesh number, *i.e.* the aperture of a 40 mesh sieve was exactly half that of a 20 mesh sieve.

Four fractions were obtained:

(1) between 10 and 20 mesh;
(2) between 30 and 40 mesh;
(3) between 50 and 70 mesh;
(4) between 70 and 100 mesh.

Compared with uniform spheres, each of these fractions contains grains varying widely in size, but they have a range of grain size similar to that of the fractions obtained when carrying out a mechanical analysis.

It can be seen from the mesh numbers that no grain in (3) differs from any in (2) by a factor exceeding 5:2. If, therefore, the conclusions concerning the mixing of uniform spheres can be applied to sand grains, mixing these fractions should not lead to a reduction in pore space. On the other hand, each grain in (1) differs from any grain in (4) by a factor exceeding 5:2, and when these fractions are mixed there should be a decrease in pore space.

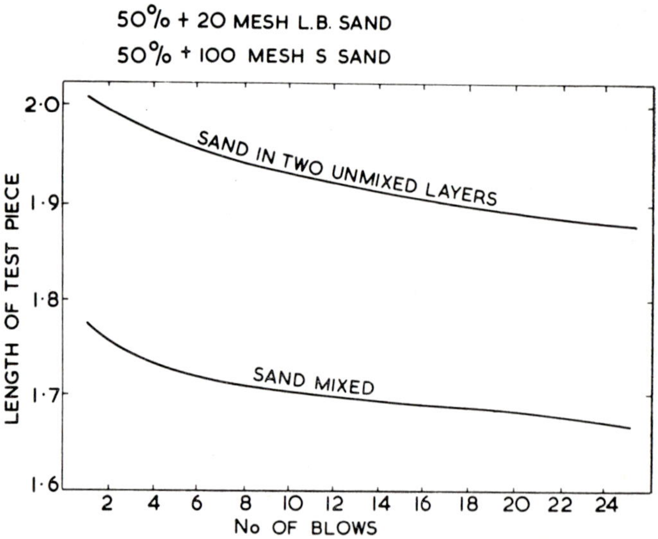

FIG. 28. Effect of mixing on length of test piece (50% + 20 mesh sand, 50% + 100 mesh sand).

To test this, each of the fractions was moistened with 1% of water. 75 g of (4) was placed in the standard test piece container, and 75 g of (1) placed on this without mixing. The sand was then rammed on the standard rammer and the length of the test piece measured after each blow, taking a bearing from the top of the rammer rod. Ramming was continued until 500 blows had been given, but as the results showed that the pattern of movement could be seen in sufficient detail much earlier, subsequent tests were restricted to 25 blows.

The sand was removed from the container, mixed and returned and the test repeated.

The lengths of the test pieces are shown in Fig. 28.

The length of the test piece after three blows of the rammer was a little under 2 in, and the pore space $43\frac{1}{2}\%$, when the sands were unmixed. When the sands were mixed, the length of the test piece was 1·75 in and pore space 31 %. As a result of mixing, pore space fell by $12\frac{1}{2}\%$ of the volume of the test piece, or nearly 25 % of the volume of the pores.

The test was repeated with fractions (2) and (3), and it was found that both before and after mixing, the curves relating length of test piece to number of blows of rammer were very close to the upper curve of Fig. 28. These results are in line with those calculated for uniform spheres.

The permeability of sands

The qualitative effect on the permeabiliy of uniform spheres of adding larger uniform spheres was derived from a consideration of the changes in the number and size of pores, always provided that the ratio of the diameters did not exceed 5:2. In this case, the permeability of the mixture must be higher than that of the smaller spheres, but there is no indication of the extent of the increase.

When the ratio exceeds 5:2, it was not possible to obtain even a qualitative result, since two changes were involved, one tending to raise and one to reduce permeability.

The quantitative effect can, however, be measured using the four fractions of sand separated by sieving.

Mixtures of (2) and (3) were made in all proportions from 0–100 % in steps of 10 %, and the permeability and length of a 150 g test piece measured. The results are shown in Fig. 29.

The length of the test piece changed only slightly throughout. The mixed sands always gave a slightly shorter test piece, but this small change is to be expected since (a) the sand does not pack in close-packing formation and there are oversize pores, and (b) sieving is such a slow operation that it was not possible to remove all the particles which should have passed +70 mesh.

The first addition of coarser sand raised permeability as had been calculated, but the rate of increase was small until the coarser sand formed half of the mixture, after which it rose rapidly.

The test was then repeated using sands (1) and (4) and the test results are shown in Fig. 30.

The picture here was completely different. The first addition of coarse sand produced a marked reduction in the length of the test piece, and therefore in pore space, and a reduction in permeability. As the proportion of coarse sand increased, both pore space and permeability fell to a minimum and then increased.

Minimum pore space was reached with 60–70 % coarse sand present. Minimum permeability was obtained with 40 % coarse sand and at this stage permeability had been halved.

The effect on permeability of adding a coarse to a fine uniform sand therefore depends on the ratio of the diameters of the two components. If the diameter of the coarse is less than two and a half times that of the fine, there is an immediate but very small increase in permeability, which continues progressively, but still slowly, until equal amounts of the two sands are present, after which it increases at a much faster rate. If the diameter of the coarse is more than two and a half times that of the fine,

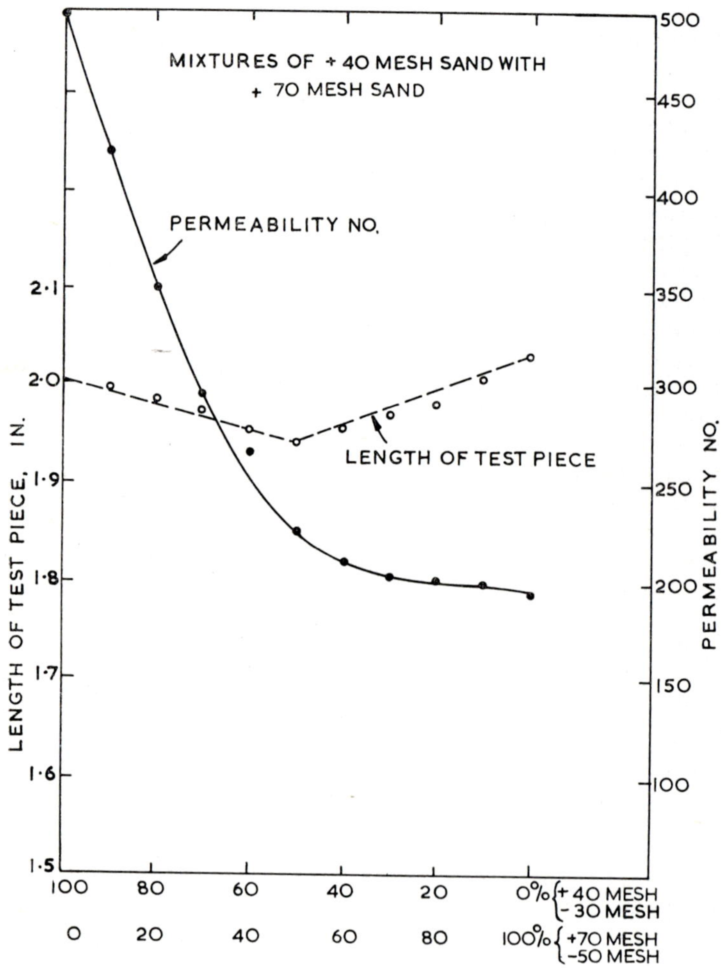

FIG. 29. Permeability and length of test piece (+40 mesh and +70 mesh sand).

addition of coarse to fine results in an immediate and significant fall in permeability to a minimum, followed by a very rapid increase.

Should the difference in size between the two components be increased, the point at which permeability is minimal, moves to the left, *i.e.* in the direction of a larger proportion of coarse sand. Minimum permeability decreases.

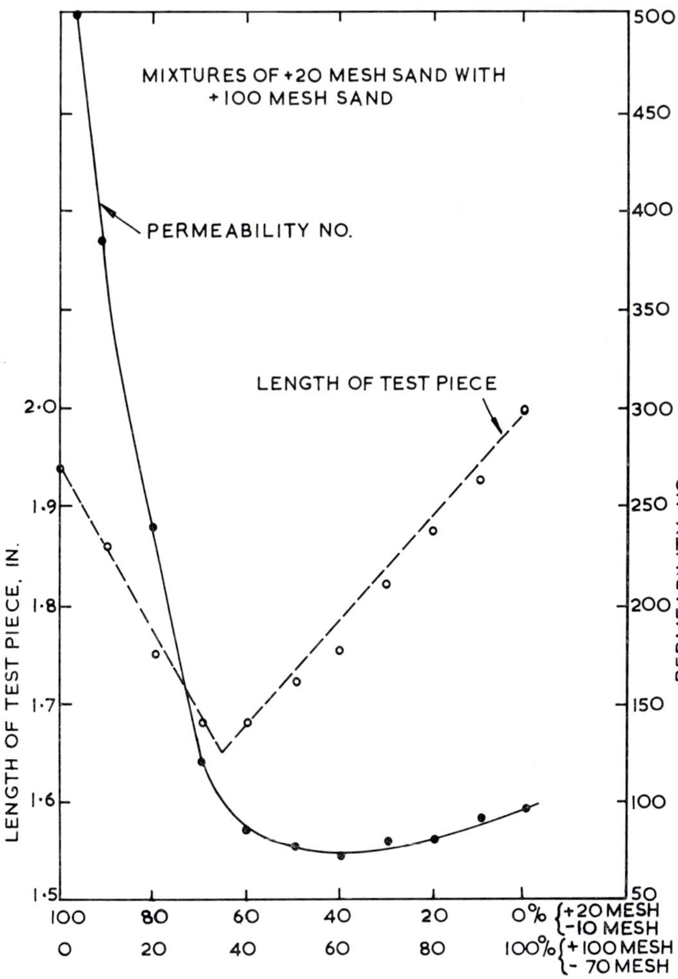

FIG. 30. Permeability and length of test piece (+20 mesh and +100 mesh sand).

CONTROL OF PERMEABILITY

Examination of the curves in Figs. 29 and 30 shows that it is impossible to raise the permeability of a sand of uniform grain size to a significant extent by adding a coarser sand unless the second sand forms more than half of the mixture. In Fig. 29 the finer sand had a permeability of 195, the mixture of equal parts of fine and coarse a permeability of 220. This is not a significant increase — it would, for example, be completely masked by an increase of a few points in mould hardness. On the other hand, the addition of smaller quantities of a finer sand reduces permeability appreciably.

For all practical purposes, therefore, the permeability of a uniform sand cannot be raised by additions, but only reduced. On this account, the sand selected for a system should have a permeability appreciably higher than that needed, and this permeability should then be reduced by the addition of a finer sand. Variations in permeability such as frequently occur in a clay-bonded sand system as a result of an increase in fines resulting from the breakdown of coal dust, or a decrease in fines due to losses at the shake-out, can then be taken care of by decreasing or increasing the amount of the finer sand. It is not necessary for the finer sand to be of uniform grain size and it may be more convenient to use one of the finer red sands rather than a fine silica sand. They are frequently cheaper and also enable the clay addition to be reduced.

Raising the permeability of non-uniform sands

Practically all the natural sands in this country have grains which cover a wide range in size and most have a low permeability. By adding a uniform sand, it is possible to reduce the proportion of fines by dilution and thus raise the permeability. As the more uniform sands are invariably clay-free,

Table 1. *Mechanical analysis of silica sands*

Sieve aperture (μ)	S1	S2	S3	S4	S5
+1003	0·1	—	—	—	0·2
+ 699	35·7	—	1·0	—	0·5
+ 251	39·3	0·2	1·0	0·7	2·4
+ 152	13·2	1·6	25·2	7·1	17·8
+ 104	1·3	34·2	63·8	81·9	57·6
+ 76	0·2	52·7	7·6	10·2	21·3
+ 53	—	6·9	0·5	0·2	4·4
− 53	—	1·8	0·6	0·3	1·0
Clay	—	2·6	0·3	0·1	0·8
Permeability number	500	75	160	170	150

Table 2. *Mechanical analysis of natural sands*

Sieve aperture (μ)	N1	N2	N3	N4	N5
+1003	—	—	—	9·3	—
+ 699	—	—	—	7·2	0·2
+ 251	1·3	6·9	0·6	15·5	0·4
+ 152	5·5	21·8	0·6	15·0	4·4
+ 104	18·6	33·4	2·4	16·3	32·1
+ 76	36·5	19·2	11·8	6·4	24·5
+ 53	11·6	6·3	39·4	1·3	10·2
− 53	16·1	6·1	29·4	3·9	7·6
Clay	10·4	6·3	15·8	14·9	20·6
Permeability number	17	60	15	60	14

this practice also reduces clay content and can only be carried on to a limited extent.

The mechanical analyses and permeabilites of a number of silica sands are shown in Table 1, and of natural sands in Table 2.

It is necessary to know the mechanical analysis of the natural sand before deciding on which silica sand is likely to be most satisfactory. If the silica sand is so coarse that its addition increases by two the number of sieves on which an appreciable amount of sand is found, permeability will be reduced rather than increased.

For example, the permeability of sand N1 would be increased by an addition of S3, S4 or S5, and particularly by S4, but would be reduced by S1. The permeability of N3 would not be raised by the addition of any of the silica sands except S2, but the permeability of N5 which has the same permeability as N3 would be raised by an addition of any sand but S1.

When the grains are uniform, it is possible to derive the approximate permeability from the mechanical analysis, but when the grains are spread over a wide range this is impossible. For example, N2 and N4 have the same permeability, but completely different grain size distribution.

Fines and permeability in the foundry

The increase in permeability which results from the addition of a suitable uniform sand to one with a wide range of grain size is due to a reduction in the proportion of finer grades by dilution. A more effective method is to remove some of the fines themselves, since removal of half the fines has the same effect as adding sufficient uniform sand to double the amount of sand in use. This removal of fines is carried on sometimes deliberately, sometimes unavoidably in many foundries.

Any kind of equipment in which dry sand is handled, can be adapted for removal of fines, or 'desilting' as it is frequently called. Vigorous disturbance of dry sand raises these finer particles into the atmosphere, where they remain suspended for some time and can be withdrawn by means of a current of air. The type of cooler in which the sand falls *via* a series of baffles from the top of a tower, is frequently fitted with means for drawing air through the sand in the last two stages of its descent, and large amounts of fines can be removed in this way. At one time 'desilters' were fashionable. In its simplest form the desilter consisted of a box into which dry sand was brought on one belt to fall onto a second belt which carried it away. The dust which arose was withdrawn by a current of air. Too enthusiastic removal can be expensive and should be unnecessary. Coal dust has a lower specific gravity than silica and the largest particles of coal removed are larger than the largest particles of silica, and it may be necessary to renew the coal dust after each cycle. If an operation of this kind is needed to obtain the required permeability, the base sand is unsuitable and should be replaced.

In most mechanised foundries there is a considerable loss of fines, particularly at the shake-out and the sand preparation plant, the amount lost increasing as the proportion of sand dried out before the shake-out increases. There is usually a much smaller loss in a hand moulding shop, or any shop where the moulds are knocked out and the sand wet down *in situ*, since the dry sand is not treated in such a way as to cause the fine grains to become airborne.

The effect on permeability can be important and the difference between the mechanised and non-mechanised shops can be large and, at least quantitatively, unpredictable. In one foundry identical castings were made on a mechanised plant and by hand moulders, using the same natural sand and maintaining, as far as possible, the same amount of the same grade of coal dust in the sand. In both, the new sand had a permeability of 25–30, reduced to 18–20 by the addition of fine coal dust. At the end of 2 months, the permeability in the hand moulding shop had fallen to 9, but it was necessary to return some of the dust from the collectors in order to prevent permeability in the mechanised section rising above 30.

Fines are always produced when molten metal is poured into moulds containing coal dust, and in the floor moulding section conditions favoured the retention of fines. The method of working was to make a few moulds, pour, shake out and 'recondition' the sand by means of a shovel. The energy required for this last operation increases as the amount of dry sand increases, and on this account the moulders did not allow the castings to stay in the sand any longer than necessary. This, combined with the careful method of shaking out, kept airborne dust to a minimum and there was a continued build-up of fines.

In the mechanised section the average time that the castings remained in the mould was considerably longer, since the speed of the conveyor had to be adjusted to allow adequate cooling time for the thickest casting poured nearest to the shake-out. More sand dried out and there was more vigorous displacement at the shake-out and during subsequent handling, with the result that not only was coal and coke removed, but also the finer sand grains. Permeability rose until it was higher than that of the new sand in spite of the addition of superfine coal dust.

Cases of this kind in a single foundry are rare, but it must be recognised that the permeability of a prepared foundry moulding sand depends not only on that of the new sand and any additions made to it, but also on the method of operation. When it is necessary to maintain a close control over permeability, the use of a sand with a permeability higher than necessary, with a much finer sand as an adjustable control, is the most satisfactory method of operation.

SURFACE FINISH

Roughness

There is no universally accepted method of measuring the roughness or smoothness of an unmachined casting. Indeed, surface finish is usually decided by the reaction of the purchaser and no quantitative measurement which ignores appearance is likely to be helpful. The most useful method is by comparison with a chart on which are shown castings of gradually increasing roughness. Such a chart has been issued by the American Society for Metals[4] and this is quite widely used. There is no difficulty in producing a suitable series by making a number of castings in sands of gradually increasing grain size.

The relation between sand properties and surface finish has been studied by Nicolau[5] and by Gittus.[6] Gittus made castings, one face of which was in contact with the plane face of a standard test piece prepared from the sand under test. Surface roughness was compared by measuring the resistance of air flow between this face of the casting and a smooth plate. Both observers found that finish improved as permeability decreased.

In a series of tests in which other metals in addition to cast iron were considered, an AFS committee[7] found that surface finish improves as grain fineness number increases, *i.e.* as the sand becomes finer. In a second part of the investigation it was found that when the castings are shot-blasted, surface finish is very much influenced by the grade of shot used and a coarse shot can spoil the surface of a casting which in the as-cast condition is good.

There is a general agreement that, as the permeability of the sand decreases, surface finish improves, always provided that permeability is not so low that there are surface defects on the casting.

Effect of coal dust

If coal dust is present in the sand the appearance of the casting is improved, although it may not be possible to find increased smoothness by any method of measurement. This is discussed in Chapter 6. There is usually a slight difference in colour and the blue skin obtained when coal dust is present in the sand seems more acceptable to buyers of light castings, although this skin is removed by vigorous shot-blasting.

Definition and surface detail

It is possible to have a casting which has a smooth surface, but is not acceptable because of a failure to reproduce fine detail. There are two possible reasons for this failure, the most frequent being insufficient aeration of the sand, although occasionally the presence of a few large grains may give trouble.

When the sand is insufficiently aerated, large aggregates are present which behave not as single large grains, as they are not hard, but as balls of a material with the consistency of, say, plasticine. When the mould is rammed they deform to some extent but do not flatten out completely. This leaves a hole in the mould and a corresponding protuberance on the casting. Should the aggregate fall on a part of the pattern where there is any fine detail, such detail may be completely masked.

This is the result of faulty sand preparation and is discussed more fully in Chapter 7. However, in some sands, single large grains are present and these, too, can impair the appearance of castings. They are not so large as the aggregates mentioned above, but are completely rigid and may have a serious effect. On a flat or gently rounded surface, fine grains usually pack round such coarse grains and finish is not seriously affected, but where any fine detail is present, a single coarse grain can block off the entrance of metal and the detail cannot be reproduced. For example, if the casting carries a name or part number, a single large grain may block off the top loop of a letter A or a figure 4, spoiling the appearance of the casting — and the better the general surface, the more do imperfections of this kind stand out.

The presence of a few large grains does not increase permeability — it may even cause a reduction — and defects of this kind are outside the general surface finish/permeability relationship.

Factors independent of sand

Metal temperature and the method of pouring employed can influence surface finish and this aspect should be examined before making changes in the sand.

In general, surface roughness increases as metal temperature increases, but reproduction of detail deteriorates as metal temperature falls. These are foreseeable effects, but the effect of high temperature can also be reinforced by the method of pouring.

When a casting is filled from the top or side and there is no choke between the pouring basin and the entry to the casting, metal may enter the mould cavity at a high velocity and exert a corresponding velocity pressure in the mould cavity, which is added to the static pressure due to the head of metal. The effect of the combined pressure may be to cause swelling or metal penetration. This is discussed in Chapter 9.

EXPANSION ON HEATING

The expansion of a sand depends on its mineralogical composition and on the degree of compaction.

The main constituent is quartz. The expansion of a single crystal of quartz differs along its three principal axes, but in a sand, orientation of the grains is entirely random and an 'average' figure is obtained. The expansion of quartz was determined by Sosman[8] and is reproduced in Fig. 31.

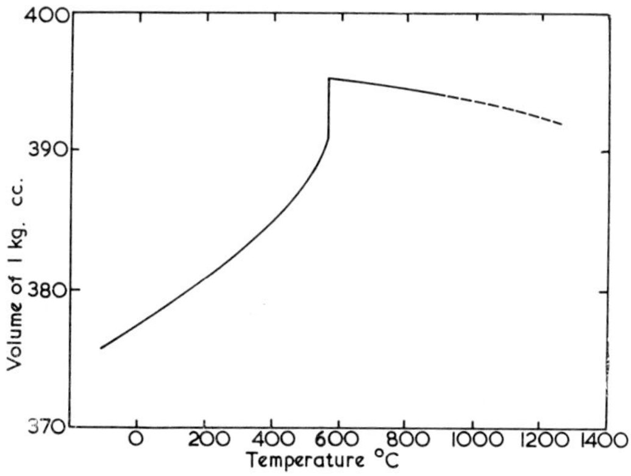

FIG. 31. Expansion of quartz (after Sosman).

Until the temperature reaches 573°C there is a regular, but high, expansion. At that temperature there is a change in the crystal structure from α to β quartz which is accompanied by an immediate expansion. β quartz itself contracts slightly as the temperature increases.

When the sand is not rammed hard, part of the expansion can be accommodated in the sand itself, but hard-rammed sand has an expansion exactly the same as the quartz rock used by Sosman. Clay contracts on heating, but in the proportion in which it is present in the moulding sand, has no effect on the expansion of hard-rammed sand.

Felspar has a lower coefficient of expansion than quartz, and a red Bunter sand, in which about 20% of the quartz grains are replaced by felspar, has a correspondingly lower expansion than a pure quartz sand. These factors are discussed in Chapter 9.

Size distribution and expansion

Provided that the grains of a sand consist only of quartz and the sand is rammed as hard as possible, the linear expansion is 1·3%. This is independent of the grading of the sand, but there is some evidence, and even more belief, that the size distribution of the sand has a pronounced influence on the incidence of defects due to expansion, particularly scabbing and finning, and that sands with very uniform grain size are to be avoided.

Perhaps this is most frequently expressed in America, where for many foundrymen the use of a 'four screen' sand, *i.e.* a sand in which the grains are spread over four consecutive sieves in their testing series, is an article of faith.

It has not always been so. Moldenke,[9] whose book published in 1917 is still worth reading, expressed the opinion that the 'ideal' moulding sand consisted entirely of spherical grains, all exactly the same size, bonded with the thinnest possible layer of the fattest clay available. This was a natural conclusion to draw at a time when no synthetic sands were in use, and the available natural sands had a low permeability which, as Moldenke realised, was due to the presence of fine grains with the coarser. When synthetic sands came into use and Moldenke's 'ideal' came nearer, certain difficulties arose and eventually the four-screen sand was adopted.

The four-screen sand should not be regarded as sacrosanct. It is a compromise between the uniform sand with its attendant ill-defined difficulties, and the widely spread grains with low permeability. Where a low permeability is needed to obtain a good surface finish, a sand spread on six or seven sieves may be found — even in America.[10]

The reason for the difference in behaviour is said to be that with a uniform sand, all the grains in the surface layer of the mould heat up at the same rate and reach 573°C, where they receive the shock of the very rapid expansion due to the crystal change together. With a wider spread of grain size, the small grains warm up more quickly than the larger ones and are contracting slightly when the larger grains reach 573°C. Although overall expansion is not changed, the shock at the change point is reduced. It is difficult to show the different type of expansion, but the effects have been established. However, very uniform clay-bonded sands are more brittle when dried than are sands with a wider range of grain size. This factor is of greater importance than any effect on the way in which the sand expands.

There is no reason why a three- or five-screen sand should not be as satisfactory as a four-screen sand — indeed, whether a sand is found on four- or five-screens is frequently dependent on the arbitrary selection of sieves. In the series used by the AFS every alternate sieve is taken from the full ASTM series. If the other half of the series is used, and this would comply with the International Standard for test sieves, some four-screen sands may become three- or five-screen.

STRENGTH

Grain size distribution

When a sand is milled with, say, 5% bentonite, the grains are coated with a layer of damp clay, the thickness of which depends on the total surface area of the grains. Coarse uniform grains have a thicker coating than fine uniform grains, and a somewhat higher strength.

If, however, the grains have a wide range of size, the strength is appreciably higher. If two sands, one with a diameter of 500 μ and one a diameter of 104 μ, are milled with the same amount of clay and tested, the coarser sand has a slightly higher strength. If, now, the two sands are milled together the mixed sand has an appreciably higher strength than either of the components.

The grain distribution desirable for any particular purpose is controlled by factors other than that of obtaining the highest strength from a given amount of clay, and there is nothing to be gained from an attempt to choose a sand which permits a minimum clay consumption. However, this effect of grain distribution does underline the need for always using the same sand when testing the bonding properties of a clay by using it to prepare a synthetic sand and testing that sand.

Grain surface

The grain surface has a slight effect on the strength of a sand bonded with a given amount of clay. This was first reported by Holmes,[11] who referred to the 'fixed' and 'mobile' bond in a moulding sand. 'Fixed bond' is not altogether an informative term. Red sand grains, coated with iron oxide, give a rather higher strength than clean silica grains of the same size and having the same clay content, and a similar effect can be obtained by precipitating hydrated iron oxide on the sand grains and heating strongly. However, this effect can also be obtained by a deep etch with hydrofluoric acid and it is likely that the deciding factor is a rough surface which affords a better 'grip' for the clay. The practical importance of this behaviour is small.

MOISTURE REQUIREMENT

The amount of water required to produce a mixture which can be used for moulding is controlled mainly by the amount and kind of clay, but the total surface area of the grains has a minor influence.

If, for example, a 500 μ sand and a 76 μ sand are milled with the same proportions of clay and water, the coarser sand with the smaller surface area will appear to the touch to have the higher moisture content.

The effect is not of great importance in the foundry, since it tends to lose itself among the greater effects of dead clay, but has a limited importance in the field of sand testing. If, for example, the strengths of a very coarse and a very fine sand with the same clay and water contents are measured, the coarse sand may have a slightly lower strength. If, however, the strength is measured over a range of moisture content it will be found that the coarser sand gives the higher strength and needs less water.

FLOWABILITY

Flowability is a term which has several definitions, and several methods of measuring it have been suggested. The test methods do not always give the same result or even place a series of sands in the same order. The *AFS Foundry Sand Handbook* simply gives a list of five references to work by different investigators and leaves it at that. In view of the present confusion, this is an excellent way of treating the subject in a publication dedicated to the describing of standard methods of testing.

However, some aspects of flowability are clear. If, for example, one of two sands gives a more uniform and denser mould than the other with the same ramming procedure, that sand may be said to have the higher flowability.

If this definition is accepted, then grain shape has an influence on flowability, since a very angular sand will in the above conditions produce a softer mould with wider hardness variations. In this case, the deciding factor is the internal friction which is higher with angular grains.

'SANDS' OTHER THAN QUARTZ

Quartz has a number of advantages as a base for a moulding sand, which have resulted in its almost universal use. Its melting point, the first essential, is 1710°C which is well above the pouring temperature of all the common metals, and large deposits of uniform quality and suitable grain size are widely distributed and easily won, so that it is relatively cheap. It is not wet by molten metal and penetration between the grains in the finished mould occurs only under considerable pressure, whilst it does not react readily with the molten metals at the temperatures at which they are poured. Heating quartz to the temperatures reached by the surface of moulds has little effect on it and sand can be re-used a number of times, the limit usually being set, not by failure of the quartz grains, but by the adverse effects of decomposed bonding material. In the steel foundry, grains near the metal–mould interface may crack, particularly if there has been some metal penetration, and a limited conversion of quartz to crystobalite or tridymite has been observed, but the amount rendered worthless in this way is small and has no influence on the consumption of material.

On the other hand, quartz has a high and irregular expansion on heating which may be a cause of casting defects unless appropriate precautions are taken. It reacts with and is wet by many metallic oxides and it was shown by Hoar[12] that steel readily penetrates a sand mould in the presence of even small traces of ferrous oxide; its reaction with manganous oxide may cause difficulty when casting manganese steel.

In consequence granular materials other than quartz are used to advantage under suitable conditions. Four such are used on a considerable scale:

Chamotte or 'Compo';
zircon sand;
olivine;
chromite.

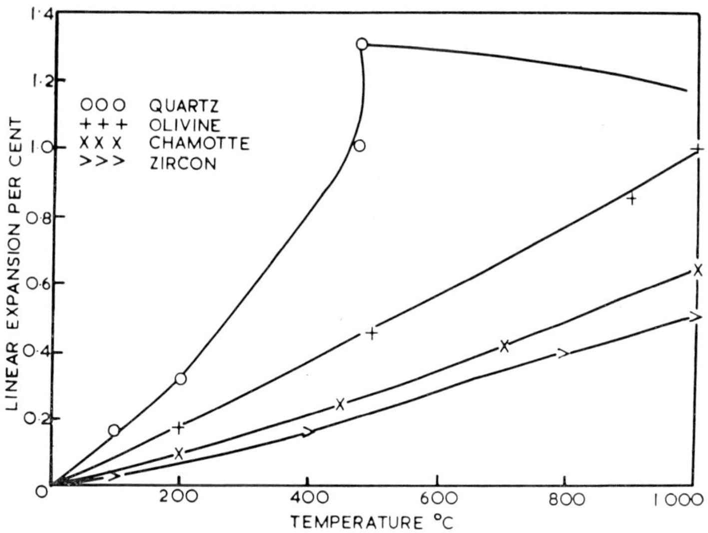

FIG. 32. Expansion of moulding materials.

In addition, attempts have been made to produce carbon in the form of hard granules of suitable size, but these have had only partial success.

All the alternative materials have one property in common — they have a lower expansion than quartz or at least are free from the rapid initial expansion up to 573°C and the further immediate expansion which accompanies the α–β quartz change. These expansions are shown in Fig. 32.

In addition, each of the materials differs in properties other than expansion, and these properties are useful under certain conditions.

Chamotte or 'Compo'

'Steel moulders' composition' or 'Compo' was formally a mixture of refractories based on ground crucibles, firebricks and any other suitable materials, with fireclay as a bond. The exact composition was frequently a secret, although if the many secrets which were formerly around in iron foundries can be used as a guide, the identical formula was treasured by many refractories makers and foundries.

Nowadays the 'Compo' consists of ground firebricks or burnt fireclay, with raw fireclay as a bond. It is used here and on the Continent, where it is known as Chamotte, for making moulds for large steel castings.

The great advantage of 'Compo' is that the various defects due to expansion are almost completely eliminated, and a corollary of this is that castings can be poured much more slowly than would be possible in a sand mould, with a consequent smaller risk of mould washing.

'Compo' has been successfully used in iron foundries for preparing moulds, and especially cores, for large castings, but the practice is not widespread.

Zircon sand

Zirconium silicate is found in very small quantities in many sands, but the amount is so small that it has no measurable influence on their properties. It is found in commercial quantities in several widely separated places — Australia, California, India — in the form of grains similar in size and distribution to a rather fine silica sand. Owing to the limited number of sources of supply and the distances over which it has to be carried, the cost is high.

The cost factor has limited the extent to which it is used, and it is confined to fields where its particular properties are of outstanding use or where maximum reclamation is possible.

In addition to its low expansion, zircon has the advantage that it is even more refractory than silica and does not so readily combine with iron oxide. It is sometimes used in steel foundries as a facing sand in areas where scabs or metal penetration are likely if the face of the mould is based on silica. In some cases it is possible to obtain with an uncoated zircon mould, a finish approaching that which can be obtained with a silica sand mould with a coating.

In iron foundries, zircon is little used except by a few foundries specialising in shell moulding. In this case, no sand other than zircon is used and the only binder is organic, so that a very large proportion of the used sand can be recovered by a simple burning operation. A small amount only of new sand is required for each cycle of operations and the discrepancy in cost between zircon and silica is greatly reduced.

The specific gravity of zircon is 4·5 and that of silica is 2·65. The cost per ton does not therefore give a true comparison between total cost of the two materials, since for a given mould the same bulk volume of material is needed and the weight of zircon involved approaches twice the weight of silica. A similar consideration must be taken into account when arriving at the cost of bonding material. A zircon shell with 2% resin is as strong as a silica shell with $3\frac{1}{2}$%, but as the zircon shell is correspondingly heavier, the weight of resin needed for each shell is the same.

Olivine

Olivine is a magnesium silicate in which varying proportions of magnesia are replaced by ferrous oxide. The refractoriness of the material falls off as the proportion of iron oxide increases and not all deposits of olivine are sufficiently refractory for foundry use. There are large deposits in Norway of suitable composition, which are used on a considerable scale in steel foundries in Scandinavia. During the last few years attempts have been made to popularise in America the use of olivine from the State of Washington.

The material does not occur as grains of the size of silica sand grains, but is found as a massive rock which has to be crushed before use. In consequence all grains are sharply angular. The specific gravity is 3·3–3·4 and it is therefore about 25% heavier than silica sand.

One of the arguments in favour of the use of olivine in foundries was that the dust, if inhaled, could not cause silicosis. The incidence of this disease in iron foundries is so low that the material really has to stand on its technical merits.

The expansion of olivine with increase in temperature from 0°C to 1000°C does not differ greatly from that of silica over the same temperature range. This expansion takes place at an almost uniform rate over the whole temperature range, whereas that of silica is complete by the time the temperature reaches 573°C. Before the expansion of olivine approaches that of silica, the temperature has reached the point at which the clay begins to soften and the hot deformation of the sand begins to increase and continues to exceed its expansion. Moulds in which silica sand is replaced by olivine are completely free from expansion scabs.

Olivine is widely used for moulds for casting high manganese steel and is said to give a better strip from these than can be obtained with any other mould material. It is not, however, widely used in this country.

Chromite sand

This has been used to a relatively small extent in steel foundries, but has not yet been adopted on a considerable scale in this country. It is used with success in South Africa and Switzerland. It has the low expansion and low reactivity desirable in the alternative to silica and is suitable for use with most kinds of binder.

Carbon sand

Provided that it can be isolated from oxygen when its temperature is high, carbon is perhaps the most satisfactory of all refractory materials. It has a low expansion, and low reactivity, and is not wet by molten metal. In the form of machined graphite block moulds, it is used with excellent results as moulds for steel wheels in America,[13] a process on which several patents were taken out.

In 1956 a process[14] for preparing hard granules of carbon similar in size to sand was developed. The behaviour of a green clay-bonded sand and of a hot box resin bonded sand[15] in which silica was replaced by carbon sand were examined by Nicholas and Roberts.[15] Similar comparisons with silicate, dry clay and air-setting bonds have been described by Elliott.[16]

Defects due to expansion — scabbing on green sand moulds and finning in cores with organic binders — were eliminated or greatly reduced. The conductivity of carbon is appreciably higher than that of silica and the tendency to chill was increased, but the most serious difficulties found were those due to the presence of more than 5% sulphur in the sand.

If the material is to be used on a large scale, both the cost and the sulphur content must be reduced.

REFERENCES

1. Boswell, P. G. H. *British Resources of Refractory Sands*, p. 13.
2. White, H. F. and Walton, S. P. (1937). *Amer. Ceramic Soc.*, May.
3. Sheehan, J. J. (1938, 1939). *Proc. Inst. British Foundrymen*, **32,** pp. 43–46.
4. (1955). *AMERICAN SOCIETY FOR METALS. Metal Progress*, **68,** p. 83.
5. Nicolau, P. (1938). *Bulletin de la Société d'Encouragement pour l'Industrie Nationale*, **137,** Nov.–Dec., pp. 423–431.
6. Gittus, J. (1933–1955). *BCIRA J. Res. Devel.*, **5,** pp. 376–389.
7. (1949). *AFS SAND COMMITTEE* 8*H. Amer. Foundryman*, Sept., p. 51.
8. Sosman, R. B. *The Properties of Silica.* Chemical Catalog Co., New York.
9. Moldenke, R. (1917). *Principles of Ironfounding.* McGraw-Hill.
10. Silk, T. T. (1949). *Amer. Foundryman*, Sept., p. 51.
11. Holmes, C. W. H. (1922). *J. Iron Steel Inst.*, **116,** pp. 61–87; (1923). See also *Foundry Trade J.*, **28,** pp. 296–301.
12. Hoar, T. P. (1950). *Iron Steel Inst. J.*, **166,** pp. 1–17.
13. British patent no. 721 059.
14. British patent no. 798 249.
15. Nicholas, K. E. L. and Roberts, W. R. (1962). *BCIRA J.*, **10,** Sept., pp. 557–567.
16. Elliott, J (1961). *Modern Castings*, **40,** Aug., pp. 31–33.

CHAPTER 4

Clays

THE FUNCTION OF CLAY IN A MOULDING SAND

The function of clay is to hold together the compacted sand grains from the time ramming is complete until after the casting has solidified. The properties of the compacted sand must be such that it retains its shape (1) while the pattern or mould is rapped or vibrated, (2) when the support of the pattern is withdrawn, (3) when the stream of metal is entering the mould, and (4) when the mould is full of liquid metal. In addition, since one of the advantages of clay-bonded sand is that it can be re-used with the addition of only a small proportion of new sand, it must be possible to break down the mould after casting without too much difficulty, so that the sand can pass round the return sand system and be remilled with water, to produce a uniform material free from aggregates of dry sand.

The properties concerned are not independent and the requirements are sometimes directly opposed to each other. This limits freedom of choice of sand properties. For example, it becomes easier to strip a mould from a pattern as deformation increases, and at the same time resistance to collapse under its own weight or the stress imposed by the molten metal increases with increasing green strength. However, as toughness, which is proportional to the product of green strength and deformation increases, it becomes more difficult to ram the sand and hardness differences between different parts of a mould increase. Increasing green strength and toughness must therefore eventually lead to difficulty in ramming and to loss of accuracy and soundness in the castings.

Properties dependent on the clay are:

green strength;
toughness;
deformation;
adhesion;
dry strength;
'wet' strength.

Green strength

As a rule, the green strength of a sand is lower than is needed to keep the mould in shape and undamaged, unless the mould is small. The transverse load in the centre of the mould in a large box is high, and that due to the upward pressure of metal may be still higher, and it is usually necessary to reinforce the sand in some way.

The method used with moulds of moderate size is to place bars in the box, but in the case of large moulds, particularly when these are hand-rammed using an unmilled sand, additional aids such as pothooks, lifters, gaggers and rods may be needed, just to prevent collapse of the mould when the support of the pattern is withdrawn. Indeed, an appreciable part of the skill of a hand-moulder making large moulds lies in his ability to produce a mould which can be handled without resorting to an excessive use of such reinforcement.

By the use of a sand with a high green strength, allied to hard ramming, it is possible to produce moulds of moderate size which can be handled without collapse even when there are no bars in the box. In a mechanised plant, this has the advantage that after pouring, mould and casting can be pushed out of the box, a procedure which eliminates noise at the shake-out, but it is necessary to place weights on the top of the mould to prevent rupture due to the upward pressure of metal.

The range of green strength in prepared sands is from 3 to 4 lb in^{-2} in the case of unmilled sands and backing sands, to occasionally more than 20 lb in^{-2} in sands for high pressure moulding.

Toughness

The resistance of a mould to failure when being handled and during and immediately after pouring therefore increases as green strength increases. It was shown in Chapter 1 that separation of mould from pattern becomes easier as deformation increases. Production of a mould should therefore become easier as toughness, which is proportional to the product of green strength and deformation, increases.

However, the energy required to compact sand and hardness differences within a mould increase as toughness increases, and the maximum toughness permissible depends on the method of moulding employed. A hand-moulder producing large moulds frequently uses an unmilled sand with a shatter index of 30–40, or a thin layer of a tougher facing sand with a backing of weak sand, in order to reduce the effort required in ramming, although this practice increases the need for reinforcing the mould and backing it up with weights. With conventional squeeze and jolt–squeeze machines, a shatter index of 60–80, depending on the contour of the pattern, is usual. With the more effective methods of compaction — simultaneous jolt–squeeze and high pressure squeeze, a shatter index approaching 100 is frequent.

Deformation

Although this property is rarely measured, it must be maintained at a minimum, and this depends on the shape of the pattern and on its surface finish. Clearly, it is easier to obtain a good strip from a pattern for a thin plate than from one with a deep vertical side with little taper, but nevertheless there are indications that variations in the deformation of prepared sands are less than those in green strength and toughness. Invariably, when sands with a high green strength are used, the shatter index is high, whereas when green strength is low, a sand with a low shatter index can be, and is, used.

Dry strength

When moulds are dried before being used, a certain minimum dry strength is needed to prevent damage during handling, and particularly during coring up. Even when moulds are used in the green state, the first entry of metal dries out the surface of the mould near the ingate and the strength of the dry portion must be high enough to resist erosion by the metal stream.

An unnecessarily high dry strength may be a nuisance, since it leads to difficulty in breaking down the mould at the shake-out and at a later stage much good sand may be lost down the tailing chute at the screen. However, deliberate control of dry strength is rare. Usually the amount and kind of clay used depend on a number of other factors and when these have been decided the dry strength just happens. Occasionally, recommendations as to a minimum dry strength have been made and 50 lb in^{-2} seems to have been a nice round figure to choose, but sands with appreciably lower strength are being used successfully. When very large dry sand or loam moulds are made, dry strength may be several hundreds of pounds per square inch.

Sand adhesion — the use of parting powders and liquids

One property of sands which has not been measured, and has so far proved incapable of measurement, is sand adhesion.

When a rammed sand is broken, failure invariably takes place in the clay layer and not at the junction between clay and sand grains. The forces holding the clay on to the sand grains exceed those joining one clay particle to another.

Indeed, the forces which hold clay particles together are in general less than those which exist between clay and any other kind of solid, even one as smooth as glass. Piper[1] showed that there is a relation between the green strength of a clay-bonded sand and the tensile strength of the moist clay. Moreover, he obtained the tensile strength by using the clay to join together two glass strips and measuring the force needed to separate them, when failure invariably occurred in the clay layer and not at the clay–glass interface. The same result is obtained if the glass is replaced by metal.

It should follow that when a clay-bonded mould is separated from a pattern, failure also occurs in the clay layer, either in the clay adjacent to the pattern itself or elsewhere in the mould. In the one case, a layer of clay would be left on the pattern, and in the other, part of the mould would be left behind. For some reason, this does not always happen and, indeed, the evidence is somewhat conflicting when the thin film present on sand grains is considered in place of the much thicker layer used by Piper. On the other hand, when a rammed sand is broken, failure invariably occurs in the clay layer, which implies that the only relevant factor is the strength of the clay, and on the other hand the 'static bond' described by Holmes (see Chapter 3) indicates that the surface of the grains has at least a contributory effect.

So far as the mould-maker is concerned, the position is clear — a clean separation between mould and pattern is by no means certain unless appropriate steps are taken. Originally a layer of 'parting powder' was dusted on to the pattern. This consisted of a finely ground powder made waterproof by a coating of wax. The preferred powder was silica flour, but its use was forbidden by the 'Parting Powder Regulations' of 1950 and ground limestone replaced it everywhere. For some time there has been a turn to parting liquids which consist of a paraffin in which a small amount, usually 2–3%, of stearic or oleic acid is dissolved. The chief advantage of using the liquid parting is that less frequent applications to the pattern are needed.

Wet strength

The term 'wet strength' was first used by Patterson and Boenisch.[2] When a mould face is heated, moisture evaporates and some of it condenses in the sand behind the mould face, forming a weak wet layer behind a thin stronger dried layer. The wet strength is the tensile strength of the wet layer and Boenisch devised a test for measuring the property.

To measure wet strength, a disc heated to 300°C is placed against one plane face of a cylindrical test piece and allowed to stand for a short time to allow evaporation and condensation to take place. A tensile stress is applied and failure occurs in the wet layer.

Boenisch showed that there is a relation between wet strength and expansion scabs and that the incidence of the defect is lowest when wet strength is highest. This is discussed more fully in Chapter 9.

THE EFFECT OF WATER ON CLAY AND CLAY-BONDED SAND

A dry, powdered clay has no adhesive properties and a mixture of sand and clay is, from the point of view of the mould-maker, just a mixture of two inert powders differing in grain size. It is not possible to prepare even the standard test piece from such a mixture as it has no green strength.

If water is added in small increments, eventually it becomes just possible to form a test piece, which has a low green strength. If, now, the moisture is raised gradually, green strength rises rapidly and then falls almost, but not quite, as quickly. With further increase in the amount of water, green strength falls, but more slowly. This can be seen in any of the curves in Fig. 33.

At low moisture contents, the test piece is extremely brittle and it is difficult even to strip the test piece from the box undamaged. A good strip can be obtained by using a split container, but even so there is usually a wide spread of test results and on this account it is not possible to compare the green strength of sands by comparing strength at the peaks.

The shatter index of the sand is low, even when the moisture content is sufficient to give maximum green strength, but continues to increase as moisture content increases and green strength falls. This combination of increasing toughness and falling green strength can only result from a rapid increase in deformation. As a result, the sand becomes less brittle and mould-making becomes easier as moisture content increases.

Eventually, with continuing increase in moisture content, shatter index begins to fall, but only slowly at first. At the same time there is also a steady fall in green strength and deformation changes only slowly. However, separation of mould from pattern becomes more difficult, not because of increased brittleness, but because of a decrease in strength and an increase in adhesion between clay and pattern. Even a liberal application of parting powder or liquid is unable to prevent part of the sand sticking to the pattern.

The range of moisture content over which a sand is neither too brittle to strip without cracking, nor too wet to separate cleanly, is generally known as the moisture range. It is not possible to measure this property precisely since both end-points are ill-defined, but wide differences can be seen readily if the full green strength/moisture curves are examined.

Dry strength increases with increasing moisture content up to an indefinite maximum and then remains fairly constant. The curve relating dry strength and moisture content is almost a straight line over the range of moisture used in the foundry, except where loam moulders are concerned.

The optimum moisture content

It would be very convenient to be able to define the moulding properties of different sands by means of their green strength and shatter index at a single moisture content, instead of using graphs similar to those shown in Figs. 33 and 34. It is not possible to do this by using a fixed moisture content for all sands, since the moisture required if a strongly bonded red sand is to be suitable for mould-making would be far too high for a weakly bonded synthetic sand. Nor is it possible to use a fixed clay/water ratio since different clays react very differently to changes in water content.

Some observers quote sand properties at 'optimum moisture content', although there is frequently no indication as to how this moisture content has been decided upon. Others regard 'optimum' moisture content, or, in the nomenclature adopted in much American literature, moisture at 'temper', as that at which the bulk density after ramming is at a minimum. This corresponds roughly with the moisture content which gives maximum shatter index. However, the curves relating moisture content with either bulk density or shatter index are very flat near the maxima and there is considerable uncertainty as to the optimum moisture content.

Other observers appear to assume that 'optimum' moisture content is that most suitable if the sand is to be used for green sand moulding and some such assumption is necessary if for any reason it is necessary to evaluate a sand by tests at a single moisture content. In such cases the sand has to be judged by 'feel', which is itself an uncertain method even for a single observer; in addition a high pressure moulder, a maker of flat castings on a plain squeeze machine and a loose pattern moulder may have appreciably different ideas as to the appropriate moisture for a given sand.

Whatever the basis for deciding the 'optimum' level of moisture content, there is therefore considerable uncertainty as to the appropriate figure for any particular sand. Moreover, green strength varies fairly rapidly as moisture content at this level changes, and strength at the 'optimum' is not a very exact method for comparing the bonding properties of sands and clays. Permeability, however, changes only slowly as moisture changes and comparisons are valid.

TYPES OF CLAY AND THEIR PROPERTIES

The term 'clay' is applied to a group if minerals which have been formed as a result of the slow decay of other silica and, usually, alumina bearing minerals. The composition of the clay depends on that of the parent mineral and on the conditions under which it has decomposed and can vary widely. There may also be wide differences between the physical properties of clays of similar composition and derived from similar parent minerals, and on this account the chemical analysis of a clay gives little information on its value except as a refractory, the melting point increasing as the proportion of silica decreases. Many clays contain free silica in an extremely fine state of division and this may appear in a chemical analysis as a constituent of the clay particle itself instead of as a diluent.

One feature common to all clays is their extremely fine state of division.

Individual particles are too small to be seen by means of an optical microscope and Hast,[3] in an attempt to view the structure of clay by means of an electron microscope, used a magnification of 100 000.

Modern ideas as to the structure of clays, however, really stem from the work of Pauling[4] who used X-ray methods to determine the arrangement of atoms in the clay structure. He was followed by other investigators, and Grim[5,6,7] was particularly concerned with the behaviour of clays as bonding materials in foundry sand.

As a result of these investigations it was shown that:

1. Clays consist of small plate-like particles with a thickness much less than length and breadth.
2. The plates are built up of alternate sheets of silica and alumina with combined water. In some clays there are equal numbers of silica and alumina sheets and in others two silica sheets to each alumina sheet.
3. There may be substitution of one atom by another, for example replacement of aluminium by magnesium on a small or on a large scale. On a small scale, such substitution has an important effect on the properties of the clay only if it produces a change in the clay structure.

Clays have been divided into a number of groups according to their structure and composition. There is a considerable number of such groups, but as many clay deposits are mixtures, there is on occasion some doubt whether a particular clay belongs to a distinct group or is a mixture of others. There are usually some differences in chemical composition between clays in the same group obtained from different deposits.

Only a few of these groups or types of clay mineral are of importance in the foundry, because they are either used in the preparation of synthetic sands or are present in naturally bonded sand. The important clays are:

montmorillonites;
kaolinites;
illites.

Montmorillonites

In the montmorillonites there are two silica sheets to each alumina sheet, the number of silicon atoms being approximately twice the number of aluminium atoms and the basic composition is $Al_2O_3.4SiO_2.H_2O$.

In all clays in this group, a small proportion of the aluminium is replaced by magnesium. If the change were such that 2MgO replaced Al_2O_3 there would not be any great change in clay properties, but when two atoms of magnesium, requiring two atoms of oxygen, replace two of aluminium with three atoms of oxygen, there is an oxygen atom in excess and a charge on the particle which can be satisfied by any cation, for example sodium, calcium, magnesium. The kind of cation has an important influence on the properties of the clay and in defining a montmorillonite the name of the cation is added as a prefix, *e.g.* sodium, calcium or magnesium montmorillonite.

Base exchange

It is relatively easy to replace one cation by another and the cations are described as 'exchangeable bases'. If a sodium salt in solution is added to a calcium montmorillonite, part of the calcium is replaced by sodium; if the calcium salt formed by this reaction is insoluble, for example when sodium carbonate is added to a calcium montmorillonite with the formation of insoluble calcium carbonate, replacement of calcium by sodium continues until all the calcium has been replaced or all the sodium used up. Some of the sodium montmorillonite used in the foundry is prepared in this way from a calcium montmorillonite. It is not possible to determine the quantity of exchangeable bases by means of a straightforward chemical analysis, since the same element may be present in the main structure and as the exchangeable base.

Replacement of calcium by sodium as the exchangeable base is described by some observers as 'activation' and a clay in which all the calcium has been so replaced is described as 'fully activated'.

Base exchange capacity

The extent to which exchangeable bases are present is known as the 'base exchange capacity' and is expressed as meq, *i.e.* thousands of the equivalent weight of the element concerned per 100 g clay.

The base exchange capacity of montmorillonites from different sources varies somewhat, but is usually around 100. A substantially lower figure indicates that a clay other than montmorillonite or perhaps some non-clay material is present.

The influence of the exchangeable base

The exchangeable base has some influence on all the properties of a clay, but is not the sole factor, and replacing the bases of a number of, for example, calcium montmorillonites by sodium would not produce a group of identical sodium montmorillonites. Wyoming bentonite, a naturally occurring montmorillonite, mostly sodium, has a very high dry strength, but this is not true of all sodium montmorillonites, and many of the sodium montmorillonites produced by base exchange have much lower dry strengths. On the other hand, some properties, notably the reaction of the clay to changes in moisture content, do appear to depend on the nature of the base to a great extent.

Swelling clays

Water can enter between the plates of some sodium, but not calcium, montmorillonites causing the clay to swell. There does not seem to be any close relation between swelling index and bonding properties, but swelling is of importance in the preparation of slurries — in the foundry as a carrier for blacking and other coatings, and in the oil industry as a base for drilling muds.

Swelling index was at one time used as an index of the purity of Wyoming bentonite, at that time the only montmorillonite used to any extent in foundries. The test was made by shaking 100 cm^2 water with 4 g of the clay and 0·2 g light magnesium oxide (this hastened the change). After prolonged and vigorous shaking, a thixotropic gel which, after a few minutes at rest, solidified so that the vessel could be inverted without displacing the gel. The presence of opaque water on the surface of the gel was a quantitative indication of any diluting clay present. Various modifications of the test, such as adding the clay in small increments with continuous shaking, have been proposed.

Sources of montmorillonite

This clay, which is a decomposition product of a volcanic glass, is widespread and deposits occur in most countries. In general, naturally occurring sodium montmorillonites are less widely distributed than those with calcium as the exchangeable base.

The first deposit from which clay was extracted for foundry use was that at Wyoming in USA. This is still the largest supply in the world. The exchangeable base is mainly sodium with a little calcium. There are also deposits of sodium montmorillonite in North Africa and samples have been obtained from New Zealand.

Calcium montmorillonites are of wider occurrence. In America 'southern' bentonite, a term used to distinguish the clay from the 'western' or Wyoming bentonite, has calcium as the exchangeable base. There are deposits in France, Germany, Greece and Italy, many of which are treated with soda for sale to foundries. In Britain the Fuller's earths, which were used at least as early as the Roman occupation for cleaning and decolouring, were first used in the foundry at about 1940. Since the deposits are frequently associated with chalk, the exchangeable base is wholly calcium and most foundries use the clay after soda treatment.

Kaolinites

In the kaolinites there are equal numbers of silica and alumina sheets and equal numbers of silicon and aluminium atoms. The basic composition is $Al_2O_3.2SiO_2.2H_2O$. Owing to the higher ratio Al_2O_3/SiO_2, the melting and softening points are much higher than those of the montmorillonites, and the clays are much used for the manufacture of refractories. Indeed, in America they are usually known as fireclays, a term which has a more restricted meaning in Britain.

In the kaolinites, exchangeable bases are not present to a significant extent, but differences between individual clays, particularly in their effectiveness for bonding sand grains, are even greater than those between montmorillonites. China clay, the purest of kaolinites, has such poor bonding properties that it has never been seriously considered as a possible binder in synthetic sands, but some of the ball clays, which are only a little less pure, are quite effective — 7% gives a green strength similar to that obtained with 5% montmorillonite.

The use of kaolinites is tending to decrease. At one time they were used on a considerable scale, mainly because Wyoming bentonite, the only montmorillonite then available, had such a high dry strength that there were difficulties in the sand-handling system. However, the clay in many naturally bonded sands contains kaolinite, since it is a decomposition product of felspathic material present in the granite from which most of our native sands were derived.

Illites

The illites — the name was coined by Grim[7] who examined a number of deposits in the state of Illinois — are alternatively described as 'secondary mica' clays, since they are decomposition products of mica. Like the montmorillonites, they have two silica sheets for each alumina sheet, but some of the silicon atoms are replaced by aluminium and the resulting charge balanced by potassium which is not exchangeable. Some aluminium is replaced by magnesium, but on a smaller scale than in the montmorillonites and base exchange capacity of the illites lies between the low value of the kaolinites and the high one of the montmorillonites.

Illites are not used in synthetic sands in this country and there is little information available on their bonding properties. Some illite is present in the clay in most natural sands.

Clay in naturally-bonded sands

Natural sands have been formed under widely different conditions and it is therefore unlikely that all contain clay of the same composition. When the clay has been formed by the decomposition of the felspars and micas present in granite, it seems likely that kaolinite and illite will be present, but no mineralogical examinations appear to have been made.

There have been chemical analysis of the clay in some sands, particularly the red Bunter sands. Skerl[8] analysed the fraction finer than 10 μ from a number of natural sands and found considerable differences in composition.

The following is Skerl's analysis of this fraction from one Bunter sand:

Compound	% Composition
SiO_2	53·76
Al_2O_3	25·39
Fe_2O_3	6·94
TiO_2	0·62
CaO	0·34
MgO	2·06
Alkalies	3·17
Loss on ignition	7·77

The silica and alumina correspond to $Al_2O_3.3{\cdot}6SiO_2$, but as some free silica is also present, the actual proportion present in the clay mineral is less than that represented in this formula. Such a composition is characteristic of a mixture of kaolinite and illite. The combined water (7·7%) indicates that illite (combined water 5%) is present in greater proportion than kaolinite (combined water 14%).

The clay from Erith loam, an estuarine sand, had higher silica and lower alumina contents than even illite and montmorillonite and on the whole, the evidence as to the constitution of the clay in the natural sands is far from clear. Moreover, the reaction of these clays to changes in moisture content is very different from that of any other clays.

CLAYS AND THE PROPERTIES OF SYNTHETIC SANDS

When comparing the properties of synthetic sands containing different kinds of clay, it is desirable to adjust the proportion of each clay used in such a way that the green strength and toughness are similar. If 5% of one clay produces a sand with a green strength of 10 lb in^{-2}, and 5% of a second a green strength of 5 lb in^{-2}, the dry strength of the two sands has no relevance in the foundry. In order to produce satisfactory moulds a certain green strength is needed and the amount of clay has to be adjusted so as to give that strength, and it is the dry strength of that sand which is important. Green strength tests on two sand/clay mixtures indicate the relative quantities of the two clays which the foundry will need to add.

The properties to be measured are green strength, toughness and dry strength; from the changes in green strength with change in moisture content, the moisture ranges can be compared, and these may give an indication of the wet strength. There is some evidence, unfortunately contradictory, that casting soundness is influenced by the type of clay used.

It is also desirable to compare the rate of breakdown of the clays at high temperatures, since this affects both the amount of clay which must be added to maintain a constant proportion, and the amount of dead clay produced, which in turn influences the amount of used sand which must be discarded.

To compare the bonding properties of different types of clay, Erith silica sand was milled with three types of clay in turn — Wyoming bentonite which is mainly sodium montmorillonite, Cyprus bentonite which is a calcium montmorillonite, and a ball clay which is a kaolinite. Tests were carried out over a range of moisture content in order to show the moisture range. Green strength, shatter index and dry strength were measured.

Green strength

The green strengths are shown in Fig. 33. Owing to the weaker bonding properties of the ball clay, 7% was added, compared with 5% of the bentonites.

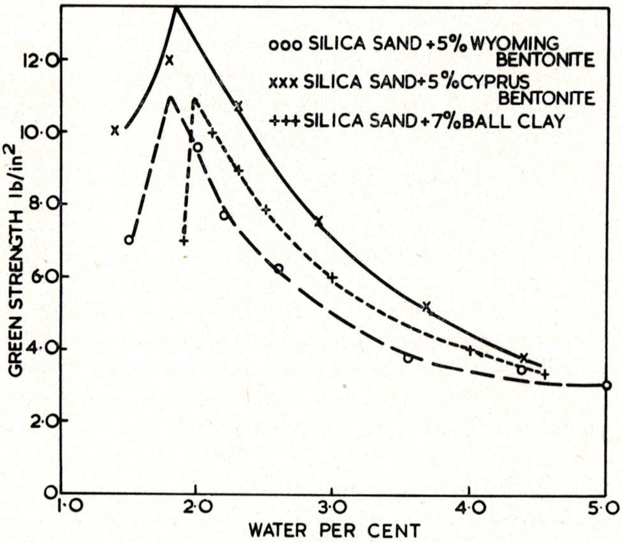

FIG. 33. Green strength of synthetic sands.

The calcium montmorillonite gives the highest green strength with the sodium montmorillonite only a little lower. This relationship holds for some but not all montmorillonites and treating a Fuller's earth with sodium carbonate, to replace calcium by sodium, has only a slight effect on strength.

The Dorset ball clay, although the most effective of a large number of kaolinites which has been examined, gives appreciably lower green strength and it was necessary to raise the amount used to 7% to obtain the strength given by 5% montmorillonite.

Shatter index

The shatter indices of the three sands are shown in Fig. 34.

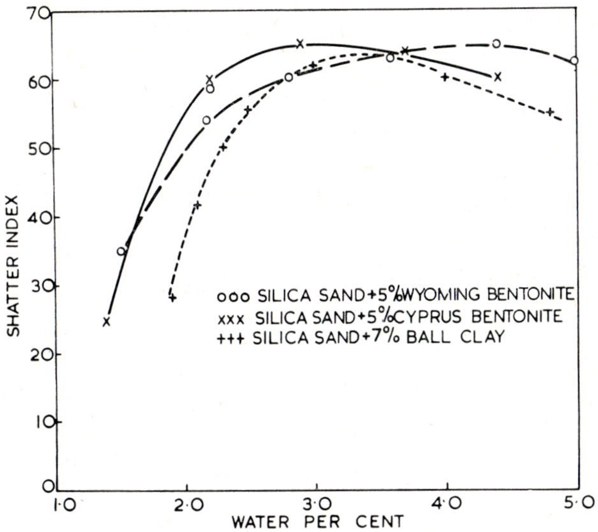

FIG. 34. Shatter index of synthetic sands.

The three sands have similar shatter indices. However, since the green strength of the sodium montmorillonite is lower than that of the calcium montmorillonite, its deformation must be higher and, in marginal cases, its stripping properties better. The difference is less than that which would follow a change of 0·2% in water content.

Dry strength

The dry strengths of the three sands are shown in Fig. 35.

The differences in dry strength are pronounced. However, the exchangeable cation is not the sole controlling factor and the very high dry strength of Wyoming bentonite is partly due to some other cause which is not known. In general, replacing calcium by sodium either raises the dry strength of a montmorillonite slightly or leaves it unchanged. A number of attempts to obtain dry strength equal to that given by Wyoming bentonite by treating calcium montmorillonite with sodium carbonate have failed. On the other hand, some natural sodium montmorillonites, notably a small sample obtained from New Zealand, resemble Wyoming bentonite in all

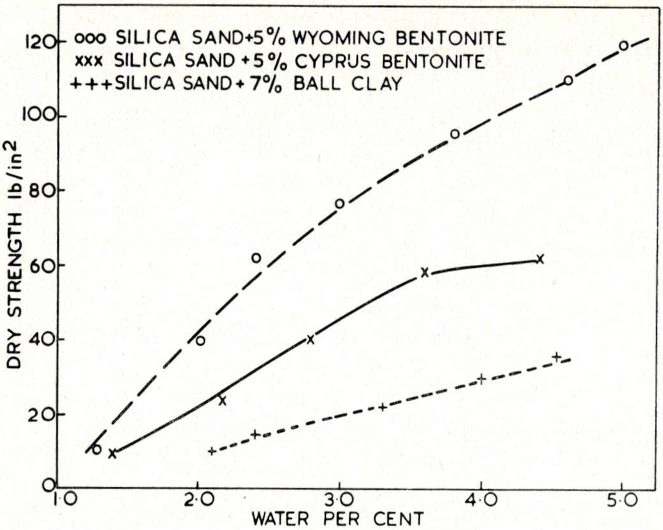

FIG. 35. Dry strength of synthetic sands.

their bonding properties. All sands bonded with kaolinites have low dry strength.

Dry strength of mixture of clays

It was pointed out by Dunbeck[9] that when a mixture of clays of different kinds is used, green strength is as calculated, but dry strength may be higher.

In his experiments Dunbeck prepared two sands of equal green strength, one bonded with Wyoming bentonite and the second with Ohio fireclay, a kaolinite. It was necessary to use 8% of the fireclay to obtain a green strength equal to that obtained with 5% bentonite. The dry strength of the sand bonded with the fireclay was much the lower of the two.

He then prepared a series of sands, gradually replacing the bentonite by the fireclay, using in all cases 1·6% fireclay in place of 1·0% bentonite. In the absence of any abnormal behaviour, it could be calculated that as the proportion of fireclay increased, green strength should remain constant but dry strength should fall rapidly. The expected green strengths were obtained, but dry strength of the mixtures was invariably higher than that calculated.

Similar results are obtained when the Ohio fireclay is replaced by other kaolinites. In a series of similar tests using Wyoming bentonite and Kernick clay, a plastic kaolinite from Cornwall, it was found that dry strength remained constant until 80% of the bond was supplied by the Kernick clay.

The green and dry strengths of sands containing the two clays and also a mixture of the two were then determined over a range of moisture content. The three sands were of the following composition:

Erith silica sand + 5% Wyoming bentonite;
Erith silica sand + 8% Kernick clay;
Erith silica sand + 1% Wyoming bentonite + 6% Kernick clay.

Green strength is shown in Fig. 36 and dry strength in Fig. 37.

The green strength curves are what would be expected, but the addition of a small proportion of Wyoming bentonite to a low dry strength kaolinite raises its dry strength to the level of that of bentonite.

It has not been established whether all sodium montmorillonites, including those prepared by cation exchange, behave in the same way,

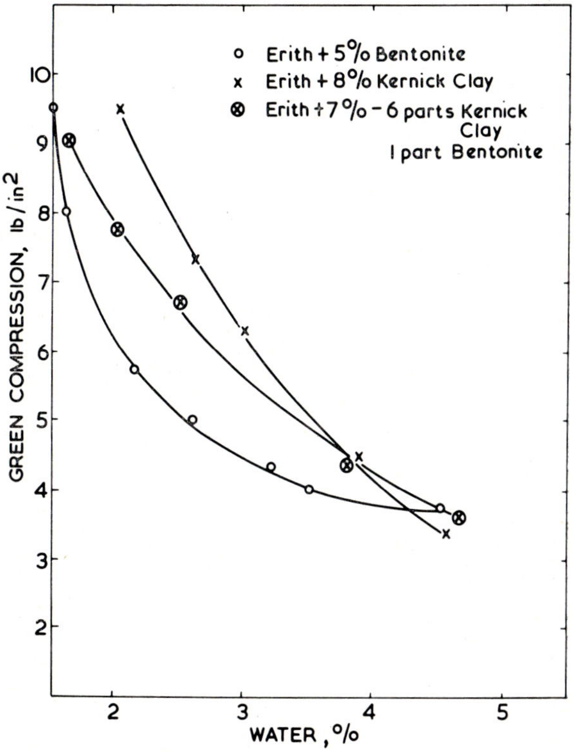

FIG. 36. Green strength of mixed clays.

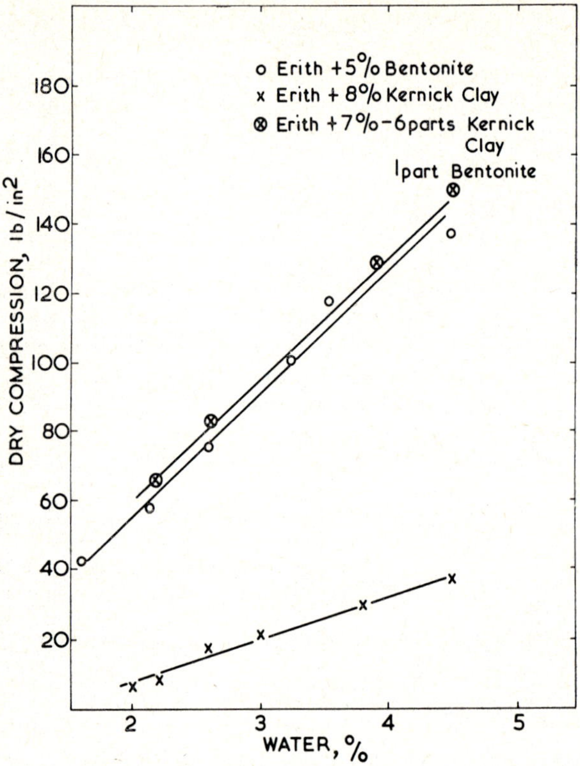

FIG. 37. Dry strength of mixed clays.

since for reasons still unknown, Wyoming bentonite has a dry strength which is unusually high, even among sodium montmorillonites. However, trial should be made before any attempt to control dry strength by mixing clays is made.

The moisture range of sands

The curves in Figs. 33 and 36 show the rate of change in green strength with change in moisture content and it is clear that the sand bonded with the sodium montmorillonite has a gentler slope. It might be expected that 7% of kaolinite would require more water and be less sensitive to changes in water content than a sand containing 5% of any other clay, but this is not true and the sand containing sodium montmorillonite has the wider moisture range.

The difference is most marked at the highest moisture content. In this region increasing water content has a very small effect on the green strength of the sand containing sodium montmorillonite, but reduces the green strength of other sands appreciably. Other sodium montmorillonites behave in the same way as the Wyoming bentonite used in the above tests.

Hofmann[10] pointed out that the very slow change in green strength when moisture content increases to a high level, is responsible for the higher wet strength of the sodium montmorillonites and the greater resistance to scabbing when these clays are used in a moulding sand.

Clay and soundness of castings

The kind of clay used in a mould has an influence on the soundness of the castings produced, but the relation is not a simple one.

Nicholas[11] produced spherical castings 3 in diameter in moulds bonded with (a) a sodium montmorillonite prepared by base exchange, (b) a Wyoming bentonite, mainly sodium but with a little calcium, and (c) a calcium montmorillonite. He found that the soundest castings were obtained using sands bonded with clay in which all the cations were sodium, and the least sound castings were obtained when the cations were all calcium. Intermediate results were obtained with the Wyoming bentonite.

On the other hand, Hofmann[10] and Toriels and Wallace[12] obtained the soundest castings in moulds bonded with calcium montmorillonites.

A test of this kind is not particularly difficult to carry out, but obtaining similar sand properties is less easy. The relevant properties are toughness, which controls hardness differences in different parts of the mould, and green strength, which influences mould hardness. The curves shown in Figs. 33 and 34 show that there is no difficulty in obtaining the same shatter index or the same green strength when using two different kinds of clay, but that it is almost impossible to obtain both. Whichever is fixed, there must be differences in mould properties. In addition, the three observers used very different castings in their tests, and close agreement could not be expected unless the influence of clay was very marked.

It can be concluded that the influence on soundness of the type of clay used is small.

pH, exchangeable cation and sand properties

Calcium montmorillonites are slightly acid and sodium montmorillonites alkaline. The addition of an acid to a sodium montmorillonite or an alkali to a calcium montmorillonite may therefore produce a considerable change in properties. There is some evidence that some of the additions made to moulding sand, or their decomposition products, may change pH. Measurement of pH has been described in Chapter 1.

Several investigators have examined the relation between the pH of sands, their physical properties and the incidence of scabbing. The results vary widely, depending on the nature of the sand used.

Booth[13] added sodium carbonate to a number of foundry sands and found that both green and dry strength could be related to pH. In one case raising pH from 5 to 10 gave a sevenfold increase in dry strength. In a discussion[14] on Booth's work, several foundrymen reported that the addition of sodium carbonate to their sand resulted in a complete elimination of scabbing. On the other hand, Houseman[15] varied the pH of new and used sands between 5 and 10·5 by additions of acetic acid and sodium carbonate, and found little change in sand properties or in the incidence of surface defects on 300 lb steel castings. Roberts[16] found that small additions of alkali or acid had a pronounced effect on pH, but little effect on sand properties and that larger additions produced little further change in pH, but had a considerable effect on sand properties.

The relation between pH and sand properties is fairly clear provided only small changes are involved:

1. When the clay present is a calcium montmorillonite, raising the pH from below to above 7 by the addition of sodium carbonate produces a small change in green strength and shatter index. There may or may not be a pronounced increase in dry strength. The change of greatest importance is, however, a reduction in the fall in green strength with very high moisture contents. This leads to a reduction in scabbing (see Chapter 10).
2. The change in sand behaviour is due to the change in the exchangeable cation and not specifically to pH. The same pH could be obtained by adding potassium or ammonium in place of sodium carbonate, but the change in sand properties would be different.
3. Small changes in the pH of sands containing clays with only small base exchange capacity have a negligible effect.

THE CLAY GRADE OF NATURAL SANDS

Information as to the precise nature of the clays present in natural sands is scanty and their properties are in some respects very different from those to be expected from a mixture of kaolinite and illite, which some observers regard as their most likely composition. The most notable difference is in the moisture range.

With any one clay, an increase in the amount present in a sand requires a corresponding increase in moisture content and in the moisture range. However, it can be seen in Fig. 33 that the nature of the clay is also important, since the sand with 7% kaolinite has an appreciably narrower range than the sand with 5% Wyoming bentonite.

The moisture range of some natural sands is far wider than that of any synthetic sand, but in order to establish that the difference is due to the clay only, it is necessary to eliminate any possible effect of the difference in the size distribution and surface of the sand grains.

Piper[1] removed the clay from a red Bunter sand, and after drying and grinding, used it to prepare a synthetic sand. He compared this with other sands obtained by adding Wyoming bentonite, Colbond, a clay at one time widely used and thought to be a kaolinite (although Grimshaw and Roberts[17] showed that it was probably a halloysite), and a china clay. The relation between green strength and moisture content of the four sands is shown in Fig. 38.

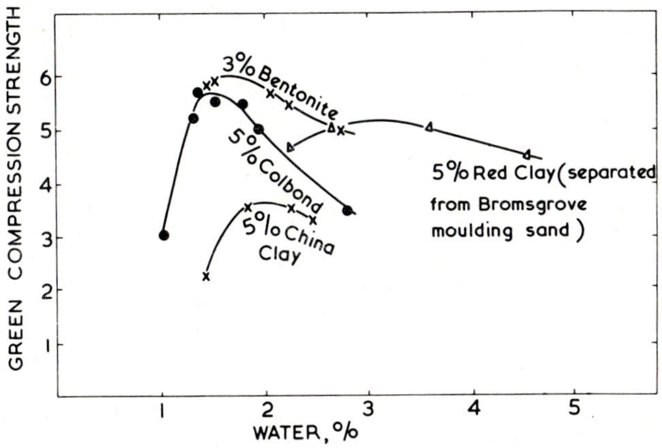

FIG. 38. Relation between green strength and moisture content (after Piper).

Piper used a method of preparing his test pieces differing from the present standard method in that the sand was rammed to a fixed green density and the shape of the curves was slightly different from those obtained by the standard method. Moreover, the process of obtaining clay from a suspension in water is lengthy and laborious and the amount obtained was insufficient to carry out enough tests at low moisture contents to show the impressive, but largely irrelevant peaks. Nevertheless, Piper's curves show that the clay from the natural sand has a wider moisture range than Wyoming bentonite. Since its bonding power is less, and it has to be present in larger quantity, the difference in practice is even greater.

It does not follow that the clays in all natural sands have properties similar to those of the clay in the red Bunter sand and this, indeed, is unlikely. No other observer appears to have emulated Piper's industry in collecting the clay from a natural sand. It would be less laborious to make the comparison by removing the clay from a large volume of natural sand and using the residual sand to prepare the series of synthetic sands.

The effect of iron oxide on clay

One obvious difference between the clay in the red sands and the clays used to prepare synthetic sands is the amount of iron oxide present, which usually amounts to between 6 and 10% of the clay grade. Boswell, at a time when no standard methods of testing were available, was able to carry out only mechanical and chemical analyses and, observing that iron oxide was a significant constituent of many natural sands, deduced that it contributed materially to the bonding properties.

In order to obtain more precise information on this point, a large number of red and yellow ochres — iron oxides produced by natural processes somewhat similar to that which has produced the iron oxide in sand — was collected and added to a silica sand to prepare a series of synthetic sands. Ochres containing less than 60% of the oxide were rejected as clay was usually present.

Tests on these sands showed that:

(1) the bonding properties of the red oxides are negligible;
(2) the yellow oxides give a little green strength — about 3 lb in^{-2} with 20% oxide.

Dry strength of the sand containing yellow oxide is very high, greatly exceeding the 400 lb in^{-2} which was the limit of the apparatus then available. If, however, the test piece was dried at a temperature exceeding 200°C, the yellow colour was replaced by red, and dry strength fell almost to 0. If, therefore, yellow iron oxide has any appreciable influence on the properties of a sand, there should be a considerable fall in dry strength when the drying temperature is raised from below to above 200°C.

Only one such sand, and that from the Continent, has been found and there the fall in dry strength was only by 20%.

It can be concluded that apart from its effect on colour, iron oxide exerts little more influence on the physical properties of sand than any other inert powder of the same size.

THE EFFECT OF HEAT ON CLAY

As the temperature of a moist clay is raised, a series of changes takes place. Some are reversible and others not.

A change may occur over a range of temperature and may not be complete at any temperature within that range except the highest. The rate of change increases with increase in temperature. The changes are:

1. Water is removed — or absorbed — whenever the clay is allowed to stand in air at room temperature. If the temperature is raised to 100°C all the 'free' water is removed. This is the water which is estimated as 'moisture content'. The change is reversible.

Removal of free water produces an increase in strength.

2. At a slightly higher temperature a further small amount of water is removed. This water is bound to the clay particles more firmly than is the 'free' water.[17,18] The change is reversible.

Removal of this 'bound' water has little effect on clay properties and there is, for example, little change in dry strength with drying temperatures between 100 and 300°C.

3. At a still higher temperature, the combined water is removed and the layer structure of the clay breaks down. This change is irreversible and the plasticity of the clay cannot be restored by the addition of water. The clay is then described as dead clay.

The temperature at which this takes place depends on the kind of clay concerned. Breakdown of some clays begins at a temperature below 500°C, but is rarely complete. Montmorillonites are not greatly affected at this temperature, but break down completely at 600–650°C. The rate of breakdown increases with increase in temperature.

4. If the temperature is raised still further, rearrangement of the constituents of the clay takes place with the formation of new compounds. This change is accompanied by an appreciable increase in strength. It is slow at temperatures below 700°C. If the clay is maintained at a temperature a little below that at which the least refractory constituents begin to fuse, there is a considerable increase in strength. Several observers have measured the 'hot strength' of clay-bonded sands heated to various temperatures. Usually, strength has been measured on a standard test piece which has been placed in a furnace at the appropriate temperature and allowed to stand for 20 min to reach a uniform temperature throughout. Under these conditions 'hot strength' may exceed room temperature dry strength by a factor as high as 5 or 6. Moreover, much of the test piece has been at temperature for a shorter time and it is possible that still higher strengths may be obtained if the soaking time is prolonged.

This change is comparatively slow. If the test piece is placed in a dielectric heater such that the high temperature is reached in 2 min or so, the 'hot strength' differs little from the cold dry strength.

5. At a still higher temperature, fusion commences and strength falls rapidly.

The high hot strength is of little significance in the foundry. The face of a mould is, in practice, heated so rapidly that the fall in strength due to incipient fusion occurs before the high hot strength has time to develop.

The life of clays

When molten metal is poured into a mould, part of the sand is heated to a temperature at which the irreversible changes of loss of combined water and breakdown of structure take place. To maintain moulding properties, fresh clay must be added, but even this practice may fail after a time since the 'dead' clay has an adverse influence.

The more readily a clay breaks down at high temperatures, the more quickly the sand loses its desirable properties and the greater the need for remedial measures. The 'life' or 'durability' of a clay is of considerable practical importance.

Three methods of investigating the breakdown of clays have been used — differential thermal analysis, casting tests and heating tests.

Differential thermal analysis

When an inert substance is heated under uniform conditions, temperature rises smoothly. If, however, a change in constitution takes place, this is usually accompanied by an absorption of heat (endothermic) or a giving out of heat (exothermic). With an endothermic change, the rate of rise in temperature is retarded; with an exothermic change it is accelerated. If an inert substance and a clay are heated side by side and temperature differences shown by means of thermocouples connected through a sensitive galvanometer, then the temperatures at which changes in clay structure take place are shown by a dip or a peak in the curve relating actual temperature with temperature difference.

Davison and White[19] examined a number of clays using this method. They found that kaolinites had an endothermic peak at 550°C, and illites at 520°C. Montmorillonites varied, some having a peak at 680°C and others one at 520°C and one at 660°C. The temperatures are approximate and there are small differences between individual clays of the same kind.

The results indicate the temperature at which the combined water is removed. This is highest with the montmorillonites and lowest with the illites. Evidence as to the rate of breakdown is less well marked.

Casting tests

Casberg[20] and Piper[1] used a method in which the clay to be tested was used to prepare a synthetic sand, which in turn was used to make a mould for a simple bar casting. The casting was poured and allowed to cool, after which the sand was re-milled with water and re-tested. This series of operations was repeated a number of times.

Piper rammed up a mould for a cylindrical bar 1 in diameter and 8 in long, using a tube 3 in diameter and 9 in long as a moulding box. He found that dry strength gave a clearer indication of 'life' than could be obtained from green strength.

Two synthetic sands were tested, bonded with a montmorillonite and a clay thought to be a kaolinite but which may have been a halloysite. He found that the dry strength of the sand containing montmorillonite fell from 45 to 22 lb in^{-2} after 14 cycles; in the same time the dry strength of the sand containing halloysite fell from 22 to 2·5 lb in^{-2}.

Heating tests

Making a series of castings and reclaiming all the sand after each operation is a tedious operation and other observers have compared clays by heating to various temperatures and measuring their bonding properties before and after treatment.

Davison and White[19] used a modification of a test devised by Henderson. A test piece is heated from one end under standard conditions such that the total time of heating is 90 min and the final temperature 820°C at the hot face and 57°C at the cold face. The test is repeated for a total number of 8 cycles, green and dry strengths being measured after each. Only montmorillonites were tested and wide variations were found; sodium montmorillonites in general had a longer life than calcium montmorillonites.

Hofmann[10] heated sands bonded respectively with a sodium montmorillonite, a kaolinite and an illite to 700°C in steps of 100°C, re-milling and testing the sand after each cycle. He found that the illite commenced to break down at 400°C and broke down rapidly, the kaolinite at 490°C with a rather slower rate of breakdown, and the montmorillonite at 650°C with an intermediate rate of breakdown.

Exact comparison between the results obtained by different observers is not possible, owing to the different methods of testing employed, but it is clear that montmorillonites have a longer life than kaolinites.

THE EFFECT OF DEAD CLAY ON THE PROPERTIES OF CLAY-BONDED SAND

When metal is poured into a sand mould, some clay is heated to a temperature at which it breaks down and some dead clay is always formed. Part of this clay may be removed with sand adhering to the casting, but some remains behind and in a sequence of operations the amount of dead clay will build up at a gradually decreasing rate until the amount withdrawn with the casting equals the amount produced.

There is no way of removing the dead clay from a sand without removing at the same time any active clay and other fine-grained material. Moreover, it is not possible to estimate the amount of dead clay in a sand which also contains active clay, with any degree of precision. Various methods have been described, usually depending on estimation of the total clay content and the active clay content and obtaining dead clay by difference. Estimation of active clay depends on its capacity for absorbing certain substances from solution, the earlier tests making use of various dyes and the more recent tests nickel salts. The results are not very satisfactory with montmorillonites and worthless with other clays. In the iron foundry there is the added complication that the clay grade usually contains, in addition to active and dead clay, finely divided coke and coal in amounts which are difficult to estimate exactly. In consequence, the method is not to be recommended for inclusion in control tests.

The effect of dead clay

In order to determine the effects of dead clay on the physical properties of a moulding sand, a series of sands was produced in which the active clay content was constant and the amount of dead clay was raised in steps. The difficulty in estimating the amount of dead clay was avoided by heating a sand with known clay content until breakdown of the clay was complete.

A synthetic sand was prepared from Erith silica sand and 5% Wyoming bentonite, and green strength, shatter index and dry strength were measured over a range of moisture content. Each sand tested was prepared from fresh sand and clay so that a constant milling time could be maintained.

The whole of the sand was heated to 800°C for 1 h and it was observed that, after this treatment, the clay adhered to the sand grains so strongly that little was found when the standard clay estimation was carried out. On milling with water the sand had the properties of a damp silica and was too weak for testing by the standard procedure. However, milling separated most of the clay from the sand grains.

This sand was then milled with 5% new bentonite as before and the series of tests repeated. Each of the sands tested contained 5% active bentonite and the dead clay from the previous addition. The actual amount of dead clay was rather less than 5% since the combined water had been removed.

The whole of this sand was heated to 800°C and the tests repeated with the addition of a further 5% of new bentonite. The procedure was repeated a third time.

In this way four sands were produced, each containing 5% bentonite with the dead clay from 0, 5, 10 and 15% bentonite. The green strengths, shatter indices and dry strengths are shown in Figs. 39–41.

One feature is common to all sets of graphs. As the amount of dead clay increases, the curves move to the right, *i.e.* in the direction of higher moisture content. Dead clay has a very large surface area, of the same order as that of active clay, and absorbs a correspondingly large amount of water. Apart from this, dry strength curves are not greatly affected and need not be considered further.

As the proportion of dead clay increases, green strength rises and shatter index falls. It was shown in Chapter 1 that an increase in green strength at constant shatter index leads to an increase in brittleness and a deterioration in moulding properties, and that a fall in shatter index at constant

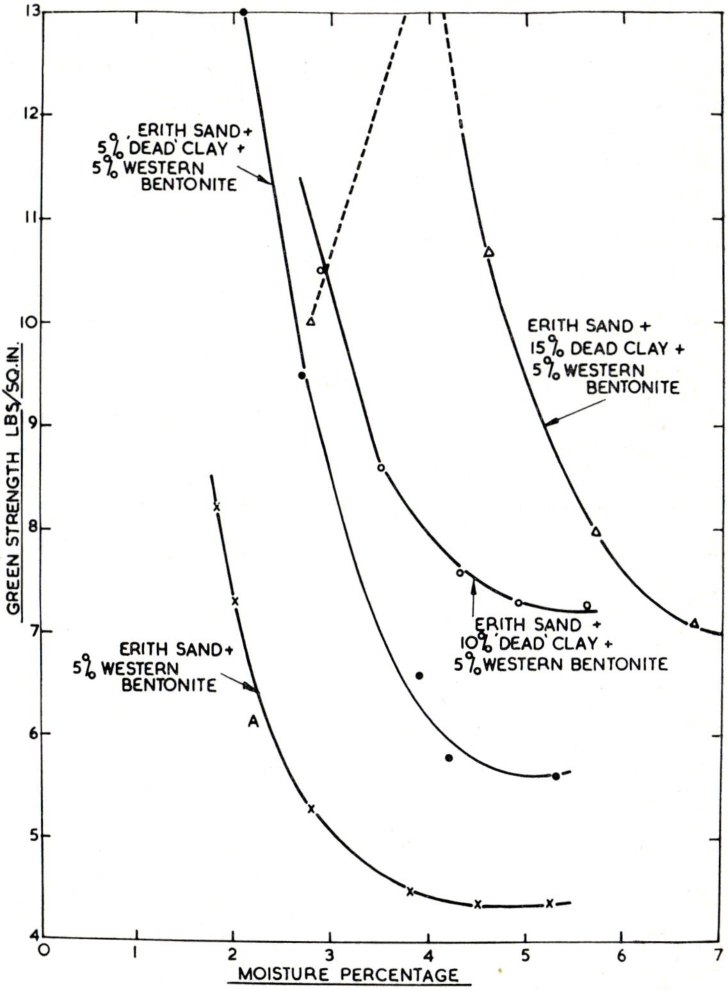

FIG. 39. Effect of dead clay on green strength.

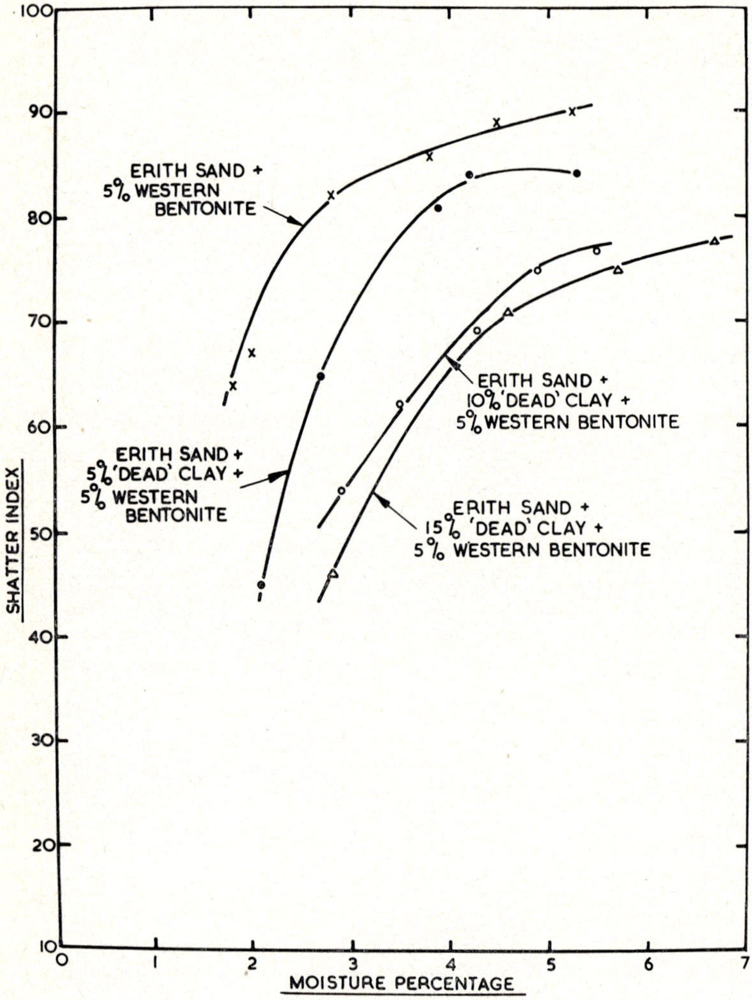

FIG. 40. Effect of dead clay on shatter index.

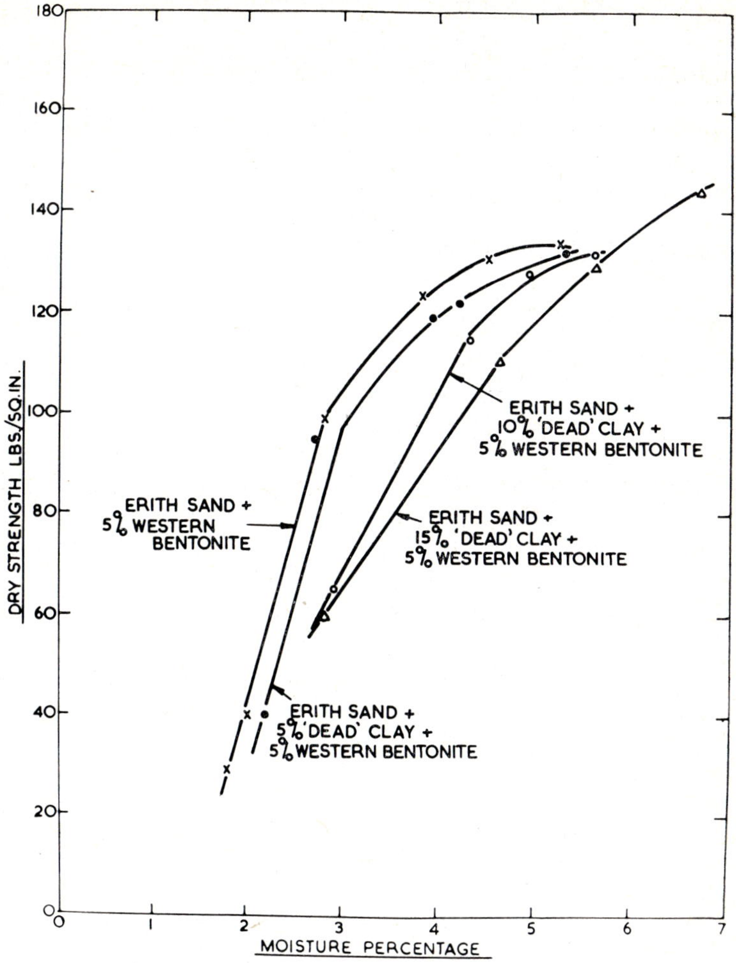

FIG. 41. Effect of dead clay on dry strength.

green strength has a similar effect. The same deterioration must occur to an enhanced degree when an increase in green strength and fall in shatter index occur together.

Dead clay and sand control in the foundry

Under most foundry conditions, the sand from the shake-out contains both dead and active clay in unknown proportions and these proportions can vary widely under different conditions. The physical properties which are most affected by dead clay are the green strength and shatter index and these are the tests used to control sand.

As a result of the loss of the bonding properties when a mould is filled with molten metal, the sand from the shake-out has both a lower green strength and a lower shatter index than the original sand. Addition of new clay to restore the original proportion of active clay would result in an increase in green strength and a fall in shatter index; but since it is not possible to measure the amounts of active and dead clay present, the foundry is limited to maintaining either a constant green strength or a constant shatter index. It is not possible to do both at the same time by varying the clay addition.

Suppose that it is decided to maintain a constant green strength. Dead clay raises green strength and to offset this it will be necessary to reduce the amount of active clay. But reducing the active clay will lower still further the shatter index, already reduced by the presence of dead clay, and to maintain a constant green strength it will be necessary to make a change which will reduce shatter index appreciably.

If, on the other hand, it is decided to maintain a constant shatter index, clay will have to be raised to compensate for the effect of dead clay in lowering shatter index. Green strength, already raised by the presence of dead clay, will increase still further.

In either case, the sand becomes brittle and it becomes difficult to make satisfactory moulds.

The effect of the dead clay and of the alternative methods of attempting control are illustrated in the following table which shows actual results:

	Original sand	Used sand	Constant clay	Constant green strength	Constant shatter index
Green strength (lb in^{-2})	10	7	12	10	14
Shatter index	75	60	70	65	75

The sand had been used for some time with satisfactory results and its properties were as shown in the first column. At this time heavier castings were introduced, and an average sample of sand from the shake-out is illustrated in the second column. The third column contains hypothetical figures, showing the kind of result which would have been obtained if it had been possible to obtain a constant amount of active clay.

The fourth column shows the test results when the amount of clay added was just sufficient to restore the original green strength. Shatter index has fallen from 75 to 65. The last column shows the test results when the clay added was sufficient to maintain a constant shatter index. Green strength has risen from 10 to 14 lb in^{-2}.

The rate of build-up of dead clay and the level reached, before the amount produced by each cycle of operations is balanced by the amount discarded with the sand adhering to the castings, depend on the kind of castings being made. When the castings have a large weight/surface ratio, and there is also a high metal weight/sand weight ratio, a relatively long time from pouring to shake-out, and an absence of cores with organic binders, build-up of dead clay is rapid. In a foundry with a system on which only flywheels and similar castings were made, a shatter index of 80 was maintained and, as a result of the increase in dead clay, green strength rose from 10 to 20 lb in^{-2} at which stage moulding became impossible.

When the castings are relatively thin and the metal weight/sand weight ratio is low, the amount of dead clay produced is small, the build-up slow and the final level low. Should cores with organic binders be present, some of the bond is destroyed by heat and clay-free sand enters the system, reducing still further the build-up of dead clay. In the foundry making the flywheels described above, a second system producing crankcases was in operation and with the shatter index of the sand maintained at 80, green strength showed no tendency to rise above 10 lb in^{-2} over a long period.

Controlling the build-up of dead clay

Should the amount of dead clay increase to such an extent that moulding becomes difficult, the only practicable method of reducing it is to withdraw some sand from the system and replace it with sand free from dead clay. In this respect sand obtained from the breakdown of cores with organic binders is quite as satisfactory as new sand.

Observation in a number of foundries has indicated that if the amount of new sand entering a system is equal to 10% by weight of the total metal poured, no undue increase in dead clay is likely.

Occasionally, dead clay is returned to a sand system unnecessarily. When castings are shaken out, a certain amount of sand adheres to them. If the castings are light, much of the clay in this sand is active and to discard the whole would be wasteful, but when the castings are 1 in or more in thickness, the adherent sand contains only dead clay. When the castings are allowed to cool, much of this sand becomes detached. In appearance, it is white in colour and resembles silica sand, and since the dead clay is firmly attached, shaking with water produces no evidence of the presence of anything other than sand grains. The simplest way of getting rid of this apparently good silica sand may be to dump it on the shake-out grid. Sooner or later, in order to dilute the dead clay, a similar amount of new sand must be added, the whole bonded with clay, and half of it discarded to prevent an increase in the amount of sand in circulation. The apparently clean, white sand adhering to thick castings should always be discarded.

Neutralising the effect of dead clay

The addition of any material which influences green strength and shatter index in a direction the reverse of that due to dead clay should neutralise the harmful effects.

Water seems an obvious choice, provided it is not present in quantity sufficient to be harmful in other directions. The moisture content in sand for green sand work is almost invariably a little below that which gives maximum shatter index and above that which gives maximum green strength, and an increase in water exceeding that called for by the moisture need of the dead clay, tends to offset the increase in green strength and fall in shatter index due to the dead clay. This procedure is occasionally used, often inadvertently, and is sometimes successful, provided the amounts involved are small. It is not satisfactory when a large increase in moisture content is needed. In one foundry where this method was used, the water content in a synthetic sand had to be raised steadily from 4 to 9%. The sand did not feel unduly wet, there was no marked increase in blowing and stripping was fairly satisfactory, but the surface of the castings was rough and there was considerable swelling, and the practice had to be abandoned. The reason for this behaviour is not clear.

A less unsatisfactory method is to reduce the addition of clay sufficiently to maintain a constant green strength. This reduces the already low shatter index. However, the shatter index can be raised without increasing green strength by adding a small amount of dextrin or one of the other green binders used in oil–sand mixtures and the properties of the sand can be maintained. Very frequent tests on sand properties are needed, since a small amount of organic binder produces a large increase in shatter index, but the greatest difficulty arises if there is a hold-up in the plant, and the mould is raised to a higher temperature than usual before reaching the shake-out. The organic binders break down more readily than the clays, there is a sharp reduction in the amount present in the sand from the shake-out and at the appropriate time a rapid fall in the shatter index of the sand. It is not easy to foresee the exact time when the sand deficient in binder reaches the mill and to make the necessary adjustments. The method has been used successfully in 'rescue' operations, but it is safer, and in the long run cheaper, to avoid the need for such measures by adding an adequate amount of new sand.

SAND TREATMENT AND THE STRENGTH OF MOIST CLAY

The properties of moist clay depend not only on the kind of clay and its moisture content, but also on the treatment it has received. If a little water is introduced into a dry powdered clay without any form of rubbing action, the only change is from a dry to a damp powder, but as soon as any work is performed on the powder some plasticity is developed. Performing any form of work, from squeezing in the hand to throwing against a wall, initiates the change, but the method most widely used is rubbing under heavy rollers. The properties of a moist clay depend not only on composition, but also on the treatment it has received.

This factor is important in any industry which uses a stiff clay. For example, some fireclays, although moist, have little plasticity as mined and are milled by the brickmakers before being used to make firebricks.

Much of the effect of the treatment is lost when the clay dries out and if it is to be regained, the original process must be repeated. This factor is of considerable importance in the foundry industry where clay-bonded sand has to be re-used after a proportion has dried out and the method of preparation has an important influence on the properties of the prepared sand. This is discussed in Chapter 7.

REFERENCES

1. Piper, G. H. (1938–1939). *Proc. Inst. British Foundrymen*, **32,** pp. 33–41.
2. Patterson, W. and Boenisch, D. (1961). *Giesserei*, **28,** April 6, pp. 157–166.
3. Hast, N. (1947). *Nature*, **159,** March 16, p. 354.
4. Pauling, L. (1929). *J. Amer. Chem. Soc.*, **51,** p. 1010.
5. Grim, R. E., Bray, R. H. and Bradley, W. F. (1936). *Trans. Amer. Foundrymen's Assoc.*, **44,** Oct., p. 211.
6. Grim, R. E. and Rowland, R. A. (1940). *Trans. Amer. Foundrymen's Assoc.*, **48,** Sept., p. 211.
7. Grim, R. E. and Cuthbert, W. (1947). *Illinois State Geological Report.* Investigation 102.
8. Skerl, J. G. A. (1931). *Inst. British Foundrymen*, report to conference. June, p. 195.
9. Dunbeck, N. J. (1941). *Proc. Inst. British Foundrymen*, **34,** p. 17. See also: (1941). *Trans. Amer. Foundrymen's Assoc.*, **49,** p. 141.
10. Hofmann, F. (1966). *Cast Metals Res. J.*, **2,** Dec., p. 4.
11. Nicholas, K. E. L. (1964). *BCIRA J.*, **12,** p. 313.
12. Torriello, L. I. and Wallace, J. F. (1963). *Modern Castings*, **44,** No. 2, p. 1024.
13. Booth, B. H. (1949). *Trans. Amer. Foundrymen's Soc.*, **57,** pp. 210–220.
14. Booth, B. H. (1954). *Amer. Foundryman*, **26,** Sept., pp. 34–39.
15. Houseman, D. H. (1958). *British Foundryman*, **51,** March, p. 194.
16. Roberts, W. R. (1961). *BCIRA J.*, **9,** p. 327.
17. Grimshaw, R. W. and Roberts, A. L. (1944–1945). *Proc. Inst. British Foundrymen*, **38,** p. A55.
18. Grim, R. E. *Clay Mineralogy*. Ch. 8. McGraw-Hill Book Co., Ltd., London.
19. Davison, S. and White, J. (1953). *Proc. Inst. British Foundrymen*, **46,** p. A86.
20. Casberg, C. H. and Schubert, C. E. (1956). University of Illinois, Experimental Station Bulletin, Apr.

CHAPTER 5

Natural and Synthetic Sands

The relative advantages of natural and synthetic sands were at one time the subject of considerable argument, each having its devotees, but, largely as a result of experience, the most suitable kind of sand for each particular set of conditions has been adopted. Given a description of the kind of castings being made and the methods of production in use, it is now possible to say whether the foundry concerned will use a natural or a synthetic sand and, to be wrong less than once in ten guesses.

To a considerable extent, comparisons between natural and synthetic sands differ in different countries and certain undesirable or desirable properties which are of major importance may be present in one country but not in another. The first survey of resources of foundry sands in Britain was undertaken by Boswell for the Ministry of Supply and published in 1918. Its main objective was to find a supply of sand in Britain which could be used by steel founders in place of the Belgian and French sands which had previously been used almost exclusively by British steel founders who made castings in green sand moulds. With the outbreak of war in 1914, the steel founding industry was in serious difficulty and had to adopt all manner of makeshifts. No large new source of sand suitable for steel founding was uncovered, and at the end of the war there was a general return to imported sand, but nevertheless the exercise was well worth while. Boswell's methods of testing were confined to mechanical and chemical analyses and his approach was frequently that of a geologist rather than a foundryman, but he did survey the more important deposits and point out likely sources not then being worked. His tests were sufficient to indicate the kind of sand available in each locality, the extent of available resources and even the kind of variation to be expected in individual quarries. Many of the quarries examined by Boswell have ceased operation for various reasons — all the suitable sand has been removed or, more frequently, is covered by roads and buildings, or the change from canal to road transport has made the approach to one site more difficult and that to another easier

— but nevertheless his memoirs still have some usefulness as a source of information on sand deposits.

Apart from its value as a catalogue, Boswell's book performed a service in that it stressed the need for separate examination of the clay and sand grades and for regarding the sand as a mixture of different grains rather than a single entity. At that time 'advanced' sand suppliers went so far as to publish chemical analyses of their products, although the bulk analysis of a mixture of sand and clay is a singularly useless piece of information. Boswell made it quite clear that not only should the sand and clay be analysed separately, but that an analysis of the sand grains could only be useful if it was used as an aid to determine the proportions of the different minerals, quartz and felspar for instance, which were present.

DIFFERENCES BETWEEN NATURAL AND SYNTHETIC SANDS

(1) The sand grains

The sand grains have an influence on the properties of moulding mixtures which depends on their grain size and distribution and on their mineralogical composition, and since there are well-marked differences in both respects between the sands with a natural clay bond and the clay-free sands, there are corresponding differences between their properties.

Composition

The clay-free silica sands, with few exceptions consist mainly of quartz, the total silica varying from 95% in the Congleton sands to more than 99% in Leighton Buzzard, Reigate and Erith sands.

Supplies of bonded sands in which the sand grains contain such a high proportion of quartz are lacking, and various other minerals are usually present. The red sands of which the known deposits are enormous and spread from Worcestershire to Lancashire and Yorkshire, contain felspar which usually varies between 15 and 20%. The rotten rocks found in Scotland and Durham contain a smaller proportion of felspar. The estuarine sands such as those at Erith and Southampton are relatively free from felspar, but contain appreciable quantities of glauconite.

The felspars are essentially silicates of aluminium and either potassium, sodium or calcium, potash felspar predominating in the red sands. The glauconites are rather complicated silicates of aluminium and potash, with magnesium and iron usually present.

Quartz melts at about 1700°C and the felspars and glauconites at a little over 1300°C. The natural sands therefore contain such a large proportion of minerals which fuse at a temperature some 200°C below that at which steel is poured that they cannot be used in the steel foundry except as minor constituents of moulding mixtures. Exceptionally, some of the rotten rocks with the smallest proportion of felspar are used in steel foundries but they are usually mixed with more refractory sands in order to reduce still further the proportion of fusible grains.

In the iron foundry where the pouring temperature is lower, the presence of felspar is not harmful and castings weighing many tons are made in red sand moulds without any sign of fusion of sand grains and resulting metal penetration.

The felspars have a coefficient of expansion which is much lower than that of quartz and in consequence, the red sands are less prone to expansion defects than synthetic sands. The difference is not so great that it has a decisive influence on the choice of sand and the only definite effect of composition is the almost complete exclusion of native natural sands from the steel foundry.

This characteristic of the natural sands is not necessarily true in all countries. Some Belgian and French sands consist of grains with silica approaching 99% and clay and natural sands containing high proportions of quartz are found elsewhere. Some Belgian sand is still imported for use in the steel foundry but the practice is declining.

Grain size and distribution

The silica sands are coarser and have grains of more uniform size than the natural sands. The division is quite sharp and the finest silica sand obtainable has a permeability higher than 50, whereas few natural sands with a clay content high enough to enable them to be used without the addition of a bonding clay, have a permeability much higher than 40. The much finer silica sands, such as those used for shell moulding, are not as a rule mined as such, but are separated from a sand with a wide range of grain size. As the coarser fraction has to be sold in competition with as-quarried silica sand, the fine fraction, usually only a minor proportion of the total, has to bear the whole cost of the treatment and is costly. Very fine, clay-free silica sands are scarce in all countries, but there are large deposits of high permeability natural sands notably in France and Belgium.

In view of the differences in permeability between natural and synthetic sands, the decision as to which kind of sand to use is frequently decided by this factor.

(2) The clays

The most important difference between the clays present in the natural sand and the bonding clays available for preparing synthetic sand is the difference in moisture range. There are appreciable differences in bonding power — no clay present in natural sands has the powerful bonding properties of the montmorillonites — and in dry strength, and these influence behaviour in the foundry, but they have little impact on the decision to use or not to use a natural sand. All the clays in the natural sands have moderate bonding properties, appreciably lower than those of the montmorillonites and similar to those of the better kaolinites, and give dry strengths in between those of the swelling sodium montmorillonites and those of the kaolinites, and similar to those of the calcium montmorillonites. There are few cases in which the dry strength of the natural sands is not adequate without being excessive, although there are many where it is advisable to avoid the use in synthetic sands of clays with very high or very low dry strengths.

The moisture range is more important, particularly to hand moulders. Several factors contribute. The sand may be mixed by hand or simply passed through some form of disintegrator after a preliminary hand mixing, in which case the precise moisture control required with a synthetic sand cannot be achieved. Even when a milled facing sand is supplied, the demand may be so small that the mill is in operation once or twice a day only, and a prepared sand with a wide moisture range is less likely to dry out and become unusable. Patching is more difficult when the moisture range of the sand is narrow. When a large patch is necessary the usual procedure is to throw water on the defective part of the mould, apply new sand and finish with a trowel. When the sand has a narrow moisture range it is more difficult to control the amount of water added so that the mould is sufficiently moist for the patch to adhere, yet not so wet that the sand begins to run. Even under the most favourable conditions, patching should be avoided wherever possible, but when only an occasional casting is required, the cost of the pattern may be an appreciable part of the cost of the casting. Pattern making is frequently skimped and the pattern kept in service long after its useful life is over. Where the patterns are of good quality and a reasonable rate of delivery of sand with controlled moisture content can be arranged, a jobbing foundry can use a synthetic sand without difficulty, but these favourable conditions are often missing.

A supply of a clay with moisture range as wide as that of the clay in the natural sands would be useful to many users of synthetic sand. The rate of air drying would be halved and hot sand would not be such a nuisance since one of the main objections to it is that the surface, and particularly the edges, of a warm mould may dry out before coring up is complete, in which case the core setter may dislodge sand from the friable surface. In addition, since the amount of water removed from a mould depends on the metal/sand ratio, the proportion removed from a high moisture mould would be lower than from a low moisture mould and when using a clay with a wide moisture range, a smaller proportion of the sand would dry out and less milling would be necessary.

(3) Other factors

Apart from the differences in composition and physical properties between the sand and clays in the natural sands on the one hand, and the silica sands and clays available for preparing synthetic sands on the other, the method of preparation has an influence on the usefulness of the sand in the foundry for a number of reasons.

(*a*) *Ease of control*

When using a synthetic sand, the sand grains can be selected so as to give any permeability required and the amount of clay can be adjusted so as to give any green strength likely to be required in the foundry without affecting permeability. By choosing the appropriate kind of clay, dry strength can also be controlled over a wide range.

Control of this kind is not possible when using any British natural sand. With each kind of sand, clay content decreases as permeability increases. In the case of the red sands, the clay content is low and the sand too weak for many purposes when permeability is above 50, and the strongest sands have a permeability below 25. The relation is not an exact one and, on the whole, sands in the Midlands have a higher clay content than Lancashire sands with the same permeability. Erith loam shows a similar relationship even though the sands are obtained from the same quarry, the upper layers consisting of a rather weak sand with a permeability of 40–50, underneath which are successive layers with descending permeability and increasing clay content.

(*b*) *Uniformity of product*

Supplies of silica sand are on the whole extremely reliable. This is in part due to the nature of the deposits and in part to the very tight specifications necessary for the glass makers who frequently draw their supplies from the same quarries. To satisfy these, it is necessary to have very precise knowledge of any variations in the quarry, and the foundries benefit.

Many natural sand deposits are more variable, and in the case of the red sands there is no constant pattern of variation. Occasionally there may be a decrease in permeability as depth increases, but just as frequently the closest sand is on top with open sand beneath it, or sand with a moderate permeability may be in between layers of higher or lower permeability. As few of the sand suppliers are equipped for sand testing the possibility of variation is always there. Some quarries supply two or even three grades of sand differing in permeability, but this practice in itself can lead to loss of uniformity. The profit on sand is low and it is not economically possible to dig sand out and dump it elsewhere, a proceeding which would be necessary if the main demand were for the kind of sand found only at the foot of the workings. More uniform supplies can usually be obtained from a quarry where only one grade is produced provided that where there is a considerable variation down the working face, the whole is treated as a single unit, sand being taken from all levels and mixed. Sand which has been through a disintegrator at the quarry, in addition to being more suitable for use in the foundry, is usually more uniform than untreated sand. The usual practice is to draw sand from each level in turn and transfer to a heap by the disintegrators, thus ensuring some degree of mixing. After treatment, the sand is transferred to a second store and further mixing takes place here. Where this practice is followed, very uniform supplies can be obtained.

Economy in operation

When molten metal is poured into a sand mould, the clay in part of the mould loses its bonding power and must be replaced. When using a synthetic sand, it is unnecessary to add more new sand than is required to prevent too high a build-up of dead clay, which is a very small amount. Sufficient clay is then added to satisfy this new sand and to replace the clay which has lost its bonding power. Total requirements are small. When a natural sand is used, a much larger amount of new sand, depending on its clay content, is required. If, for example, 9% clay is required to provide suitable moulding properties and the new sand contains only 10%, destruction of 1%, or one-ninth of the clay by heat at each cycle of use would eventually make it necessary to use a moulding mixture containing 50% new sand.

In view of the differences in the clay content of natural sands from different sources and of the great differences in the time and temperature of heating in the foundry, as well as the different stabilities at high temperature of the various kinds of bonding clay, it is not possible to give a comparison between the consumption of natural and of synthetic sand for the same purpose. However, consumption of synthetic sand is usually much lower. At the same time there is less used sand to discard, and the cost of removing and dumping is less.

Cost

Although the consumption of natural sand is much higher than that of a synthetic sand, the cost is not necessarily higher and in some places may well be lower. In districts remote from their source of supply, the cost of transport is greater than that of the sand at the quarry — a foundry in Falkirk drawing sand from the Midlands may find it three or four times as great — and this has an overriding effect. In many areas, and particularly in the Midlands, clay-bonded sand is usually cheaper than silica sand and when the cost of clay for the synthetic sand is added, the difference is wide and frequently more than balances the differences in consumption.

CHOICE OF SAND

When a choice has to be made between synthetic and natural sand, many foundries find that their freedom to choose turns out to be extremely limited, confined by operating conditions or by the sands available. If the foundry industry is divided up into broad sections, it can be seen that frequently the kind of sand to be used chooses itself.

(1) Heavily cored castings

When the moulds contain a considerable total weight of cores bonded with an organic binder, much of the binder decomposes in the interval between pouring and shaking out, and the core collapses. A large part of the sand from these cores enters the moulding sand at the shake-out and in many cases supplies all the new sand required. It does not contain any clay and if its clay content is to be built up by the addition of a natural sand, a very large amount of new sand will have to be added. In practice, the only practical means of maintaining the system is by adding clay and thus preparing a synthetic sand.

(2) Light castings

Surface finish is particularly important, while there is no need for a very high permeability. A large majority of foundries producing castings of this kind use natural sand.

(3) The jobbing foundry

In the foundry making only a few castings of one kind and frequently having to use patterns of poor quality, the narrow moisture range of the synthetic sands is a source of serious trouble, particularly to hand moulders. A great majority of foundries in this class use natural sands. It is possible to use synthetic sands even under these conditions, but usually the arrangements for distributing prepared sand have to be modified, so that it does not have to stand for a long time before being used. At the same time, the moulder has to modify his technique if any patching has to be done.

When the patterns are of good quality and little or no patching is required, there is greater freedom of choice and other factors become important. If, for example, ramming is by slinger instead of by hand, the moulds are harder and a higher permeability sand is desirable to offset the reduction caused by an increase in ramming density. At the same time the sand is used up more rapidly and can be stored in a closed hopper after it has been prepared, under which conditions objections to a narrow moisture range no longer hold true. A synthetic sand is therefore used quite frequently.

(4) Mechanised production of castings with few cores

The need to take care of the influx of sand from the cores is no longer present, and there is no necessity to use a synthetic sand on this account. The need to ram moulds hard in order to obtain completely sound and accurate castings, is being increasingly recognised and, except in the light castings' foundries, synthetic sands are preferable to take care of permeability.

(5) High pressure moulding

Not only is it necessary to use a sand of high permeability, but a sand with high strength and toughness is needed in order to avoid difficulty in stripping. It is frequently necessary to reinforce the clay with organic binder and a synthetic sand is necessary.

(6) Dry sand moulding

When the moulds are dried and contain no coal dust, little gas is evolved during pouring and, although the castings are large, a high permeability is not necessary. Usually the moulds are coated with blacking and a high permeability does not result in metal penetration. Permeability is not a particularly important factor under these conditions. Hand moulders generally prefer a natural sand, particularly when any patching has to be done, but with repetition work, good patterns and any mechanical form of ramming either kind of sand is satisfactory. The choice of sand is then governed by cost or convenience of supply, or by some factor depending on the method of operation. If, for example, de-coring is carried out by means of a hydroblast, a large volume of sand from which the clay has been removed is obtained, and the most economical procedure is to return this to the foundry for use in a synthetic sand. Some new sand is needed and this may be either a silica sand or a natural sand. In the latter case the sand grains are first used with their natural clay bond and later, after recovery from the hydroblast, with a bentonite or other suitable clay.

SEMI-SYNTHETIC SANDS

Usually, the most important factor to be considered when making a decision as to whether a synthetic or a natural sand is to be used, is the permeability required and if that permeability greatly exceeds 40, a synthetic sand is essential. However, there are in existence large deposits of red sand with permeability exceeding 60 and sometimes exceeding 200, which contain only a small amount of clay. The possibility of using these sands in synthetic moulding mixtures is attractive since they require a smaller addition of clay than is needed by the clay-free silica sands.

Sands of this kind were at one time extensively used for preparing pig-beds, but with the introduction of casting machines, this outlet has gone. Large quantities were also used by the building industry, but here also there has been a falling off in demand.

Deposits of this weakly bonded, high permeability red sand are widespread. Most of the sand quarried in Yorkshire and used as a moulding sand, has a permeability number of about 60 and a clay content too low for it to be used by many foundries. There are also deposits with permeability exceeding 100 and only about 2–3% clay. Sands with permeability of 60–100 and a somewhat higher clay content are quarried in Nottinghamshire and still more open sands are available. There are also deposits with permeability in the 100–150 range within 20 miles of Birmingham.

The mechanical analysis and approximate permeabilities of a number of such sands are given in Table 1.

Permeabilities were measured on the sands after milling. The weakest sands with the lowest clay content pack more readily than the stronger sands and therefore have somewhat lower permeabilities than would be expected from the mechanical analysis. If to each sand there were added sufficient clay to bring green strengths and shatter indices up to a common level, the relative permeabilities would be changed. Under these conditions, for example, Ryton sand would have a permeability slightly lower, rather than higher, than Stourbridge sand. The differences are not, however, of great importance in the foundry, especially when coal dust is added to the sand. The addition of 6% of fine coal dust would reduce permeability by about one half, but its influence would be different on sands with different mechanical analyses, as would be expected from the effect on permeability of mixing sands of different grain size (see Chapter 3).

Open, weakly bonded sands of the above kind are widely distributed over the areas where the stronger red sands are found, and the quantity available is for practical purposes unlimited. When they have been worked, it has usually been for the building trade, the requirements for which are in some respects very different from those called for by foundries. For example, many deposits contain appreciable proportions of coarse grains which may not be harmful in concrete but would be a nuisance in a moulding sand, and such deposits should be avoided unless it is possible to arrange working of the quarry in such a way as to exclude them. In many cases, a similar procedure has to be followed when quarrying some deposits of the more usual type of red sand, in order to exclude certain layers or pockets which are either too coarse and weak, or composed almost entirely of clay.

The mineralogical composition of the red sand grains is very similar, whatever their size and size distribution. Felspar to the extent of 15–20% is present, which makes red sands useless in the steel foundry but not in the iron foundry.

Given a sand with suitable grain size, the addition of an appropriate amount of clay produces a semi-synthetic sand which has all the advantages of a fully synthetic sand but not all its disadvantages. Permeability and green and dry strength can still be controlled independently of each other, and in addition, the properties required can be obtained with the addition of a smaller amount of clay than would be needed for a fully

Table 1. *Mechanical analysis of high permeability red sands*

Aperture (μ)	Ryton, Warks.	Stourbridge, Worcs.	Ormskirk, Lancs.	Warsop, Notts.	Mansfield, Notts.	Selby, Yorks.	Pollington, Yorks.
+699 (22 mesh)	0·6	1·3	0·4	0·2	2·6	—	0·2
+500 (30 mesh)	0·7	1·6	1·2	1·2	5·4	0·4	0·3
+353 (44 mesh)	4·0	6·2	4·8	5·6	25·6	1·4	5·2
+251 (60 mesh)	21·0	23·7	14·0	16·5	30·4	22·0	32·7
+152 (100 mesh)	42·3	35·3	32·8	38·6	21·0	53·0	27·3
+104 (150 mesh)	12·9	22·2	20·2	19·2	6·2	15·1	18·9
+ 76 (200 mesh)	4·9	6·3	13·4	7·0	1·8	3·4	4·8
− 76 (−200 mesh)	5·8	2·2	6·4	4·1	1·4	1·7	3·4
Clay	7·8	1·2	7·0	7·2	5·6	2·6	7·2
Permeability number	120	120	63	90	190	170	100

synthetic sand. The moisture range, although narrower than that of a completely natural sand, is still sufficiently wider than that of a fully synthetic sand to reduce difficulties on its account to small proportions. The lower coefficient of expansion reduces the risk of scabbing.

The use of semi-synthetic sands of the above kind has been adopted on a considerable scale. Few of the sands in Yorkshire contain sufficient clay to enable the sand to be used for moulding. At one time many users added the stronger Mansfield sand in order to improve bonding properties, but mixtures of this kind have a permeability very near to that of the finer sand, and control of permeability is limited. The addition of a small amount of clay in place of a strong natural sand is finding increasing favour. Quite large quantities are worked at Warsop and considerable quantities at Ryton.

A further increase in this practice is likely, particularly in jobbing foundries that have adopted some form of mechanical ramming for castings of medium size, since the harder ramming will call for the use of sands of higher permeability.

PRESENT AND FUTURE PATTERN OF CONSUMPTION

Exact figures concerning the use of natural and synthetic clay-bonded sand are not available, but it is possible to obtain a rough comparison between the amounts of each used at the present time.

Consumption of silica sand for all purposes is appreciably higher than that of natural sand, but much of the silica sand is used in the production of oil–sand cores, in the various air-setting, shell and hot box processes and in the CO_2 process, and although sand used for one of these purposes may later be used in a synthetic clay-bonded system, the amount involved is a small proportion of the total. On the whole, indications are that the amount of naturally bonded sand is similar to the amount of synthetic clay-bonded sand used in foundries.

This is understandable. Provided that the method of mould making and kind of castings being made are such that the physical properties of the available natural sand are suitable, control of a natural sand system is simpler. It is necessary to control the moisture content and the amount of coal dust, but beyond this all that is needed is to see that the amount of new sand added is sufficient to maintain the amount of active clay and prevent an undue build-up of dead clay. Since the natural sands contain only a small amount of clay in excess of that needed to maintain moulding properties, additions must be fairly large and restriction of the build-up of dead clay is automatic. The addition of an excessive amount of new sand does not cause any serious difficulty, and where a facing sand is used, it may contain as much as 50% new sand, and all that is needed to maintain the system is to add more new sand than the minimum required to maintain moulding properties. This greatly simplifies sand control. Moisture content is likely to vary just as widely, whatever kind of sand is used, and quite frequent testing is needed. Coal dust is also influenced by factors quite independent of the kind of sand used. Tests for permeability and particularly for green strength and shatter index, however, need not be made with the frequency required when using a synthetic sand. This is discussed in more detail in Chapter 7.

CHAPTER 6

Organic Additions to Clay-bonded Sand

In the iron foundry, moulds are not usually made from clay-bonded sands alone, and one or more of a wide range of organic materials is added. The nature of the addition depends on whether the mould is to be cast in the green state or after drying.

When and why the additions were first made is a matter for conjecture, but coal dust was certainly being added to moulding sand a hundred years ago. The use of fibrous materials goes back much further. Benvenuto Cellini published an autobiography in the sixteenth century in which he described his sand preparation, and an English translation in two volumes was published in London in about 1900, but the edition was limited and it is difficult to obtain a copy. In referring to the preparation of sand to be used in the casting of statuary, Cellini said that his 'moulding clay, a lean, not a fat clay, was mixed with horse manure and cloth shreddings and beaten with an iron bar two inches thick.' In view of the present-day value of a casting made by Cellini, devotees of horse manure can claim with some justification that they are in good company.

Although there is no information as to why the addition was made, it is not difficult to appreciate the reasoning which could have led to the use of fibrous materials in clays or heavily bonded sands to prevent cracking on drying, since straw was added to clay bricks made on the banks of the Nile before Exodus, and hair or other fibrous material has for a long time been used for the same purpose in plaster.

It is less easy to find the reason which first prompted adding coal dust to moulds which are to be poured in the green state. Some books and papers published before 1914 state that the coal dust burns out, leaving voids in the sand and so improves permeability, but since the coal dust first swells and frequently does not 'burn out' until after the casting has solidified, this explanation cannot be sustained. Moreover, coal dust is not used to reduce blowing and eliminates misrunning only under rather special conditions. Perhaps the first use of coal dust was accidental.

PURPOSE OF THE ADDITION

The main objects of additions are (1) to improve surface finish, (2) to prevent scabbing, or (3) to influence the bonding properties of the clay.

The same addition may be used to improve surface finish and to prevent scabbing, but under severe conditions a second material may be needed. For example, coal dust is used mainly to improve surface finish and is also used more widely than any other material to prevent scabbing, but it is sometimes necessary to add a second material, such as wood flour, when the design of the casting is such that the risk of scabbing is high. All the materials added to improve surface finish are reducing agents; those added solely in order to prevent scabbing are fibrous. These are used widely in the iron foundry, but very rarely in the steel foundry. On the other hand, the additions used to influence the bonding properties of the clay are used widely in the steel foundry, but rarely in the iron foundry except when very tough sands, such as those used in high pressure moulding, are required. They are used everywhere as green binders in conventional oil–sand cores.

The number of materials suitable for improving surface finish or modifying the bonding properties of the clay is limited and they are well known, but sources of fibre are much wider and occasionally an isolated foundry may be using a material which is not generally known. Some fibres are used nationally and others have a mainly local significance.

The following list includes most of the materials used, but cannot claim to be complete:

Reducing agents: coal dust
pitch
fuel oil

Fibres: wood flour
peat
straw
chaff
spent hops
horse or cow hair
shoddy
horse manure

Binders: dextrin
starch
molasses
sulphite lye

REDUCING AGENTS

(1) Coal dust

There is a definite improvement in the surface finish of castings when coal dust is added to the moulding sand. A reason for this improvement which has been put forward frequently, is that when coal dust is present, there is present a layer of gas at the mould surface which prevents contact between mould and metal, but it is difficult to understand how such a layer could persist. The improvement in finish is obtained even when coal dust free from very fine particles, say those finer than 100 μ (150 mesh), is used, in which case the coal dust does not coat the sand grains but appears as discrete particles. In this case, the pressure due to the gas must be not less than the pressure due to the column of metal between the appropriate point on the mould and the top of the metal in the pouring basin. Gas pressure must not exceed metal pressure or gas will bubble through the metal itself. It is difficult to imagine a mechanism whereby the gas pressure at each and every point in the mould will automatically adjust itself to balance the pressure due to the metal column, whether the height of that column is 2 in or 2 ft. This conception, together with that concerning the increase in permeability when coal dust 'burns out', must be discarded.

Surface finish

When a casting is poured into an undried mould containing sand and clay only, it may be found that in certain areas a layer of sand is so firmly fixed to the casting that it cannot be removed by vigorous brushing with a wire brush. It can be shown that the sand is cemented onto the casting by a thin layer consisting of silicates, mostly those of iron and aluminium.

Sand adherence of this kind can, and in most cases probably does, develop after the casting itself is solid. If a fairly thick casting, for example a bar 2 in diameter, is poured and is removed from the mould immediately after it has solidified, a considerable number of sand grains can be seen adhering to it. If this bar is then plunged for half its depth into a sand containing coal dust, leaving half exposed to the air, and allowed to cool, then it will be found that that part of the bar which has been buried in the sand can be cleaned by a light brushing, leaving a blue skin. The part of the bar exposed to the air cannot be cleaned in this way and although the adherent sand can usually be removed by shot-blasting, the appearance of the bar is inferior.

When molten metal is poured into an undried mould containing sand and moist clay only, the gas present in the pores of the mould is a mixture of air and steam which reacts readily with iron at high temperatures. The exact nature of the reaction products when the iron is liquid and contains 2% or so of silicon and 3% of carbon is uncertain, but at temperatures not too far below the melting point of the metal, iron oxide is formed. The oxide, which is not only liquid itself at this stage, but is also chemically active, reacts easily with clay and sand to form mixed silicates with still lower fusion point. When the casting cools, these silicates finally solidify, binding the sand grains to the casting.

By drying the mould, and so removing the free water which is partly responsible for the formation of iron oxide, the reaction can be lessened, but it cannot be entirely suppressed since oxygen is still available in reduced amount from the air in the mould pores and the combined water in the clay.

When coal dust is added to the sand, strongly reducing decomposition products fill the mould cavity and the formation of iron oxide is suppressed completely.

The mechanism was shown clearly by Dietert *et al.*,[1] who embedded steel pins in cylinders of clay-bonded sand which were heated in a tube furnace to temperatures well below the fusion temperature of the pins. Various gases were passed through the tube and the sand test pieces throughout a test and until the furnace had been allowed to cool to room temperature.

When an oxidising gas — air, steam or carbon dioxide — was used, the pin disappeared entirely, being replaced by slag which had penetrated deeply into the sand. When the oxidising gas was replaced by a neutral gas, such as nitrogen, slag formation was restricted to such an extent that there was only a thin layer which caused the sand to adhere to the casting. When a strongly reducing gas such as coal gas or hydrogen was used, the pin was unaffected.

Dieterts' tests reflected what happens in a mould with considerable accuracy, but also some exaggeration. When an oxidising gas was used, large amounts of iron oxide and slag were formed, as they are at the mould–metal interface when free moisture is present in the sand; with a neutral gas the amount of slag formed depended on the smaller amount of combined water in the clay, as when the mould is dried; with a reducing gas no oxide or slag is formed, as when coal dust is present in the sand.

Metal penetration

There is some evidence that under appropriate conditions coal dust reduces the extent of metal penetration into the small passages between the sand grains, but in all the experiments coal dust reduced permeability, and it may be that this factor was the cause of the reduced penetration. On the other hand, it is just possible that the reducing action of coal gas may have been a contributory factor.

Molten iron or steel does not wet silica and if a small amount of iron is poured into a flat open mould, a layer partly covering the bottom is formed, as in Fig. 42.

Surface tension at the metal/sand/air interface is high and the contact angle, θ in Fig. 42, exceeds 90°.

With cast iron the thickness of the layer is about $\frac{1}{8}$ in, and if a casting much thinner than this is required, the height from casting to pouring bush must be increased, so as to provide pressure to force metal into the narrow mould cavity. If a casting should be $\frac{1}{8}$ in thick and the amount of iron poured is only half that required to fill the casting, the result is not a casting $\frac{1}{16}$ in thick, but half a casting the full thickness.

A similar effect was seen in rainwater pipes made by spinning in sand-lined moulds. As the weight of metal poured was reduced, the thickness of the pipe was also reduced, but only up to a certain point. When it was insufficient to make a pipe $\frac{1}{16}$ in thickness, a thinner pipe was not obtained, but a spiral $\frac{1}{16}$ in thick. By increasing the speed of rotation, *i.e.* by increasing the pressure, a slightly thinner pipe could be made.

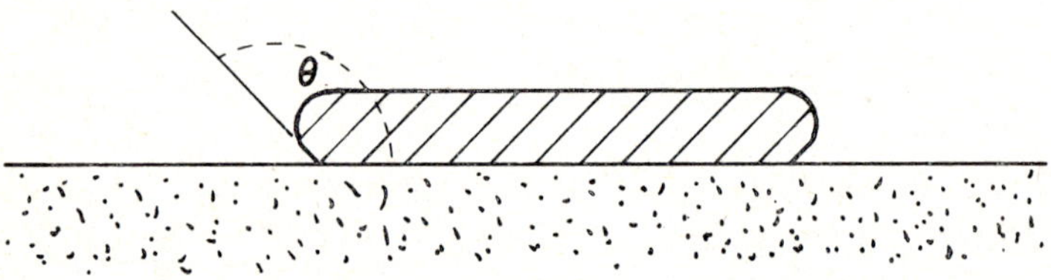

FIG. 42. Molten iron on sand surface.

It is clear that a very considerable pressure is needed to force molten iron or steel into the very narrow passages between sand grains in a mould. However, Hoar[2] showed that the presence of a very small amount of iron oxide reduces the contact angle almost to 0 and allows penetration to proceed under a very low pressure. At the temperature of molten metal, iron oxide itself is liquid and penetrates easily.

Hoar's work was carried out on low carbon steel and its relevance to cast iron, with much higher carbon and silicon contents, is doubtful.

Reduction in scabbing

Coal dust is used more widely than any other addition in order to reduce the incidence of scabbing. It is usually successful, although under particularly severe conditions it may fail.

Various reasons have been put forward — that it reduced expansion, which is not true, or that it reduced hot strength, which is true, but irrelevant — but its main effect is that it increases hot plasticity. The coal dust used contains some 30% voltaile matter and most coals of this type partially melt at a temperature of 450°C, producing a viscous liquid which persists over the range of temperature in which silica expands most rapidly.

It is not possible to show this effect in a sand test piece heated in a furnace, since there is a steep temperature gradient and the liquid formed in one layer decomposes before any is formed in a neighbouring one. At no time, therefore, is there more than a narrow annulus of sand containing liquid.

The effect has been shown qualitatively by using steel grit bonded with a clay, heating in an induction furnace and measuring deformation. The results are shown in Fig. 43.

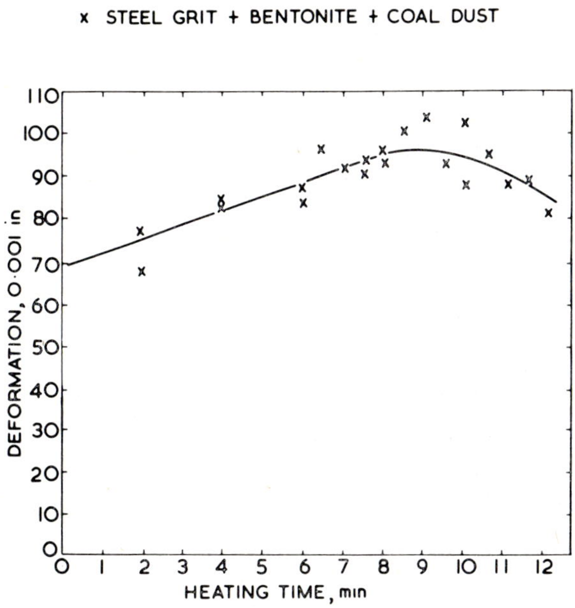

FIG. 43. Deformation of bonded grit containing coal dust.

Attempts to obtain fast heating rates resulted in disintegration of the test piece and the real nature of the effect of coal dust is not shown.

Sand test pieces can be heated more rapidly in a dielectric heater and test results are shown and discussed in Chapter 8.

Other effects of coal dust

(1) *Permeability.* Coal dust is a granular material which has no bonding properties at room temperature, and its effect on permeability is the same as that of any other granular material with the same grain size distribution. Moreover, one grain of coal dust has exactly the same effect as a grain of silica of the same size, but owing to the difference in specific gravity it would have only about half the weight. Therefore 5% coal dust will have the same effect on permeability as 10% silica with the same grading.

The size of the coal dust itself is not the only factor. When metal is poured into a mould, any coal dust which is heated to a temperature exceeding 450°C melts, swells and finally decomposes, leaving a residue of coke. When there is ample room for expansion, and this is always present in the pores of a sand, the coke produced is friable and breaks down to a fine powder — some of it is so fine that it can be found in most samples of air-borne dust collected in iron foundries. The ultimate effect of coal dust, whatever its size, is therefore to reduce permeability, provided that the coke formed by its decomposition remains in the sand.

There are therefore two factors to take into account: the effect of the coal itself, and the effect of coke.

(a) Effect of coal dust. The influence of the coal dust itself depends on the relation between its grading and that of the sand; if the coal dust is finer than the sand, permeability is reduced, but if size distributions are similar, permeability is not affected. Since coal dust forms only a minor part of the sand, it cannot raise permeability.

The reduction in permeability can be considerable. The effect on permeability over a wide range of moisture contents was measured when 6% of a commercial grade of coal dust, described as 'fine', was added to each of two sands, a coarse synthetic sand and a fine natural sand. The grading of the two sands and of the coal dust are given in Table 1.

Table 1. *Grading of sands and coal dust*

Aperture (μ)	Synthetic sand	Natural sand	Coal dust
353 (44 mesh)	1·2	0·1	—
251 (60 mesh)	23·2	2·0	3·9
152 (100 mesh)	60·2	12·9	15·6
104 (150 mesh)	7·3	32·4	16·8
76 (200 mesh)	0·7	18·3	13·2
53 (300 mesh)	0·2	12·3	14·3
−53 (−300 mesh)	0·2	7·0	36·2
Clay	7·0	15·0	

The permeabilities of the sands with and without coal dust are shown in Fig. 44.

The coal dust is much finer than the synthetic sand and reduces permeability markedly. It is spread over the same range of sieves as the natural sand, but has appreciably more finer than 53 μ. Permeability is therefore reduced, but to a much smaller extent.

It must, however, be borne in mind that the margin of safety, so far as permeability is concerned, is so much greater when using the synthetic sand that a large reduction here may be less harmful than the much smaller reduction in the permeability of the natural sand.

(b) Effect of coke. The amount of fine coke produced depends on the amount of coal dust which is heated to a temperature exceeding 450°C, but the amount which remains in the sand depends also on the way the sand is treated before it is prepared for re-use. In the foundry described in Chapter 3, in which identical castings were made by floor moulders

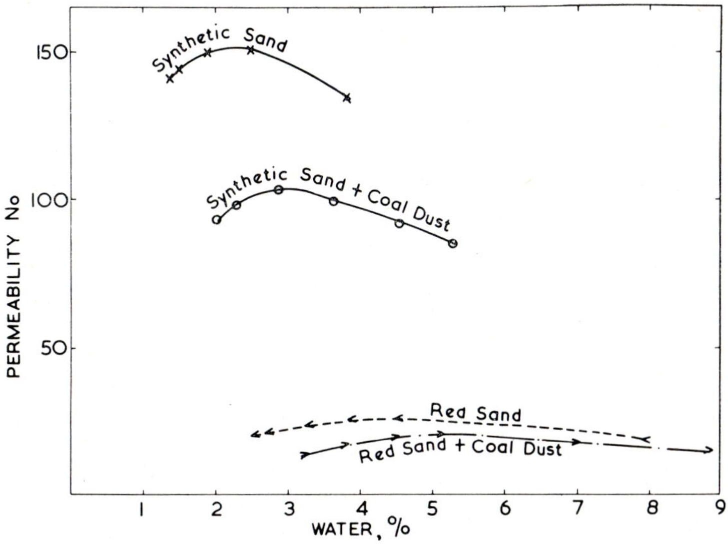

FIG. 44. Effect of coal dust on permeability.

and on a mechanised plant, the amount of fine coke produced differed a little because of the difference in time during which the castings remained in the mould, but was of the same order. The changes in permeability were very different — a steady fall in the hand moulding section and an increase in the mechanised plant.

The fall in the hand moulding section can be foreseen, although there is some uncertainty as to the rate of fall. In the mechanised section even the direction of change may be uncertain when a natural sand is used. All natural sands which occur in this country contain an appreciable amount of small sand grains and under some conditions these may be removed along with coke and the smaller particles of coal. Removal may take place anywhere in the sand system when dry sand is disturbed and the amount removed may be large, particularly when the shake-out has a down draught exhaust system so designed that a current of air is drawn through the falling sand. The permeability of the sand system may then be higher than that of the original sand before coal dust was added. Indeed, the successful use of natural sand in mechanised foundries frequently depends on the removal of sufficient fines to prevent any continuing fall in permeability.

(c) Coal dust, permeability and misrunning. Provided that metal temperature is satisfactory, the occurrence of misruns in a foundry producing thin castings is usually due to the use of a mould with low permeability. The air inside a mould cavity expands rapidly when it is heated by the incoming metal and, unless there is a ready means of escape, sets up a back pressure which may retard the entrance of metal long enough for any metal which has entered the mould to cool to such an extent that a misrun is obtained.

Table 2. *Grading of sand and coal dust*

Aperture (μ)	Sand (%)	Coal dust (%)
+500 (30 mesh)	0·2	1·2
+353 (44 mesh)	2·3	3·1
+251 (60 mesh)	30·2	9·1
+152 (100 mesh)	62·9	13·7
+104 (150 mesh)	3·1	13·6
+ 76 (200 mesh)	0·6	11·7
− 76 (−200 mesh)	0·4	47·6
Clay	0·3	—

Occasionally, however, the inclusion of coal dust in the sand has been found to eliminate misrunning, although it caused a pronounced fall in sand permeability. In one such case a bowl, 3 ft in diameter, was poured open-end down, with metal entering at the rim. In order to avoid as far as possible any undue back pressure from the expanding air inside the mould, a sand with a permeability of 200 and free from coal dust was used, but even so, misrunning was so severe that no mould was filled to within 2 in of the top, and even a whistler off the highest point in the mould cavity did not remove the defect. No further misrunning occurred when 5% of a fairly fine coal dust was added to the sand, reducing permeability to below 150.

(2) *Green strength, shatter index and moisture range.* Coal dust has no bonding properties, but it was noted in Chapter 3 that when a coarse and a fine clay-bonded sand of similar clay content were mixed, the mixture was stronger than either of the constituents.

The addition of coal dust reduces clay content by dilution and if the grading of the sand and coal were the same, strength would be reduced.

When, however, the coal dust is much finer than the sand, there is also the tendency to raise strength, which may be even more significant. The final result depends on the grading and relative proportions of the sand and coal dust.

Nicholas and Roberts,[3] investigating the relationship between mould properties and the soundness of castings, added coal dust to a silica sand and measured its effect. The grading of the sand and coal dust are given in Table 2.

The effect of 2, 4 and 6% of coal dust on green strength and shatter index are shown in Figs. 45 and 46.

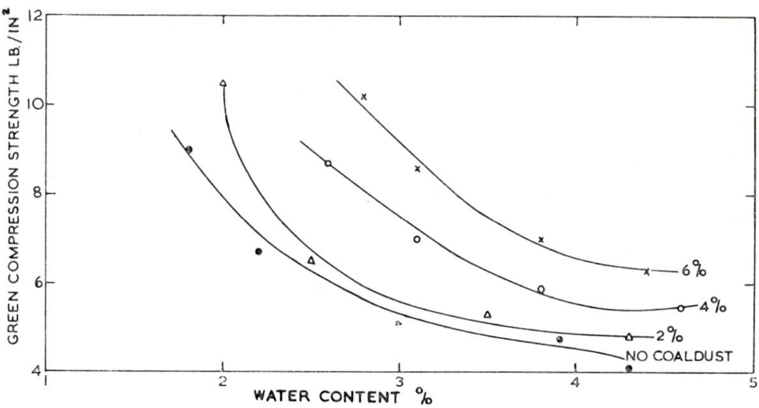

FIG. 45. Effect of coal dust on green strength (after Nicholas).

The addition of coal dust has increased the amount of water which must be used, the curves being displaced in the direction of higher moisture content. In the presence of 6% coal dust, the additional water required was about 1% and the increase was almost entirely due to the fraction finer than 76 μ (200 mesh). There was an appreciable increase in strength.

Differences in shatter index were small and no greater than would be expected from duplicate tests on the same sand. The increase in strength was therefore accompanied by a slight increase in brittleness.

(3) *Effect of decomposition products of coal dust.* When coal dust is heated, both gas and organic liquids are given off and there is some evidence that the liquids may condense in the sand and influence green and dry strength. Hird[4, 5] carried out a number of experiments with coal dust during the period 1930–1937.

In one of these, he poured iron into a mixture of silica sand and coal dust, a mixture which had negligible strength and plasticity and could not easily be shaped into a mould. The whole of the sand which had been used was mixed with more coal dust, and the sequence repeated a number of times. After ten cycles a sand was obtained which was quite suitable for making moulds of simple shape. The physical properties of the sand were not measured and it is not possible to say whether there was an appreciable development of green strength or whether the improved moulding properties were entirely due to an increase in plasticity.

Piper,[6] when carrying out his tests on the durability of clays, described in Chapter 4, found that when coal dust was present in the sand, the fall

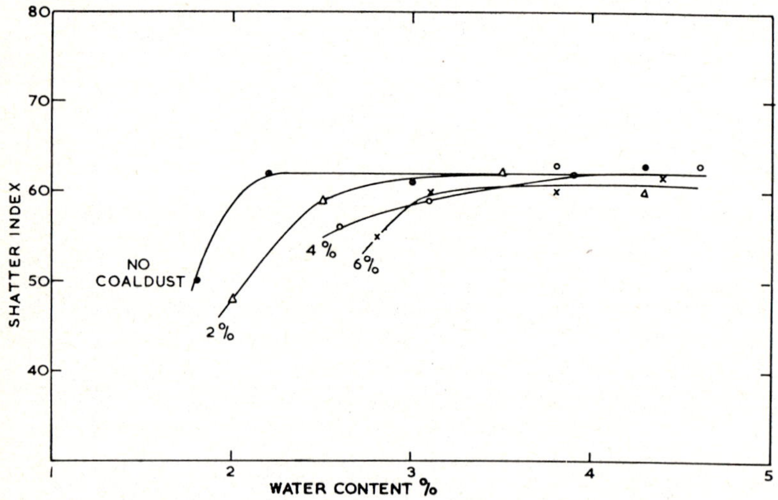

FIG. 46. Effect of coal dust on shatter index (after Nicholas).

in dry strength with continued re-use was greatly reduced. In some unpublished work carried out at the same time, Piper re-used a naturally-bonded red sand fourteen times, adding coal dust after each cycle, and found that dry strength remained constant. There was no obvious change in moulding properties.

(4) *Coal dust and soundness*. Womochel and Sigerfoos[7] studied the effect on the soundness of castings of making changes in the ramming or composition of the mould.

These observers found that adding coal dust to the moulding sand reduced both piping and internal porosity. Since coal dust also affects such properties as green strength and the amount of water which has to be used, it is difficult to decide how much of the improvement was due to the primary effect of coal dust and how much to the secondary effect resulting from the consequent change in sand properties. However, the improvement was greater than would be expected to follow the relatively small change in sand properties.

Other observers[8,9,10] have also noted the effect of certain carbonaceous materials in improving soundness.

Nicholas and Roberts[3] carried out a series of tests designed to show the effect of a number of variables, including sand composition, on the soundness and accuracy of castings made in undried moulds.

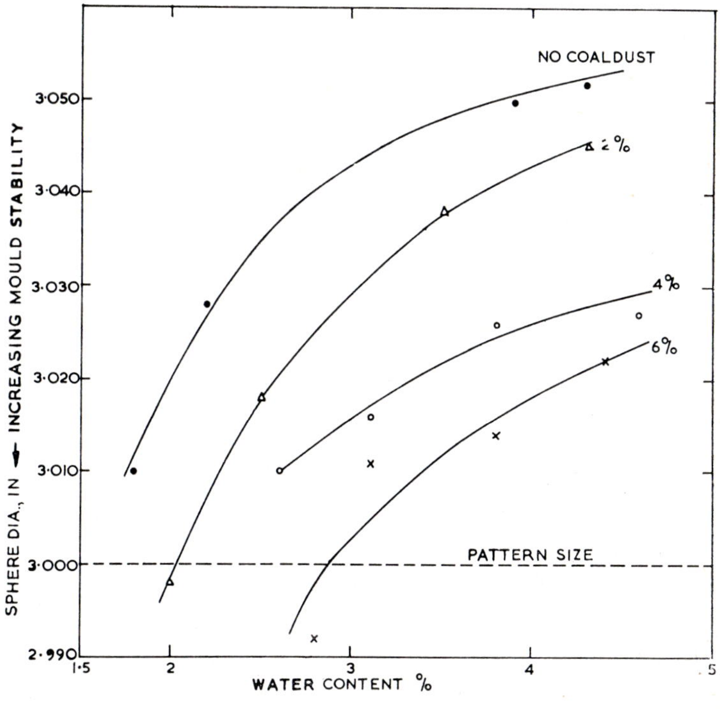

FIG. 47. Effect of coal dust on casting size (after Nicholas).

The castings made were spheres 3 in diameter and previous work by these observers had shown that there is a close relationship between the diameter of the castings, measured in three directions at right angles, and the degree of soundness or unsoundness. As the diameter of the castings increased, so did unsoundness and measurement of diameter provided a quantitative index of unsoundness.

Figure 47 shows the effect that the addition of 2, 4 and 6% coal dust to the moulding sand, has on the diameter of the spheres.

Increasing moisture content increases sphere diameter and unsoundness, but by carrying out tests over a range of moisture content, Nicholas was able to separate the effects of moisture and coal dust. The addition of 6% coal dust reduced sphere diameter by about 1% and sphere volume by 3%.

The mechanism by which additions of coal dust reduce unsoundness is not clear. The effect is not confined to coal dust and Nicholas obtained similar results with a number of other materials. These included not only carbonaceous materials such as pitch and wood flour, but also hydrated lime, 1% of which had an effect similar to that of 6% coal dust.

All the effective materials increased both green strength and mould hardness, but the reduction in porosity was much greater than could be obtained if the same change in these properties had been obtained by an increase in clay content. There does not appear to be any evidence as to the effect of coal dust with a grading similar to that of the base sand, which would not increase strength.

(5) *Coal dust and pinholes.* Under certain conditions a type of pinhole which is almost spherical in shape and is coated with a coherent layer of graphite is found just beneath the surface of iron castings. Usually pinholes of this kind are found only after prolonged shot-blasting or light machining, but in severe cases they break through to the surface.

Gittus,[11] examining the formation of such pinholes in magnesium-treated iron, found that their incidence could be greatly reduced, although not entirely suppressed, by adding coal dust to the moulding sand.

Dawson[12] extended this work to include flake graphite irons and found the incidence of pinholes could be reduced by including coal dust in the sand.

(6) *Selection of coal dust.* When choosing a coal dust, the two most important factors to be considered are its effect on surface finish and on the incidence of scabbing. The reduction in pinholes and the improved soundness reported by a number of observers are interesting and valuable, but in none of the recorded cases has there been any evidence whatsoever as to the influence of the composition or grading of the coal dust used. The increase in green strength observed when coal dust was added to certain sands was the result of a certain relationship between the size distribution of the sand and coal dust used and could be obtained at less cost by changing the grading of the sand or the clay content.

It is necessary to take into account the effect of coal dust and its decomposition products on permeability, but even in this case, the effect on selection of the kind of coal dust to be used is only marginal. Where it is cheaper or more convenient to use a sand with a permeability near the danger line, a slight increase in the size of the coal dust may be desirable, but it is safer to use a sand with an adequate safety factor as regards permeability, in order to permit the use of whatever kind of coal dust is needed to help in the production of castings which are free from scabs and have a satisfactory surface.

Two factors are important: composition and grain size distribution.

(a) Composition. The usual specification for coal dust is that the volatile matter shall be not less than 30% and the ash not more than 10%. This is generally satisfactory and only one case has been experienced where a coal dust containing 30% volatile matter did not soften in the normal manner when heated, and this led to a disastrous increase in scabbing in the foundry concerned. Most manufacturers keep to coal from the same source and thus avoid all risk.

Slight deviation from the specification, particularly in ash content, is of little importance. Owing to the reducing atmosphere inside a mould, no coal is burned to ash and the end point is coke. A coal containing 30% volatile matter produces 70% of coke and from this point of view ash content is unimportant. However, wide variations in ash content should be avoided even when the minimum volatile matter is maintained. Coal containing 30% volatile matter and 10% ash can be regarded as a mixture of 90% coal with 33% volatile matter and 10% of inert matter. If the ash content is 20% and volatile matter 30%, the mixture consists of 80% coal with 37½% volatile matter and 20% inert matter.

These may behave differently, but a considerable variation in ash content is needed before any effects become obvious in the foundry. Coal dust should not be rejected because the ash content is a little out of line.

(b) Grading. The effectiveness of coal dust depends on the evolution of a reducing gas and the presence of some liquid phase at least until the mould has been filled, and coking must not be complete until the necessary time has elapsed. The relevant time intervals increase as the size and thickness of the castings increase, and the corresponding increase in coking time can be obtained by increasing the amount of coal or its size. However, there is not complete freedom of choice. Should a coarse coal dust be used in moulds for light work, the evolution of comparatively large volumes of gas at isolated points produces depressions in the castings which mar the appearance. Fine coal dust must be used here. When fine coal dust is used in moulds for heavy castings, a very large increase in the quantity added is necessary, and there is a decrease in permeability sufficient to produce blows or 'solid' scabs. It is necessary to use a coarse coal dust which reduces the amount required and minimises the fall in permeability.

In practice, therefore, it is necessary to increase both the size and the amount of coal dust used as the size and thickness of the castings increase. At different times, attempts have been made to specify the grading appropriate for each particular section size and the following, made by Skerl[13] is typical. The recommendations are useful as a rough guide.

Section size of castings (in)	Grading of coal dust
1. $<\frac{1}{8}$	Passing 240 mesh
2. $\frac{1}{8}$–$\frac{1}{4}$	Passing 200 mesh
3. $\frac{1}{4}$–$\frac{1}{2}$	15% between 100 and 200 mesh; 85% passing 200 mesh
4. $\frac{1}{2}$–1	50% between 30 and 100 mesh; 50% between 100 and 150 mesh
5. 1–2	25% on 30 mesh; 50% between 30 and 100 mesh; 25% between 100 and 150 mesh
6. >2	50% on 30 mesh; 50% between 30 and 100 mesh

They cannot be used except as a rough guide. In order to satisfy the specification exactly, a costly sizing operation would be needed and this is neither necessary nor desirable. Even on a single casting, section size may vary and when a number of different castings has to be made, using the same sand, there are still wider variations, but only one grade of coal dust can be used. Moreover, manufacturers who produce a range of coal dusts do not as a rule quote the grading, but label their products 'superfine', 'fine', 'medium', and so on, and the size distribution of the various grades depends on the kind of grinding and sizing plant in use. It has not been possible on this account to obtain any agreement between manufacturers as to a uniform system of nomenclature, in which a particular description is always applied to a coal dust with a definite grading.

The lack of uniformity has caused occasional difficulty when changing the source of supply, since coal dusts of similar size may be given different labels by different manufacturers. Before making a change, it is desirable to carry out a sieve test on the coal dusts concerned and to rely on the results of this test rather than on the label.

Difficulties are not as a rule serious unless a range of castings varying widely in size has to be made, in which case it may be advisable to use separate facing sand for the heaviest work. When there are no such wide differences, the usual practice is to use 2–3% of a very fine grade in a foundry making very light castings and up to 10% of the coarsest grade for castings exceeding 2 in thickness and of considerable size, with appropriate modifications for castings of intermediate section.

Since the objection to an increase in the amount of fine coal dust is due to the resulting decrease in permeability, there is much greater latitude when using more permeable synthetic sands.

(7) *Control of coal dust*. In most foundry moulding mixtures, the main constituent is sand which has been used before. The proportion may occasionally be as low as 50% in some facing sands, but is usually higher

than this and may exceed 90% in unit synthetic sands. This sand contains unchanged coal dust together with coke, and due allowance must be made for the unchanged coal only, since coke does not soften and increase the plasticity of the sand at high temperature and, although it does react with any oxidising gases present in the mould, its action is much slower than that of coal gas.

The most reliable method of estimating the amount of coal dust is by measuring the volatile carbonaceous matter in the way described in Chapter 1. In a mechanised foundry in which there are long runs of very similar castings and conditions in the sand handling plant are constant, it is possible to obtain a useful control by maintaining a constant loss on ignition, but any major change in the kind of castings being made reduces the usefulness of the method.

In the absence of facilities for testing, some help can be obtained from inspection of the castings. Perhaps the most frequent method of control is to increase the coal dust whenever scabs appear and reduce it when castings misrun, a method which is usually successful, but which is applied only after some defective castings have been produced. Earlier warning can be obtained by an examination of suitable castings; the first signs of burning on near the ingate indicate a deficiency, and a shiny surface on a thin section remote from the ingate indicates an excess of coal dust. Unusually high or low metal temperature influences the appearance of the castings in the same way as a deficiency or excess of coal dust and this factor must always be taken into account.

Visual examination of the sand itself is not informative.

Control of coal dust is simpler when a facing sand is used. In this case, much of the coal dust present in the sand is sufficiently near to the hot metal for its temperature to reach the point at which volatile matter is removed, and the used sand contains only a small amount of coal dust. Moreover, the proportion of new sand used is higher than it is when using a unit sand and the two factors combine to reduce uncertainty as to the amount of coal dust introduced by way of the used sand.

(2) Pitch

The beneficial effects of coal dust are due to the formation of a viscous liquid and evolution of a reducing gas at any temperature exceeding 450°C; the most serious difficulty which accompanies its use is the fall in permeability which follows the accumulation of fine coke.

From these considerations, it appears that any other organic material which melts forming a viscous liquid and gives off a strongly reducing gas when heated may have similar effect. Moreover, if at the same time the residue left after heating is small, the effect on permeability is minimised.

Pitch is clearly a possibility. Some 80% consists of volatile carbonaceous matter, or nearly three times the amount in coal and a reduction in the amount used of the same order is to be expected. Moreover, the residue of 'coke' is only 20%, and the two factors combined would ensure that the finely divided 'coke' remaining after pitch had been used would be only about one-tenth that left by coal dust. Changes in permeability would then be negligible.

Although it was generally recognised that pitch was a possible alternative to coal dust which might have some advantages, little headway was made at first because the only means of reducing pitch to a suitably fine state of division was by grinding. Ground pitch contains quite large proportions of very fine particles which readily become air-borne and make handling the material dirty and potentially harmful. However, in the early post-war period a method was developed for producing pitch in the form of small pellets with a fairly uniform diameter, less than 1 mm, and entirely free from dust. The material was cleaner to handle than coal dust, and none of it was of such size that it could remain air-borne and be inhaled.

Once this material became available, it was used in place of coal dust on a considerable scale. At no time, however, did it seem likely that either material would ever be used to the exclusion of the other and eventually a fairly steady balance was struck. Something of this kind was to be expected. It cannot be said that pitch is 'better' than coal dust or *vice versa*, but rather that the properties of the two materials are such that one is in some circumstances more suitable and in others less suitable than the other.

In order to select the material which will give most satisfactory service it is necessary to know how their properties differ.

Effect of heat on pitch

Although there are resemblances between the behaviour of coal dust and pitch when heated, there are also differences in detail.

When coal dust is heated, part of it liquefies at a temperature of about 450°C. Decomposition of the liquid commences immediately and proceeds at an increasing rate as temperature rises until only coke remains. The grade of pitch used in foundries is sufficiently fluid to be poured at 100°C, and three changes occur — to some extent at the same time — as the temperature rises. The more volatile constituents are driven off, some constituents decompose and some polymerise to form others of higher molecular weight and higher viscosity.

At temperatures above 450°C the rate of change appears to be much slower with pitch than with coal dust. In consequence, the proportion of pitch usually found adequate in moulding sands is only about one-sixth that of coal dust needed under the same conditions, whereas the relative proportions of volatile matter indicate that it might be more than one-third.

Surface finish

The effect of pitch in preventing the formation of iron oxide and the consequent adhesion of sand to the castings is similar to that of coal dust, and from this point of view the two can be regarded as interchangeable. In two aspects, however, there are differences in surface finish.

Only one grade of pitch is available and this is appreciably coarser than the finest grades of coal dust, with the result that slight pockmarking may occur on very thin castings. The defect is most obvious when using new sand and pitch and tends to fade as the proportion of used sand increases. Any pitch which has been sufficiently close to the casting surface to be heated to a temperature of 100°C or more melts and is more uniformly distributed, and marking disappears. Nevertheless, new sand and pitch must be added and it is wise to avoid using pelletted pitch on very thin castings when appearance is important.

At the other end of the scale, sands containing pitch give a much better surface to thick castings, particularly to those which, for some reason, must be poured slowly. The surface of the casting shown in Fig. 109(b), which was produced in a mould containing 1% pitch, is superior to that which could have been obtained by the use of coal dust, whatever the grade and amount.

In less extreme conditions, equally good finish can be obtained from pitch and coal dust, but it is necessary to have the correct amount of the appropriate grade of coal dust, whereas only one grade of pitch is needed and 1% is adequate. It follows that pitch is more satisfactory than coal dust when the range of castings which has to be made in the same sand is such that more than one grade of coal dust would be needed.

Scabbing

Pitch and coal dust are equally effective in reducing scabbing. For this purpose the amount of pitch needed is about 1% and excess is harmful. This is discussed in Chapter 10.

Green strength, shatter index and moisture range

Since the pitch used is finer than the sand, it might be expected that it would affect the properties of the green sand in much the same way as they are influenced by coal dust. However, the effect is extremely small since the amount of pitch used is so small, and very fine grains are absent.

Dry strength

When a sand containing pitch is dried at a temperature exceeding 100°C, the pitch melts and resolidifies when the sand cools to room temperature.

As a result, there is a large increase in dry strength, and the addition of 1% pitch raises dry strength of most clay-bonded sands by a factor between two and five.

This may influence the behaviour of the sand in the foundry and, indeed, the dry strength factor is sometimes decisive in deciding whether to use pitch or coal dust.

When the castings are shaken out as soon as possible after pouring, the high dry strength does not develop. The hot sand, containing molten pitch, breaks up readily and is partly mixed with cold sand. No serious difficulty is found in reconditioning the sand.

If, however, the poured moulds have to be allowed to stand until cool, the pitch solidifies and the strength of the mould increases. Removing the sand from the box becomes more difficult — very much so if the size of the mould and castings is such that the whole of the sand has reached a temperature of 100°C before cooling down. When the sand has been removed from the box, considerable effort may be needed to force it through the shake-out grid and if there is a rotary screen in the system, much good sand may be rejected with the oversize and small pellets which may survive the milling operation. Should these appear on a mould face they may cause at best an unsightly mark, and at worst a scrap casting. Under these conditions coal dust should be used in preference to pitch.

Soundness

Nicholas and Roberts[3] in their experiments showing the relation between the composition of the sand and the accuracy and soundness of castings, found that the influence of pitch exceeded that of coal dust. The amount of pitch present in the sand was 2%, a higher proportion than is generally used in practice. It is uncertain whether a sand with 1% differs from a sand with 2% pitch so far as casting soundness is concerned.

Pinholes

The effect of pitch is similar to that of coal dust.

Pitch as a waterproofing agent

Pitch itself does not absorb water and under suitable conditions can increase the resistance of other materials to wetting by water.

If a sand containing as little as 1% pitch is heated to a temperature exceeding 100°C and allowed to cool, it is more difficult to re-moisten the sand uniformly than it would be in the absence of pitch. If the sand is allowed to stand in a moist atmosphere, the absorption of water and loss of dry strength are also less marked.

Under some conditions this resistance to wetting is a nuisance; under others it is a help.

When a moulder reconditions backing sand on the floor by piling it into a heap, adding water and mixing by hand or by means of an aerator, the use of pitch must be avoided. The sand is almost completely waterproof under these conditions and most of the water runs off. When the sand is milled the difficulty does not arise.

On the other hand, the presence of pitch is an advantage when moulds are skin-dried before pouring, since absorption of water on standing is prevented. A further advantage is that pitch is a viscous liquid over the range of temperature used when skin-drying, and reduces the risk of cracking the mould, even when there are steep temperature gradients in the sand. The use of skin-dried clay-bonded sands has decreased — a facing of silicate-bonded sand is frequently more convenient and equally satisfactory — but when it is used the presence of pitch is of very considerable assistance. This was reported by the team which visited America to study productivity in grey iron foundries.[14] A synthetic sand containing 3% pitch was used and the moulds passed through a tunnel where they were heated by air at 900°F (482°C), blown down on to the surface through a large number of spaced-out pipes. In the absence of pitch, the mould would certainly crack under these conditions, but in its presence there was sufficient viscous liquid present, wherever water had been removed, to prevent cracking. Halfway through the tunnel cold air was used in place of hot, and a mould for a casting a few tons in weight was ready for coring up in a little less than 1 h. The surface of the mould was extremely hard and abrasion-resisting and there was no striking back of moisture, but there was no serious difficulty at the shake-out because except for 2 in or so near the face, the mould consisted of weak green sand.

(3) Heavy oil

Heavy oil has occasionally been used in place of coal dust or pitch. It has a similar effect on surface finish and sand adherence, but it has not been used on a large scale and no detailed comparisons have been made. It is unlikely that it would be effective in eliminating scabs on large moulds with flat tops.

Heavy oil has one advantage — it widens the usable moisture range of a sand because it confers some plasticity in its own right and in its presence the sand can be moulded when the moisture content is appreciably lower than would be necessary in the absence of the oil. Other materials — dead clay and very fine coal dust, for example — also widen the moisture range, but they all require an increase in the amount of water present.

This is a useful property, but in spite of it the use of oil in place of coal dust or pitch has usually been abandoned after a period of use on account of its influence on working conditions. Some of the oil decomposes when a casting is poured and sufficient free carbon is formed for the sand to be black, and the black oily material is so unpleasant to use that moulders object to it. The smell after pouring is extremely objectionable.

Nevertheless, the use of oil may be useful as a temporary makeshift when a combination of a short moisture range, hot sand and a hot climate makes mould production particularly difficult.

FIBROUS MATERIALS

Fibrous materials are used to reduce the risk of scabbing and, to a smaller extent, cracking of moulds, particularly loam moulds, during drying. They increase the deformation the sand can undergo without cracking and have this effect in the cold as well as when hot, since their action is not due to the formation of a liquid on heating.

All the fibrous materials are more effective than coal dust or pitch in preventing scabbing, but in green sand moulds they are most frequently used to supplement the pitch or coal dust which is in any case added to improve surface finish and make cleaning easier. In dry sand and loam moulds, which are frequently given a coating of plumbago or other protective material, the fibrous materials are used without coal dust or pitch.

Although all the fibrous materials have a similar effect on the properties of the sand, they are not necessarily interchangeable. In green sand moulds only short fibres are used, partly because it is difficult to obtain a uniform distribution of long fibres by means of a short milling operation, and partly because of the occasional deleterious effect of long fibres on the surface of uncoated moulds. Longer fibres are more effective in dry sand moulds in which the mechanism of the formation of a scab is somewhat different from that in green sand moulds, partly because they hold together loose pieces of the mould, should it crack during drying or pouring. Indeed, if a sand test piece containing long fibres is broken in the compression test machine, it can be picked up by grasping the top edge, without the loss of anything beyond a few sand grains. Molten metal may enter a crack and give the fettlers a little extra work, or a few sand grains may be found embedded in the top surface of the casting, but the risk of obtaining a serious defect is small.

Pliable fibres are the most satisfactory to use. The objection to stiff fibres is that they may be bent during the mould-making operation and that some of those near the mould surface may spring and straighten, tearing some sand from the mould. This can be a nuisance, particularly when sweeping up loam by rotating a barrel on to which the loam is fed, the excess being cut off by means of a suitably located fixed knife. Should a fibre spring just before the operation is completed, a groove may be torn out of the loam and to fill this it is necessary to feed on a little more loam and turn the barrel again, with a risk that other fibres may spring elsewhere. Alternatively, the groove can be ignored and the blacking applied, but even a shallow groove can carry a layer of mud which will not dry out completely in the stove and a scrap casting may be obtained.

Of the various fibrous materials available, wood flour is one of the most widely used in green sand moulds. Peat is satisfactory under some conditions. All have an influence on the properties of the sand which must be taken into account.

Wood flour

The effect of wood flour on the behaviour of sand when it is rapidly heated and on the elimination of scabbing is discussed in Chapters 9 and 10.

The secondary effects on the other properties of the sand are, however, considerable. The addition of 1% of wood flour raises green strength a little without greatly increasing the shatter index and the sand may become more brittle and stripping may become a little more difficult. The change is small.

Dry strength is reduced and if the type and amount of clay present are such that dry strength is already rather low, sand washing may occur. It is advisable to avoid wood flour where its use would reduce dry strength to much below 50 lb in^{-2}. It is possible to offset the reduced dry strength by adding a little dextrin or similar organic binder, but the number of different constituents in a moulding sand should be kept to a minimum so as to simplify control.

One effect of wood flour is that when a certain proportion is exceeded, it becomes impossible to ram a mould properly. If a hand rammer is being used, then once the mould has been partly compacted, compression by the rammer head in one place is accompanied by an expansion elsewhere; on a squeeze machine, the sand compacted during squeezing may expand when the pressure is removed. The rammed sand feels, and to some extent behaves, like sorbo rubber. The level at which this becomes noticeable depends on the kind of wood flour in use and is usually between 2 and 3%. Since wood flour is most frequently used in conjunction with coal dust and few foundries have facilities for estimating the proportion of these when both are present, control is difficult, particularly when using a unit sand. It is advisable to restrict the amount present to 1% and to use it in a facing sand only.

Like coal dust, wood flour gives off a reducing gas when heated and since the volatile matter in wood exceeds 70%, which is rather more than double that in coal, the addition of 1% wood should permit a reduction of 2% in the amount of coal dust. In practice, the reduction must be less than that calculated on the basis of volatile matter, since the gas from wood is less strongly reducing than that from coal. During the last war, coal dust was at times in short supply and the Institute of British Foundrymen[15] published an article in which it was suggested that wood flour could help to fill the gap.

The grading of the wood flour is of some importance. Very coarse material should be avoided when the castings are small and the moulds are not dried, owing to the risk of pinholes such as are obtained with coarse coal dust. The maximum size is of less importance when the moulds are dried and some foundries get by with sawdust from the pattern shop. The very fine material obtained from a sanding machine gives variable results and is unsuitable. It is more satisfactory to buy a graded wood flour than to use any waste material from the pattern shop.

Suppliers of wood flour use a nomenclature different from that used by foundrymen when describing sand. A 60 mesh sand refers to sand which remains on a 60 mesh sieve and usually to sand which has also passed the next coarsest sieve in the series. A 60 mesh wood flour, on the other hand, passes through a 60 mesh sieve and includes all sizes finer than that.

Since very fine material is always present in wood flour, its addition is always accompanied by a fall in permeability. The charcoal formed when wood is heated in a reducing atmosphere retains the shape and size of the original wood fragments and there is no rapid disintegration such as occurs when coal dust decomposes and forms fine coke. The fall in permeability is limited to that due to the fines in the original wood flour.

Wood flour has a considerable affinity for water, and the addition of 1% to a sand raises the amount of water needed to produce a mouldable mixture by $\frac{1}{2}$–$\frac{1}{3}$%. The wood flour should not be dried before use, since dry wood absorbs water rather slowly. A sand to which the dried flour has been added may appear suitable for moulding, but the subsequent migration of $\frac{1}{2}$% water from the clay to the wood may reduce the plasticity of the sand so that it becomes unusable, although the total moisture content is unchanged.

Occasionally flour from 'hard' or 'soft' wood may be specified, but the effects of the two are so similar that it is almost impossible to deduce the kind of wood present from the physical properties of the sand or its behaviour in the foundry. Most users take 'run of mill' wood flour, specifying only the mesh number.

Peat

In many ways the effects of peat on the behaviour and properties of sand resemble those of wood flour. It is equally effective in reducing the incidence of scabbing and slightly increases green strength without raising the shatter index. Care is necessary in selecting the peat to be used, since properties change as the depth below the surface of the peat bog increases.

Peat does not impair the moulding properties of sand in the manner described when wood flour in excess of 2–3% is present. On the other hand, once it has dried out, it is even more resistant to moisture absorption than wood flour and this factor limits its usefulness.

The use of peat is by no means new. Much of the horse manure formerly used in foundries consisted mainly, in some districts at least, of peat which had been used as stable litter.

Horse manure

The value of horse manure lies in the amount of fibrous material present which may be derived from the diet of the horse or the stable litter, or both.

At one time it was by far the most widely used fibrous material and was favoured by both dry and green sand moulders. This wide use was probably due to its low price — in the days when most traffic was horse-drawn it was cheaper for stables in the centre of a town to give the material to a neighbouring foundry than cart it a considerable distance for sale to a well supplied countryside.

With the reduction in horse-drawn traffic, supplies have dwindled, but this change has probably only hastened its departure from the foundries. The digestive system of the horse had one advantage — it softened and made pliable the stiff fibres of hay and grass stalks. On the other hand, quality was variable and the effects depended on the horse's diet and on the stable litter usually present. There is no need to lament its disappearance from the foundry.

Straw, hay and grass

Both straw and hay, chopped into lengths of $\frac{1}{2}$–2 in are used to a considerable extent. They have the disadvantage that the fibres are stiff, since most are derived from the dried stalks.

Grass, green or dried, obtained from lawn mowings is much more satisfactory than either straw or hay because it consists of leaf, which is pliable, without any stalk. The limitation is solely one of supply. It can be obtained during a few months only in each year and although it can be dried and stored, it is sufficiently bulky to be a nuisance even if an adequate supply were available.

Spent hops

These are fairly widely used. As received, the hops are flat plates rather than fibres and the mill should be fitted with an effective disintegrator which will chop them up. They are then entirely satisfactory.

The moisture content of the hops as received may vary widely, but it is not necessary to measure the moisture present in each batch in order to be able to add a constant amount. Bulk density increases with increasing moisture content and although a given volume is heavier when its moisture content is higher, the extra weight is due almost entirely to the additional water. A given measure always contains approximately the same weight of dry hops and it is only necessary to determine this weight on one sample. The method is not exact, but is sufficiently accurate for control purposes.

Horse and cow hair, shoddy

These materials are satisfactory, but the supply is limited and their use is mainly confined to a few localities.

Asbestos

This has been used on an experimental scale and is very successful unless the fibres are too short.

Asbestos differs from the other fibrous materials in that it is inorganic and the amount broken down during and after casting is less than when organic materials are used. Adequate control could therefore be obtained by adding asbestos in proportion to the amount of new sand entering the system.

Nevertheless, the material has never been used as an addition for foundry sands except on an experimental scale, because of possible risk to the health of operators. Whether this risk is serious when handling sand containing 0·5% of asbestos is uncertain, but with so many alternative non-toxic materials available there is nothing to be gained by using one which carries even a suggestion of risk.

Quantity needed

The influence of the fibrous materials depends on the number of fibres present and the proportion by weight on specific gravity, thickness and length. It might be expected that with so many different materials, there would be appreciable differences in weight, but in practice this is not so. Except in the case of horse manure, which contains a variable amount of non-fibrous material, and when available is rarely measured by an instrument more accurate than a shovel, the proportion of any of the long fibres added is usually between 0·5 and 1·0%. It may be that in some cases a smaller quantity would suffice, but the very small saving in cost which would ensue does not supply the incentive needed to set an investigation going.

ORGANIC BINDERS

The organic binders used to improve the moulding properties of conventional oil–sands are also used to some extent in clay-bonded sands. At one time this practice was almost entirely confined to steel foundries, where a combination of bentonite and starch in green sands was quite usual, but as long ago as 1933 Hudson[16] described the addition of wood extract — a sulphite lye — for raising the dry strength of a clay-bonded sand in an iron foundry without increasing its green strength. On this occasion a period of several weeks had gone by without the addition of any new sand and complaints arose in the foundry, although the only change in sand properties was a fall in dry strength to 28 lb in^{-2}. However, in view of the conditions described by Hudson, the sand must have been brittle owing to an accumulation of dead clay, but as there were then no means of measuring either deformation or toughness, the deterioration in moulding properties was not shown up by the test results, although there were complaints from the foundry. The addition of sulphite lye raised the dry strength, but its most important effect was removal of the brittleness which was the cause of complaints from the moulders.

Nowadays one or other of the organic binders is frequently added t sands used for high pressure moulding. The following binders are available:

dextrin;
starch;
molasses;
sulphite lye — sodium, calcium or ammonium.

Their most important effects are those on shatter index and dry strength, whilst starch has some influence on the moisture range. There are differences between them which are important when they alone control moulding properties, and these are discussed in Chapter 10, but when the organic binder is used only to supplement the clay, the organic binders are almost interchangeable, except that starch can be used to prevent scabbing.

Shatter index and green strength
All the binders raise the shatter index, but have little effect on green strength. Their effect on shatter index is much more pronounced than that of clay and the addition of 1% of any one would raise the shatter index of most clay-bonded sands to 100. In consequence, close control of the addition is needed, or it will be difficult to obtain moulds of adequate and uniform hardness when ramming by hand, slinger or most squeeze and jolt-squeeze machines. Exceptionally, moulds rammed by high pressure squeezing are frequently denser and harder in the presence of organic binders, and sands with a shatter index of 100 are frequently used. This is discussed in Chapter 8.

Dry strength
All the binders raise dry strength and in this respect also they are more effective than the clays, the increase with a 1% addition being 50 lb in^{-2} or more.

The increase in dry strength is important in clay-bonded, but insignificant in oil–sands. The dry strength of most iron foundry green sands is in the range 50–100 lb in^{-2} and the addition of 1% of the organic green binders increases this by a factor of about 2. The baked compression strength of clay-free oil–sand cores is usually 1000 lb in^{-2} or more and the effect of the green binder is negligible.

Moisture range
The addition of dextrin, molasses or sulphite lye to a clay-bonded sand has little effect on the amount of water required to produce a mouldable mixture or on the permissible moisture range. When starch is used, the amount of water must be increased and the permissible range is made wider. This is illustrated in Fig. 48, which shows the relationship between

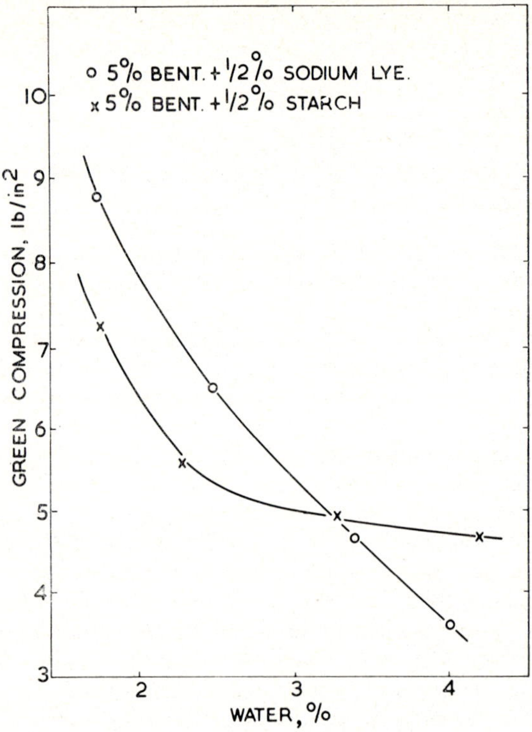

FIG. 48. Effect of starch and sulphite lye on moisture range.

green strength and moisture content for two clay-bonded sands, one containing sulphite lye and the other starch.

Scabbing

The addition of starch to a clay-bonded sand reduces the incidence of scabbing, and starch is used for this purpose in the steel foundry, but not on any great scale in the iron foundry. No detailed examination of the reasons for the reduction in scabbing seems to have been made, but the addition of starch slightly increases hot deformation, and probably reduces the extent of the fall in strength in the wet layer which forms behind the mould face when a green sand mould is poured.

In the iron foundry, the increase in shatter index which follows the addition of starch increases the difficulty in producing moulds with uniform high hardness, and unless the amount used is closely controlled, there may be an increase in unsoundness. Since coal dust, pitch and the various fibrous materials are available to prevent scabbing in the iron foundry and are usually more effective than starch, there is no need to use starch and it should be avoided wherever there is a risk of obtaining unsound castings.

Changes in the properties of sands containing starch

When a sand containing starch is allowed to stand for a considerable period, for example, overnight, there is a fall in both green strength and shatter index, although the conditions of storage may have been such that moisture content is unchanged. At the same time, the sand appears to have become wet to the touch and tends to adhere to a pattern in the way in which would be expected if the actual moisture content had been raised.

An emulsion of starch and water is viscous liquid which is not a true solution. On prolonged standing, some starch dissolves and viscosity falls, a change which also occurs when the starch is present in damp clay-bonded sand, causing the above changes in sand properties.

As a result of these changes, the control of a unit sand containing starch becomes more difficult, since the properties of the whole of the damp sand in the system change during any prolonged period of standing.

No difficulties of the above kind occur when dextrin, molasses or a sulphite lye is used in place of starch. On the other hand, these binders have a negligible effect on the moisture range and therefore on the strength of the wet layer and are less effective in reducing the incidence of scabbing.

ORGANIC MATERIALS AND THE FOUNDRY ATMOSPHERE

The foundry, not entirely without justification, is often described as dirty and this condition is attributed to the use of coal or pitch in the moulding sand. This is only partly true, since steel foundries which use neither coal nor pitch in their moulding sand are not noticeably cleaner than iron foundries.

Nevertheless, about half of extremely fine breathable air-borne dust consists of organic matter from one source or another. The slightly coarser fraction contains a similar proportion.

Much of the 'dirt' in iron foundries is due, not to coal or pitch in themselves, but to the effect of their decomposition products on sand and clay. A new sand containing coal or pitch is just a sand containing black specks, but when the sand has been in use, the whole becomes blackened by sooty decomposition products. Any dust which becomes air-borne is then just as black as coal itself. In the absence of coal or pitch, the smoke from any cores with organic binders darkens, although it does not completely blacken, the whole of the sand in a mould, and so far as appearances are concerned there is little difference between black and near-black.

The haze in foundries is in part due to the use of coal or pitch in the sand. Immediately after casting there is a copious evolution of volatile matter, but this burns producing very little smoke. After an interval, the rate of evolution slows up to such an extent that it is no longer possible to maintain combustion and the unburnt smoky compounds are evolved until the mould cools. In this respect the pitch is sometimes said to be 'worse' than coal dust, because in the proportions in which it is used it produces less, not more, volatile matter and combustion ceases earlier. Wood produces less visible smoke, but more acrid gases than either pitch or coal.

Cores with organic binders produce volatile matter after casting and this burns for a time until the rate of evolution slows up and unburnt material escapes. Less smoke is produced from the cores than from moulds containing coal or pitch, but acrid gases are a more serious problem, particularly when the cores are made from a conventional oil–sand mixture containing a green binder. The decomposition products of the carbohydrates dextrin, starch and sugar, which are used as green binders, contain an appreciable proportion of acrolein, which is most objectionable. Conditions are at their worst when heavily cored castings are poured and knocked out over the whole of the foundry floor, especially so since this method is most frequently used when the numbers of identical castings are not large enough to justify the use of hot boxes and too large to allow the use of air-setting binders, two methods which do not require the use of a green binder. So long as the casting stays in the mould, there is some restriction of the amount of fume evolved, since part of it is absorbed by the moulding sand, but if the castings have to be shaken out while hot there is an immediate surge of smoke and fume which continues until the core sand has cooled.

Getting rid of the oil–sand core would have a more beneficial effect on the atmosphere in many foundries than the elimination of coal and pitch.

REFERENCES

1. Dietert, H. W., Doelman, R. L. and Bennett, R. W. (1945). *Trans. Amer. Foundrymen's Assoc.*, June, pp. 1053–1071.
2. Hoar, T. P. (1950). *J. Iron Steel Inst.*, **166,** Sept., pp. 1–17.
3. Nicholas, K. E. L. and Roberts, W. R. (1961). *BCIRA J.*, **9,** pp. 519–536.
4. Hird, B. (1930). *Foundry Trade J.*, **47,** Sept. 2, pp. 188–190.
5. Hird, B. (1937). *Foundry Trade J.*, **51,** March 4, pp. 192–193.
6. Piper, C. H. (1938–1939). *Proc. Inst. British Foundrymen*, **32,** pp. 33–41.
7. Womochel, H. L. and Sigerfoos, C. C. (1940). *Proc. Amer. Foundrymen's Assoc.*, **48,** pp. 591–618.
8. Sanders, C. A. and Sigerfoos, C. C. (1951). *Amer. Foundryman*, **19,** Feb., pp. 19–55.
9. Wallace, J. F. and Evans, E. B. (1957). *Modern Castings*, **37,** Sept., pp. 47–49.

10. Bohan, W. F. (1951). *Amer. Foundryman*, **19,** Jan., pp. 26–29.
11. Gittus, J. (1953–1955). *BCIRA J. Res. Devel.*, **5,** pp. 594–603.
12. Dawson, J. V. (1960). *BCIRA J.*, **8**, p. 805.
13. Skerl, J. G. A. Unpublished work at BCIRA.
14. *Grey Ironfounding* (1950). *Anglo-American Council on Productivity*, pp. 70–71.
15. Use of wood dust in sand (1943–1944). *Proc. Inst. British Foundrymen*, **37,** p. B92.
16. Hudson, F. (1933). *Foundry Trade J.*, **49,** Dec. 28, p. 270.

CHAPTER 7

Sand Preparation and Control

In the foundry, moulding sand rarely consists of all new sand, other materials, for example coal dust, being frequently added, whilst in almost all cases, a proportion of sand which has been used previously is included. The amount of used sand can vary widely. A facing sand prepared from a naturally-bonded sand may contain as much as 50% new sand, but when such a large addition is made, it is usually found that all additions are made by way of the facing sand, and the backing sand consists entirely of used sand which has been reconditioned. A unit synthetic sand may contain more than 90% of used sand.

The properties of the prepared sand therefore depend, not only on those of the new materials, but also on those of the used sand, which in turn depend on such factors as the kind of castings being made, the amount of sand in the mould and the procedure after the castings have been poured.

In addition, sand properties are affected to a large extent by the method of preparation. Moreover, once the sand treatment plant has been installed, it may be very difficult to make any major changes and care should be taken to ensure that the plant, and especially its capacity, is adequate for any service it may have to perform.

SAND PREPARATION

Milling

A mill has two functions:

(1) to mix uniformly the sand, clay, water and any other materials used;

(2) to rub the moist clay in order to develop its bonding properties.

Mixing is a simple operation which is carried out quickly in modern mills. Rubbing the clay in order to develop its bonding properties may be a lengthy operation, the time and energy required depending on the condition of the clay when it reaches the mill.

If a dry, cool clay is allowed to stand in a moist atmosphere, it absorbs water, at first rapidly and then more slowly until equilibrium is reached between the moisture in the atmosphere and that in the clay. It is on this account that in the standard method for measuring dry strength, the test piece is either cooled in a desiccator or tested while still warm.

The rate of absorption of moisture increases as the area of the surface exposed to the damp atmosphere increases, and is greatest when the clay is in the form of a fine powder. If such a clay is allowed to stand in a saturated atmosphere, it absorbs water and eventually the amount of water absorbed is similar to that present in the clay in a sand suitable for green sand moulding.

In this condition the clay is a damp powder with little or no bonding properties and to develop the latter, rubbing is necessary. In the first stages, even rubbing between the hands develops a certain amount of plasticity, but to obtain the required increase in the plasticity of the clay in a large volume of clay-bonded sand, a more prolonged and intensive rubbing is necessary. When preparing a moulding mixture from dry sand, dry clay and water, the rubbing operation requires more time and more horsepower than is needed for mixing.

When a sand dries out, the effect of rubbing is lost and the operation has to be repeated before the remoistened sand is fit for re-use. The time and energy required for milling depend almost entirely on the proportion of the sand which has dried out by the time it returns to the mill.

In a light castings foundry, in which the ratio of the weight of metal to the weight of sand is low, and the time between pouring and shaking out is short, little of the sand dries out and a short milling time is adequate. On the other hand, when the castings are heavier, the metal to sand ratio higher, and the time between pouring and shaking out longer, the sand may be completely dry before it reaches the mill and longer milling is needed.

Failure to recognise this relationship between sand condition and milling time has led to difficulties in a number of foundries.

Milling capacity needed

In estimating the milling capacity needed there is a rough and ready method of calculation which is usually adequate, but which may need to be modified when very tough sand is used.

In a light castings foundry in which much of the return sand reaches the mill without much loss of water, a continuous mill is likely to be satisfactory if its power consumption is one horsepower for each hourly ton of sand treated; when the return sand is mainly dry, this horsepower should be doubled. When translating this requirement to batch mills, the proportion of 'down time' when the mill is being filled or emptied should be taken into consideration.

As the toughness of the sand increases, the minimum horsepower which is needed to keep the mill moving also increases, and the maximum amount of sand which can be dealt with in a batch mill or maintained in a continuous mill depends on this factor. If, for example, the nominal capacity of a batch mill is 10 cwt, it is quite likely that it will be capable of handling 12 cwt or more of a weak backing sand with a shatter index of 40–50, but will stall with more than 8 cwt of a sand bonded with bentonite and starch and having a shatter index of 100.

In general, the stated capacity of a batch mill can be accepted if the shatter index of the sand does not exceed 90, but caution is necessary if toughness exceeds this figure, and allowance should be made for the smaller permissible mill charge when calculating output.

With a continuous mill, the amount of sand which can usefully be maintained in the mill itself decreases as sand toughness increases and this has a corresponding effect on output.

Reducing milling time

It is possible to reduce milling time or increase the effectiveness of milling by pre-treatment of the raw materials. It is not possible to reduce the time and energy needed for rubbing clay which has dried out, but it is possible to shorten the time required to distribute and moisten any new clay, so that milling proper can begin.

When using a synthetic sand this object can be attained by adding the clay as a slurry. Should the sand be prepared from all-new materials, this method cannot be used, since in the prepared sand the moisture content will need to be about half the clay content and a mixture of two parts of clay and one of water cannot be pumped. Indeed, preparing such a mixture, which would have a consistency similar to that of plasticine, would be a lengthy operation. When, however, the bulk of the mill charge consists of returned sand, the amount of water needed is greatly in excess of the amount of clay. Moreover, in one respect, the method is self-compensating. The amount of water needed depends on how much of the sand has dried out, but if conditions are such that only a small amount of water has been removed, then little of the mould can have been raised to the temperature at which clay loses its bonding properties. The sand which requires least water also requires least clay.

There are somewhat different considerations when using a natural sand. The clay in the new sand may have developed some plasticity as a result of treatment at the quarry, and if loss of moisture in storage is avoided, this has a beneficial effect. Its extent is small when soft sands such as that from Erith or Southampton, which do not require extensive treatment before being sent to the foundry, are used, but is considerable in the case of the Bunter sands and Scottish and Durham rotten rocks, which require such treatment. For example, the red Bunter sands should be treated at the quarry in order to break down aggregates, which would not be broken down in a foundry mill and would impair surface finish without contributing anything to the bond of the sand. The operation is performed by means of a bar disintegrator which uses a large amount of energy, much of which is necessarily used in rubbing the clay, a task which the distintegrator performs with great rapidity. The saving in milling time which results is greatest when a facing sand is used, owing to the larger proportion of new sand required.

Excessive milling

There is some not very clear evidence that prolonged milling may lead to an increase in the incidence of foundry defects, especially scabs. Whatever the reason, the risk will be avoided if two horsepower for each hourly ton is not greatly exceeded. Apart from the possibility of producing defective castings, unnecessary milling is just another way of converting power to heat in a place where the heat does more harm than good.

Setting of the mill rollers

When a mill is fitted with rollers which revolve above a pan, these should be so set that they clear the pan by 1 in or more when the mill is empty. Should the rollers make contact with the pan when a layer of sand only a few grains deep is present, crushing may occur with a reduction in permeability. Maintaining this clearance has little effect on milling time. As soon as there is sufficient sand in the mill for the ploughs to pile up a layer more than 1 in thickness in front of the rollers, the full weight of the latter comes into play.

Order of additions

When using a mill with rollers rotating above a plate, the returned sand should be placed in the mill first. Should this sand be moist, the order of subsequent additions may have a slight influence on the effectiveness of milling, but does not raise any serious problems; when the returned sand is dry, the order of the additions is important, particularly when a synthetic sand is used. Two points should be kept in mind.

1. Obtaining a uniform mixture of dry materials differing in grain size is not easy. If a coarse and a fine dry sand, differing in colour, are placed in a glass beaker, it can be seen that no amount of stirring produces a uniform mixture, but if the sands are moistened, mixing is easy and rapid. If the mixed sand is dried, any form of agitation causes segregation, with the coarse grains tending to come to the top and the fine to the bottom.

If the only disadvantage of dry mixing were its lack of effectiveness, no serious harm would result, but when clay and coal dust are present, dry mixing is not only useless, but likely to cause subsequent trouble in the foundry. After a short time the smallest particles — clay, coal dust and any very fine sand grains — form a ribbon in the bottom corner of the mill which is not swept out by the ploughs, as illustrated in Fig. 49.

When water is added, either through a ring round the mill body or through inlets rotating with the ploughs, some trickles down the side of the mill turning the ribbon to mud. This may be dragged out as soon as the sand develops any appreciable plasticity, in which case no great harm will be done, but it may remain in position until the mill is discharged, in which case the sand system contains pellets which are high in clay, coal dust and water. Such pellets are likely to cause blowholes if they end

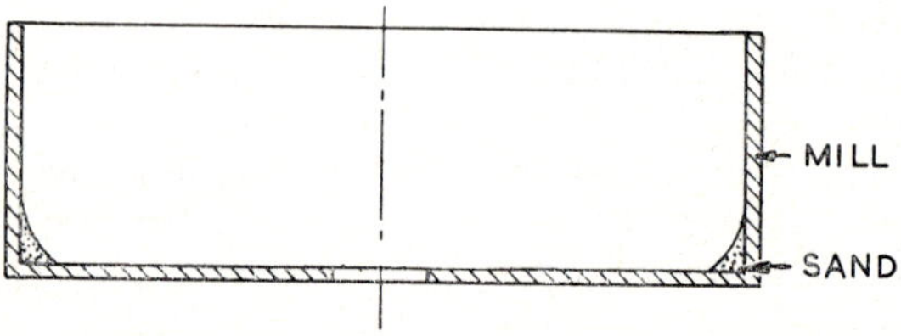

FIG. 49. Ribbon of fines undisturbed by ploughs.

up on the face of the mould. If they avoid the face of the mould first time round, they may dry out in the time between pouring and shaking out, in which case they are sufficiently hard and strong to maintain their identity until they find their way into the discard. Should they at any time be on the face of a mould, they are likely to cause a blow or a slag-like inclusion. There are so many sources of slaggy inclusions that the identification of the cause may be a lengthy operation.

2. Milling dry sand fulfils no useful purpose. Development of the bonding properties of a clay commences only when the clay is moist.

There is a further point not connected with the properties of the sand or the effectiveness of milling. Frequently the area round the sand plant is the dustiest place in the foundry and wetting the sand as soon as it enters the mill has a small, but useful, influence in reducing the amount of dust.

When preparing a synthetic sand, the returned sand, together with the new sand if this is the most convenient method, should be placed in the mill first, followed immediately by part of the water. After a short interval during which water is distributed by the mill, clay and coal dust are added. The final adjustment of the water content is made a little later.

These problems do not arise if the returned sand is moist. They are not serious in a continuous mill, when the various additions are measured out automatically and added continuously. The water should not be allowed to fall directly on to powdered clay on the surface of the incoming sand.

Selection of mill

The first and most important point to be decided is whether to use a batch or a continuous mill.

There is no difficulty here. A continuous mill is the more suitable when a steady supply of sand with the same properties is required throughout the day, but otherwise a batch mill is to be preferred. A continuous mill is not suitable for preparing a series of appreciably different sands — *e.g.* facing and backing sands, or sands for green and dry sand moulding — in sequence, since at each change it would be necessary either to stop the mill for cleaning or to produce a quantity of sand with intermediate properties.

Under suitable conditions, continuous mills have considerable advantages and greater possibilities. There is no time wasted in filling and emptying, but the most important point in their favour is that with the automatic measurement and control of water recently developed, it will be possible to produce a uniform supply of sand without the continuous intervention of an operator. He will be needed only when a change is required in the rate of feed of one or other of the constituents, either because the control tests show that the properties of the sand are getting out of line, or because it has been decided that a change in sand properties is desirable.

Continuous mills have at times had a bad reputation which was not entirely deserved. Most of the early mills of this kind were designed for use in mechanised foundries producing the large numbers of thin castings, formerly used in the manufacture of domestic equipment such as cookers. The returned sand was damp and little milling was required. Accordingly, the mills were designed to deliver a large quantity of sand using only one horsepower for each hourly ton, which pleased the foundry engineer, but when the mills were used in foundries producing heavier castings, where the returned sand was dry, the result was undermilled sand. Various devices, for example putting a retaining strip round the plate in order to increase the depth of sand and the actual dwell time in the mill, were used, but these were never satisfactory. Mills which are suitable for use when the returned sand is dry are now available.

There is no difficulty in obtaining a continuous supply of sand to the moulding machines when using a batch mill, simply by installing a surge hopper. However, the output of most of the batch mills is relatively small, and in practice, whenever a large supply of sand is needed, the choice is between one continuous mill and several batch mills. The latter can be controlled automatically, but it would be necessary either to have a control unit for each mill or to set up a switching arrangement which would be expensive. In addition, the mill would have to be so designed that filling and discharging could be completed in a few seconds.

Efficiency of mills

There is little evidence to show whether one mill is more efficient than another in translating horsepower into sand treatment. To determine this, it would be necessary to have the different mills set up in one place and compare their performance on identical sands. There is little likelihood that this will be done, owing to the cost of carrying out such a test.

The rule of one horsepower for each hourly ton of damp sand and two for each ton of dry sand is satisfactory in practice whatever type of mill is used, and it seems likely that differences in efficiency are not very great. When selecting a mill of any particular type, choice should be governed by whatever information can be gleaned from other users concerning reliability and cost of maintenance.

There are many mills which are entirely satisfactory. Those illustrated have been chosen because they illustrate the principles involved in milling.

Batch mills

A majority of batch mills have the same basic design — a fixed plate with fairly deep sides to retain the sand, one or more rollers revolving above the plate, usually with a clearance of 1 in or more, and ploughs to break up the cake of sand formed on the plate and pile it in front of the rollers. The type of mill in which the rollers revolve only on their own axes above a revolving plate is now little used.

In another type of mill, small rollers are suspended from a joint so that when the mill is not in operation they hang down vertically, but in operation they are forced out horizontally and press against the mill sides, exerting a higher pressure than is obtained when depending on the weight of the rollers.

In a third type there are no rollers, but a paddle in the base which forces the sand upwards against rubbing strips on the vertical sides. This is widely used for preparing sands with organic binders, but only occasionally for clay-bonded sands.

With all batch mills, milling time can, in theory, be controlled at will. In practice this is true only if there is ample capacity, a condition not always fulfilled. If sand is required at the rate of 10 tons h^{-1} and one batch mill with a capacity of 10 cwt has to be used, the maximum milling time, including that for filling and emptying is 3 min.

Mills with rollers revolving on the base

The majority of the mills have two rollers and two ploughs and a typical example is shown in Fig. 50. These are made in various sizes, varying from a laboratory model with a capacity of 10–20 lb to large mills with a capacity of 1 ton or more.

FIG. 50. Mill with rollers. (Courtesy of Richards Structural Steel Co.)

Different manufacturers introduce modifications. One roller, sometimes with a stirrer, may be used in place of two, whilst some mills have a number of rollers, some plain and some grooved.

Mill with rollers bearing on mill side

The only mill of this kind, the Speedmuller, was designed in America, but is now made in a number of countries. It is illustrated in Fig. 51. The upper casing has been removed.

The pressure applied by the rollers is due to centrifugal force and is much higher than that due to the weight of the rollers, which in other mills revolve on the plate. A horsepower which is very high in relation to the size of the mill can be used effectively. In consequence, milling time is very short and to take full advantage of this, the design is such that filling

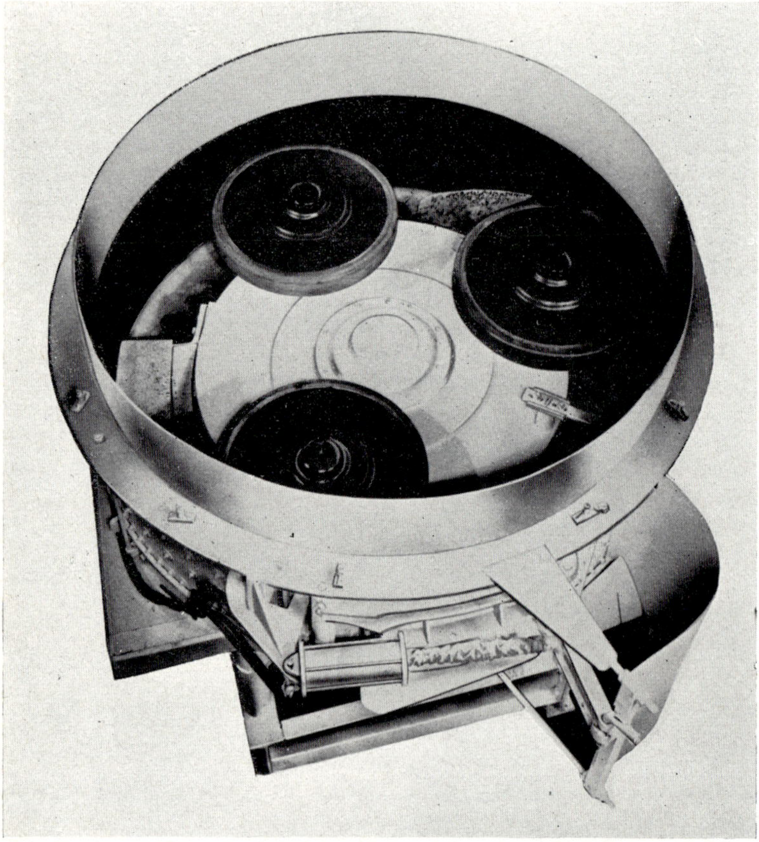

FIG. 51. Speedmuller. (Courtesy of Herbert Morris Ltd.)

and discharging require only a few seconds. The sides of the mill and the rollers are coated with rubber in order to avoid crushing of sand grains.

This mill produces far more sand in a given space than can be obtained from any other mill, batch or continuous, and since this sand is produced by a single unit, automatic control presents no difficulties of the kind likely to be met when several batch mills are used. In consequence the mill is sometimes used in preference to a continuous mill for the production of large quantities of sand of identical composition over the working day.

All the energy put into a mill eventually appears as heat, and with the high energy input of the Speedmuller, the sand can become too warm to be satisfactory if the milling time is extended beyond the minimum needed. A blower is provided, by means of which air is blown through the sand in the mill to cool it, but as the cooling effect depends only on the evaporation of water, moisture control is lost. The mill should not be used unless a timer is provided which will ensure automatic discharge after a pre-set milling time. Indeed, the provision of a cooling blower is, in itself, a warning that overheating is a possibility.

Occasionally, the provision of the blower is taken as evidence that sand can usefully be cooled in the mill and that the difficult operation of cooling the returning sand before it gets to the mill is unnecessary. This is a risky assumption. Control of moisture content is necessary and is made more difficult if an unknown quantity is removed from the sand in the mill by evaporation.

Paddle mixers

There are several types of paddle mixer, one of which is illustrated in Fig 52.

It consists of a cylindrical body, deeper in relation to its diameter than the body of the roller mills, with a rotating paddle in the bottom. This paddle forces the sand upwards round the periphery to return later round down the centre. Rubbing strips on the sides of the mixer reinforce the rubbing and mixing action, although the efficiency of the machine does not fall off seriously when these strips are worn flat, which is a feature very useful to foundries which like to put in a piece of equipment and forget maintenance, except for routine lubrication.

This type of mill is slow when used for preparing clay-bonded sand, since the rubbing action is not very intense. It is mainly used for preparing oil–sand mixtures where intensive rubbing is not needed. For this latter purpose, it is more satisfactory than mills with rollers. This is discussed in Chapter 12.

Continuous mills

Plate mills

The earlier continuous mills consisted of a revolving flat plate above which a roller revolved on its own axis. Alternatively some makers preferred a stationary plate and a rotating roller. Usually sand was fed on to the centre of the plate and discharged by means of a plough at the periphery.

This kind of mill is satisfactory when the return sand is moist but not when it is dry, and to overcome this limitation various devices have been used to put more energy into milling — replacing one large roller by a number of smaller ones and raising the pressure by means of springs, or raising the mill sides so as to increase the depth of sand on the plate, a device which made it necessary to increase the horsepower needed to keep things moving.

FIG. 52. Paddle mixer. (Courtesy of Fordath Ltd.)

By the use of these devices, it has been made possible to mill sand satisfactorily even when the returned sand is dry. When considering the possible use of a mill of this kind, it is important to obtain details of the output and power consumption, since mills designed to the original one horsepower for each hourly ton of sand specification are still made and are, indeed, quite adequate for makers of light castings. Figure 53 shows a mill that is suitable for use when the returned sand is dry.

Twin mills

A second form of continuous mill consists essentially of two roller batch mills joined together like Siamese twins. Sand is fed into the first, where it

FIG. 53. Continuous plate mill. (Courtesy of Foundry Equipment Ltd.)

FIG. 54. Twin mill. (Courtesy of August's Ltd.)

is partly milled and thrown into the second. If the exit to the second is closed, there is a continuous exchange of sand between the two. The second mill is provided with a weir over which excess of sand flows and in operation the rate of discharge is equal to the rate at which sand is fed into the first. The mill is shown in Fig. 54.

This kind of mill has some flexibility, in that the output can be increased to a limited extent simply by increasing the rate of feed, but this can only be obtained at the expense of milling intensity. When the return sand is dry, the rated output should not be exceeded.

Horizontal cylinders with rollers

There have been several attempts to mill sand by passing it repeatedly between two horizontal roller bars, rotating towards each other, but at different speeds, inside a horizontal cylindrical drum fitted with lifters. These have had only limited success.

The various difficulties were overcome by causing a roller suspended on knuckle arms to rotate against the side of the revolving drum. An agitator screw rotates in the same direction as the drum, but at a much higher speed and with its axis off centre. The interior of the mill is shown in Fig. 55.

The horsepower is between two and three for each hourly ton of nominal capacity.

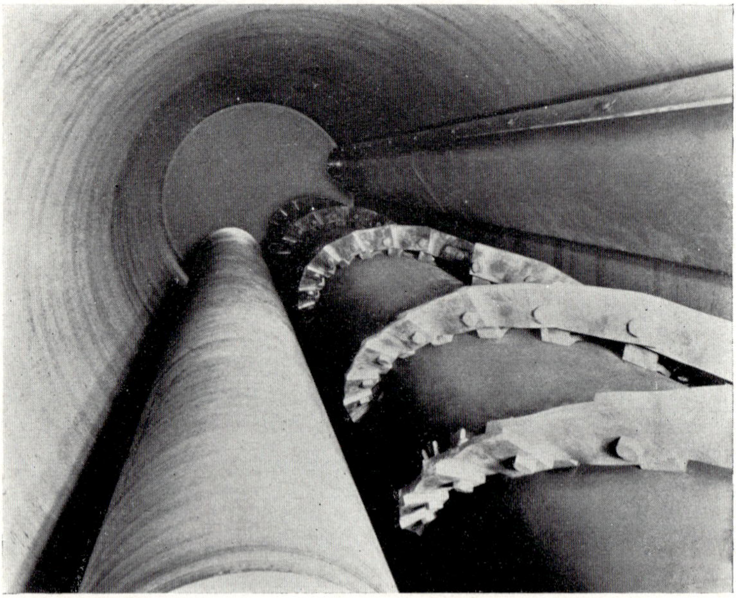

FIG. 55. Cylindrical drum mill. (Courtesy of Junkerather Maschinenbau, GmbH.)

Is milling necessary?

Quite a number of foundries operate successfully without any kind of mill. Indeed, such a practice was quite usual among hand moulders, who mixed their sand by means of a shovel, but some mechanised foundries are also able to get by without mills.

Any operation which does work on a sand develops the bonding properties of the clay. If an unmilled sand is used to make a mould on a squeeze machine and the mould is then immediately broken up and the sand disintegrated, there is an appreciable increase in both strength and shatter index. Simply passing an unmilled sand round a foundry sand system has some influence, particularly when a disintegrator is present.

The 'milling' effect depends on the amount of energy used. By far the most effective tool used on foundry sand is the bar disintegrator used at the quarry for breaking down soft sandstones such as the red Bunter sands, and this produces quite a tough sand.

Of the machines used for ramming moulds, the slinger head has the most intensive rubbing action, and most mechanised foundries which do not use a mill use a slinger for ramming. It is helpful if the castings are thin so that little sand dries out, in which case ramming has a cumulative effect, and the method is easiest to apply when the castings are shallow and no difficult lifts are involved.

Nevertheless, castings of considerable size have been made. In the report of the grey iron foundry team which visited America in 1950,[1] there is a description of a foundry in which a slinger was used to ram moulds for castings up to 3 cwt in weight in boxes 64 × 50 × 24 in. The moulds were knocked out and the sand spread out on the floor, allowed to cool, and damped down. New sand and clay were spread over the sand and the tractor slinger, with the ramming head replaced by a chute, driven through the sand leaving a neat windrow behind it. The slinger was then reversed and the head replaced, and worked through the heap of sand towing two stripping machines behind it. Half a dozen men produced about 200 moulds in a day.

It is necessary for the sand to be thoroughly mixed before it reaches the slinger. The rotor inside the head picks off a small quantity of sand and discharges it at high speed, development of the bonding properties of the clay occurring in the short time between when the sand is first cut off and when it is finally at rest in the moulding box, but each small quantity of sand is treated individually and no mixing occurs.

Disintegration

When sand is discharged from a mill, much of it is in the form of coherent aggregates, some of which may be quite large. If such a sand is rammed against a pattern, the aggregates are deformed to some degree, but not sufficiently for their outline to be lost and the mould and casting have

(a)

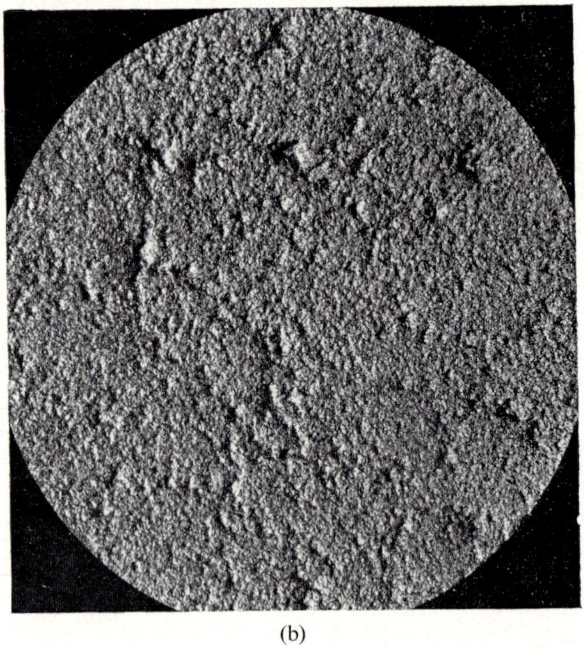

(b)

FIG. 56. (a) Mould from unsieved sand (×7·5); (b) surface of casting (×2⅝).

(a)

(b)

FIG. 57. (a) Mould from sieved sand (×7·5); (b) surface of casting (×2⅝).

rough surfaces. A disintegrator breaks up these aggregates and improves the surface of the mould and casting.

The effect on the surface of moulds and castings is shown in Figs. 56 and 57.

Figure 56(a) shows part of the surface of a mould made on a jolt–squeeze machine using sand as it came from a mill, and Fig. 56(b) shows the surface of a casting made in that mould. Figures 57 (a) and (b) show a similar mould and casting made from the same sand after passing through a sieve with $\frac{1}{8}$ in apertures.

This sieve is not a good disintegrator, but it does serve to break up the larger aggregates, and the improvement in the finish of the casting is quite plain. The illustration shows the futility of trying to improve surface finish by using a finer sand under conditions which produced the mould and casting shown in Fig. 56.

The disintegrator

The most effective method of disintegration is to throw the sand with considerable velocity through some form of obstruction such as staggered pins. However, sand is exceedingly abrasive and there is considerable wear and eventually a falling off in performance, but since the disintegrator is usually hidden away in a box at the discharge of the mill and the deterioration of casting surface is slow, the change may be overlooked until an inspector or a customer complains. There is then a demand for a disintegrator which does not wear. This may make things worse rather than better, since if the machine does not wear it does not as a rule disintegrate sand; the only way of reducing wear while maintaining performance is by using an abrasion-resisting material.

It is sometimes claimed that when sand is conveyed by means of compressed air, a disintegrator is unnecessary, but this is not always true. In one particular case where this method was used, the sand contained softer than usual aggregates which increased in number as the shatter index of the sand was raised above 70. It was not clear whether the compressed air failed to break down the aggregates or whether new ones formed at the discharge end.

Reaggregation of sand from a disintegrator

When sand which has passed through a disintegrator is subjected to any compressive force, aggregates are formed and, indeed, those in the sand coming from a mill are produced by pressing the sand between the rollers and base plate of the mill. A mould, which is simply an aggregate of a particular size and shape, is made by jolting, squeezing or throwing the sand at high velocity against a pattern. There are not many examples of jolters in a sand distribution plant, but there are several examples of throwers and squeezers. Wherever there is a column of sand, the lower layers are squeezed by the weight of the sand above, and whenever the sand falls from a higher to a lower level, it is rammed by impact.

Reaggregation takes place mostly in hoppers, and is serious when the hoppers are over the actual moulding machine. When the hopper is full, the sand is compacted by squeezing, and when nearly empty it is compacted by impact. The deeper the hopper, the greater the pressure when full and impact when empty, and very deep hoppers should not be used over a moulding machine.

The most convenient place to put a disintegrator is at the discharge from the mill and this is the location usually chosen. It is also the least effective, since there is then the whole of the sand distribution system in which aggregates may form.

It does not seem to have been possible to put a disintegrator immediately before the moulding machine, which is the place where it would be most effective. When producing castings which must have a very good surface finish and accurate reproduction of fine detail, it is frequently necessary to use a sieve — one with apertures a little larger than the largest sand grains gives the best results — and sieve the sand on to the pattern. This is a lengthy operation which slows up production, but no other method produces equally smooth castings in green sand moulds.

The extent of reaggregation is greatest when the sand has both a high clay content and a high water/clay ratio, the second factor being the more important.

TREATMENT OF USED SAND

When sand is removed from a mould which has been poured, part is hot and dry, and part is damp and relatively cool. The proportions of the two can vary widely, the amount of damp sand decreasing as the metal/sand ratio and the time between pouring and shaking out increase.

In addition, the sand contains some tramp iron, mainly from spillage during pouring, and core butts, if cores bonded with organic binders or sodium silicate are used.

Before the sand can be prepared for re-use, the extraneous material must be removed and the sand should be cooled. However, cleaning and cooling sand pose problems which can differ greatly in foundries using different methods of production.

At one extreme are the disappearing hand moulders who carry out all operations except the actual melting of the metal — ram moulds, pour, shake out and partly clean the castings, and re-condition moulding sand. Cleaning the sand is a somewhat lengthy operation. Any large fragments of iron are removed from the top of the mould before shaking out, but complete separation of small pellets is impossible, and under these conditions the sand frequently contains several per cent of iron. This does not as a rule cause serious troubles, because under the above conditions a facing sand containing a large proportion of new sand is used, reducing the proportion of iron by dilution, and the sand is sieved on to the pattern which removes all but the very small pieces. Sand cooling is not a problem. Production is slow and the sand is not re-used several times a day, which restricts heat input, and the sand is spread out over a large area which favours cooling. Finally, the use of a relatively large proportion of cold new sand reduces temperature at a time when the sand is not very hot and cooling by other means is difficult.

At the other extreme is the mechanised foundry in which all the sand from the shake-out passes over a magnetic pulley to remove tramp iron and through a screen to remove core butts and other foreign materials. Cleaning presents no difficulty, but the sand may be used several times a day, heat input is raised and the time for cooling reduced. Cooling presents a serious problem.

In between the two extremes there exists a whole variety of conditions. Moulds may be made on machines, but are knocked out on the floor, in which case the volume of sand to be dealt with is too great for the moulders to handle. Treatment of the old sand may take place on a second shift, but even so, some mechanical aids are necessary. The sand may be treated by means of a sand-cutter and left in piles by the side of the machines, a method once in great favour in America and used by a few foundries here, or a small replica of a mechanised plant may be provided — a short length of belt with a magnet and a screen feeding a mill. The whole of the sand may be treated in this way, but frequently a separate facing sand is used and only the sand needed for this relatively small quantity is dealt with, the backing sand being moistened and hand-mixed or passed through some form of disintegrator.

There are two separate problems — cleaning sand in the non-mechanised foundry and cooling in the mechanised foundry. In the first case, the problem is frequently overcome, or at least minimised, by the use of facing sand and this method is occasionally used in a mechanised foundry when there is difficulty in overcoming the cooling problem. In the latter case, it should be regarded as a temporary solution only.

The use of facing sand

When there is difficulty in controlling sand properties closely, the use of facing sand is advisable. The facing sand has more coal dust and more new sand than is present in the backing sand, and additional cost would be incurred in bringing the backing sand up to facing sand quality, although in practice, the saving in cost when using the two sands in place of a unit sand is negligible over a period. Only the sand near the mould surface is heated to the temperature at which coal dust decomposes or clay loses its bonding properties, and the sand remote from the mould surface is not affected. Once all the sand in the foundry has been made suitable for a facing, the cost of maintaining it in that condition is only marginally greater than that of maintaining separate facing and backing sands.

The real purpose of a facing sand is to keep sand which is too wet, too dry, too hot or in some other way unsatisfactory, away from the pattern during moulding and the mould face during casting. The difficulties experienced in sand control when moulds are shaken out all over the foundry floor are real, and using a separate facing sand not only enables effort to be applied to a much smaller quantity of sand, but also eliminates some of the uncertainties which call for attention in a unit sand. For example, since coal dust is added only to the facing sand which is near to the mould surface, much of it decomposes and little remains in the return sand. In preparing the next batch of facing sand, the returned sand is diluted with a large proportion of new sand, which still further reduces the amount of coal dust to the point where it can be ignored. At the same time, the large dilution with new sand prevents any serious build-up of dead clay. When using separate facing and backing sands, two of the factors which are difficult to control when there is a single sand system take care of themselves.

When a separate facing sand is used in a mechanised plant, it is in most cases for one or other of three reasons.

1. Attempts to cool the return sand have been unsuccessful and facing sand is used to keep hot sand away from the pattern. There is usually no difficulty in cooling the small amount of used sand needed in a facing, and the addition of a comparatively large amount of new sand completes the job.

2. Milling capacity is inadequate and better results can be obtained in the mould-making department by concentrating half of the available milling in a fifth of the sand to be treated.

3. A wide range of castings is being made, and a better than usual surface finish is required for one particular kind, or one is of such a design that scabbing is likely.

In general a need for a separate facing sand on a mechanised plant can be accepted as evidence that the plant is in some respects inadequate, or that the range of castings being made is unsuitably wide.

Some control of the backing sand is necessary. It should not be so wet that appreciable moisture travels to the mould face in the interval between moulding and pouring, or so dry or weak that it cannot support the facing sand in the moulding box, but these are not difficult requirements. Its permeability should not be appreciably less than that of the facing sand, but unless a coal dust free from fine grains is used, the backing sand is usually found to be more permeable.

The need for cooling

It is essential that the sand in contact with the pattern shall be cool, or difficulties such as sand sticking to the pattern or the face of the mould air drying and becoming friable, are likely to arise. When a unit sand is used, all the returned sand must therefore be cool.

In addition, hot sand containing a little moisture can cause serious difficulties in the sand plant itself. Whenever hot, moist sand is in contact with a body cooler than itself, and especially when that body is a good conductor, such as a metal, water condenses on the cooler body and some sand adheres to it, in the same way that hot sand adheres to a metal pattern. When the plant is idling, even for a short time, and small gaps in supply do occur here and there in a sand plant, the adherent sand remains firmly fixed to the metal. As time goes on, the layer of sand increases in thickness, reducing the capacity of the equipment concerned. In consequence, the buckets on elevators handling hot sand are frequently left with only half their nominal capacity, and designers of foundry plants sometimes anticipate this difficulty by arranging for sufficient excess capacity. In a storage hopper, a layer of dry sand may gradually build up on the inner surface and it is not unknown for a cylindrical hopper having as much as 6 ft diameter to contain such a thick layer of hard dry sand that the only active part is a 2 ft hole in the centre. This reduces the capacity to a ninth of its original value and, taking into account the sand in the remainder of the plant, may reduce the quantity in circulation by as much as half. From then on the same amount of metal is distributing its heat among half the amount of sand and there is a further rise in temperature. Some means of cooling the sand is therefore necessary.

Methods of cooling

Sand may be cooled by radiation, by conducting its heat to a cooler body, when the sand is moist by making use of the latent heat of evaporation, or by dilution with cold material.

Coolers are available and are in use in a number of foundries. In one of these, the sand cascades from the top of a tower fitted with baffles, air being drawn through the falling sand. The equipment is bulky.

An interesting method has been described by Zehnder.[2] The sand from the shake-out passes through a paddle mixer where the moisture content is adjusted and spread uniformly. The sand is then carried by means of a perforated belt through a chamber, where air is drawn through the moist sand. The final temperature is very little above room temperature.

Loss of heat by direct radiation from the sand is usually small. When poured moulds are allowed to stand for a long time, the loss by radiation from the sand and box eventually exceeds the gain by conduction from the hot casting, but usually, amd especially in mechanised foundries, this stage is not reached and sand temperature rises as the time between pouring and shake-out increases. There is not time for significant loss by radiation when the sand passes round the plant, because there is a free radiating surface for a short time only.

Loss by radiation is, however, sometimes significant. When sand is in contact with any body cooler than itself, it loses heat to that body by conduction, but if the lower temperature of the body is to be maintained, heat must be removed from it by radiation. Heat removed in the first place by conduction, is finally dissipated by radiation.

Heat can be transferred from sand to air by conduction, although, for the amount to be significant, air must pass through the finely divided sand. In this case, radiation plays an insignificant part and continuation of cooling depends on replacing the warmed air by fresh cool air.

Cooling resulting from the evaporation of water is usually the most important factor when the sand is hot and damp. Evaporation of 1 g water absorbs 540 cal at 100°C and 580 cal at 30°C. As the specific heat of silica is about 0·2, evaporation of 1% water reduces sand temperature by 20–25°C.

Effect of sand temperature on rate of cooling

The rate at which heat is removed from sand increases rapidly as temperature rises. When heat is lost by conduction, the rate of loss is proportional to the difference in temperature between the bodies concerned, but loss by radiation is proportional to the fourth power of the absolute temperature. The rate of evaporation of water depends on its vapour pressure and this increases with increase in temperature as shown in Table 1.

Table 1. *Vapour pressure of water*

Temperature (°C)	Pressure (mm Hg)
30	31
45	71
60	149
70	234
80	355
90	526
100	760

As a result, it is difficult to prevent loss of heat from sand, particularly damp sand with an exposed surface, when its temperature is above 90°C, but there is no significant loss when the temperature is below 45°C.

The addition of a cooler material reduces temperature, but the point at which the cool material is added is of some importance.

Should the new sand be placed on the mould before the shake-out, the average temperature of the sand at that position would be reduced and Table 1 shows that, when the sand is hot and moist, a reduction in temperature of only a few degrees reduces the rate of heat loss in the initial stages. When the amount of new sand added is only a few per cent of the total in circulation, the additional heat lost when the sand is added at the mill instead of the shake-out can influence the final temperature by only a few degrees. However, adequate cooling in the sand plant can, in many cases, be obtained only by obtaining the maximum loss of heat at each of a number of places, and a partial loss of effectiveness at the position where this loss is most rapid, may be compensated only with difficulty at a later stage.

Cooling in the sand plant

In the absence of any cooling, the temperature of the sand in the plant would rise progressively with each cycle of operations, but since some cooling always occurs, and the rate of cooling increases as temperature rises, equilibrium is eventually reached. This equilibrium temperature must be below that at which the various difficulties due to the use of hot sand become serious.

Provided that as the sand passes round a sand recovery, and preparation plant precautions are taken to ensure that the factors which favour cooling have free play, it is frequently possible to ensure that the sand at the discharge of the mill is at a suitable temperature without recourse to any special equipment for cooling.

The first essential is that the sand shall contain some moisture. In a mechanised plant — and it is only here that sand is likely to be re-used several times a day, with a corresponding effect on sand temperature — the time between pouring and shake-out is short and it is unusual for the whole of the sand to be dry. Where necessary a water spray can be used.

The second essential is that, wherever possible, the sand shall be broken up so that the maximum surface is exposed, since evaporation of water cannot take place inside a solid mass of sand.

The third essential is that any air saturated with water shall be withdrawn and replaced by unsaturated air.

When the appropriate precautions are taken, cooling takes place at a number of points in the sand system.

Shake-out

At the shake-out the sand is broken up as it falls through the grid, exposing a large surface. The hot, moisture-laden air rises and is withdrawn by the exhaust system. One limitation is that at this point the sand has not been thoroughly mixed and no moisture is present in the hottest sand.

Belts and transfer points

The surface of the sand on a belt dries out quite rapidly, but when a dry layer a few grains in thickness has been formed, little further evaporation occurs, and the rate of cooling is reduced to negligible proportions. Some help can be obtained by using ploughs or other means of turning over the sand and exposing a fresh surface.

At transfer points, the sand is broken up and a large surface is exposed. When the transfer point is enclosed in order to contain dust and fume, an adequate exhaust should be used to remove the saturated air.

Elevators

The main loss occurs at the discharge end, where the sand is broken up into individual grains and small aggregates, and falls through unsaturated air. There is some loss of heat from the elevator buckets which are continuously warmed by the incoming sand, but the amount so lost is small.

Rotary screen

Conditions can be made most favourable for cooling at this point, since sand aggregates are broken down and the maximum sand surface is exposed to the air.

In order to obtain the greatest possible cooling, the full length of the screen should be in use. In some cases, almost the whole of the sand passes through the first quarter of the sieve, apparently in order to avoid any possibility of losing good sand with the tailings. This is unnecessary. The slope of the screen and the mesh size should be such that the whole length is in use and the risk of loss eliminated, if necessary, by using a coarser mesh for the last 1–2 ft.

Steam and dust are produced at the screen and it is usually necessary to enclose and ventilate the equipment. The amount of air extracted should be sufficient to maintain a slight negative pressure in the enclosure and ensure that saturated air is being withdrawn.

The extent of cooling depends on the length of time during which it can occur, which in turn depends on the length of fall. At the rotary screen and all other places, such as transfer points, where cooling depends on contact between air and falling sand, the height of fall controls the heat loss.

Cooling in storage

The rate of loss of heat when sand is stored is extremely slow, but on the other hand the time available for cooling is comparatively long. When sand is falling at a transfer point or rotary screen, the time available is a few seconds, but in a storage hopper it may be as many hours, although there is no certainty as to how much of the cooling occurs during the actual storage and how much during the fall into the hopper.

It is safer to use two hoppers, or a hopper sectioned vertically, than a single large hopper. If one hopper is filled while another is being emptied, the possibility of building up an increasing layer of hard sand on the inside of the hopper is avoided.

Cooling in mill

In an open mill, a fresh sand surface is being continually exposed and heat is lost, mainly by the evaporation of water. On the other hand, the energy used in driving the mill eventually appears as heat, and temperature changes in the mill depend on the relation between the two factors.

In general, there is a fall in temperature when warm sand is prepared in a large open mill with shallow sides and a rate of production which is low in relation to mill size; there is a rise in temperature when the mill has deep sides and a high rate of production, particularly when the whole is enclosed.

The sand should, however, be cool when it reaches the mill, since any cooling in the mill itself is due to loss of water. This increases the difficulty in controlling the moisture content of the finished sand.

SAND CONTROL

When molten metal is poured into a sand mould, the following changes occur in the sand:

(i) some moisture is lost;
(ii) some coal dust or pitch, if present, is decomposed;
(iii) some clay is heated to the temperature at which its bonding properties are destroyed.

The carbonaceous residue from coal or pitch and the dead clay remain in the sand.

Water, clay and coal dust can be added to replace any which has been removed or decomposed, but the carbonaceous residue and dead clay remain and have a definite influence on sand properties. In consequence, it is not possible, in a mixture containing used sand, to regain the properties obtained with all new materials.

When the castings are removed and the sand is being prepared for re-use, three further changes take place:

(iv) fine material is removed as dust, and this may consist of the finer grains in the original sand or the decomposition products of coal;
(v) some sand is removed with the casting;
(vi) some new sand is added to replace that removed with the castings and to maintain the volume of sand in the system.

The extent of these changes depends on a number of factors of which the most important are (1) the ratio between the weight of metal poured and the weight of sand in the boxes and (2) the time between pouring and shaking out. It might be expected that if only one particular casting was being produced on a system, then all changes would be constant and a sand with constant properties could be obtained simply by adding a fixed proportion of water, new sand, clay and coal dust.

In practice this is not successful, because the equilibrium so obtained can be upset by any one of a number of mishaps which are likely to occur from time to time. If, for example, there is a hold-up at the shake-out, the hot castings remain in the mould for longer than usual and more water is driven off. If the amount of water added is constant, some sand with too low a moisture content reaches the moulding machines. Some bad moulds are made which are not poured, and, since the only convenient way of dealing with them is at the shake-out, a batch of sand enters the system with higher than usual moisture content, causing still more trouble at the moulding machines. In a short time the system becomes completely out of control, showing that the preliminary equilibrium was unstable.

The object of control tests is to give warning of changes in sand properties and enable remedial action to be taken before there are any harmful effects in the foundry.

Moisture control

Of all control tests, that for moisture content is by far the most important. In the first place, the moisture content in the returned sand varies more widely and more rapidly than any other property. For example, a casting would have to be left in the mould for a very long time to produce any appreciable increase in the amount of dead clay formed or coal dust decomposed, but a much shorter period has a considerable influence on the amount of water removed. In the second place, an unwanted change in moisture content can give trouble in sand distribution and mould-making and lead to an increase in rejects.

Sand distribution

Except that sand always tends to stick in hoppers, and sticking is more pronounced when moisture content is high, the most serious effect of changes in moisture is the effect on the rate of feed from devices such as the vertical cylindrical hopper with a rotating table. If the moisture content is high, the rate of feed is reduced and if it is low, the rate is increased.

Such a change can lead to inaccurate measurement of the constituents of the prepared sand, but its effect is most serious when sand must be fed at a constant rate to a machine such as a sand slinger. In one particular case, the arrival of a batch of sand with a low moisture content resulted in the immediate discharge of the whole of the contents of the hopper — about 2 tons — in a few seconds, swamping the immediate area and stopping production. Smaller changes in moisture content lead to less spectacular, but not less important, changes in the behaviour of the machine, which produce moulds of uneven hardness.

Ramming and stripping

A moderate increase in moisture content leads to some increase in toughness, which in turn leads to an increase in hardness differences between parts of the mould, which may lead to mould cavity dilation in particularly susceptible cases.

A large increase in moisture content leads to the adherence of sand to the pattern. Stripping can be improved by heating the pattern, but this should not be needed and if carried too far, can lead to the production of bad castings. The surface of the mould is raised to a temperature such that air-drying takes place, and the mould surface becomes friable, particularly at the edges of the mould cavity. Should it be necessary to insert cores, there is an increased risk that sand grains will be displaced during the operation and appear as inclusions in the castings.

Similar, and sometimes more pronounced, difficulties are found when sand is hot. In this case, moisture condenses on the cold pattern forming a layer of high moisture content at the face, and the sand behaves in the manner noted above when the sand was too moist.

Defects in casting

Since moisture content has a pronounced influence on sand properties, particularly green strength, toughness and dry strength, and also forms most of the gas evolved when metal is poured, it can be responsible for almost any defect which is due to failure of the mould. The most important are:

Low moisture. This causes (1) brittle sand and bad lifts, and (2) low dry strength leading to inclusions and erosion scabs.

High moisture. This causes (1) high and prolonged gas evolution leading to blows and misruns in thin sections and roughness and scabs in thick sections, (2) irregular packing leading to bad surface finish, and (3) high deformation and low strength, leading to swelling and unsoundness, and break-out.

Permeability, moisture content and blowholes

When molten metal is poured into a green mould, most of the gas given off is water vapour and this has to escape through the mould wall if it is not to pass through the metal. It might be expected, therefore, that there would be a direct relationship between mould permeability — as calculated from sand permeability, mould hardness and mould thickness — moisture content and the incidence of blows. In practice this is not necessarily so.

In order to demonstrate this, two sands were prepared which had the same moisture content, but widely different permeabilities. The first was a synthetic sand bonded with a calcium montmorillonite and having a permeability of about 200 and a moisture content of 5·0%. The second was a red sand containing coal dust and having a permeability of 25. Its moisture content was also 5·0%. The synthetic sand appeared to be extremely wet, and with sands of this kind 2% water is adequate, but the red sand was suitable for moulding.

Similar moulds were prepared from the two sands and poured. All castings made in the red sand moulds were free from blows, but a large proportion of those made in the synthetic sand were blown.

This was taken a stage further. The synthetic sand described in Chapter 4, in which there was so much dead clay that the moisture content had to be raised to 9% before mould-making became possible, was used to prepare a simple mould. Sand permeability was 40. The moisture content of the synthetic sand with a permeability of 200 used in the previous experiment was raised to 9% and the sand kneaded into shape, as it was too wet for ramming and stripping. The mould containing dead clay poured quietly, but that produced from all new sand and having the same moisture content and five times the permeability blew up.

This has been observed on many occasions and it follows that *if a sand contains moisture appreciably in excess of that required to form a mouldable mixture, blowing is almost certain, whatever the actual permeability and moisture content.*

With any one sand mixture, the tendency to blow increases as the moisture content increases and this increase becomes much more rapid when the moisture content passes the point at which sand sticks to a pattern. Perhaps it is even more important for the foundryman to recognise that he cannot avoid the ill effects of poor moisture control by using a sand with a very high permeability.

Methods of control of moisture content

Direct methods. In almost all foundries, moisture is controlled by 'feel', with an experienced operator to do the job. Actual measurement of moisture is carried out only to check the accuracy of the method used.

With the arrival of the methods described in Chapter 1 which give the water content within a few seconds, there is now no difficulty in controlling each batch of sand from a batch mill to within one or two tenths of 1%.

The constituents of the batch are mixed, the moisture content measured and the correct amount of water added.

A parallel procedure is possible with a continuous mill, provided that the appropriate precautions are taken. The most serious limitation is that there is no immediate prospect of producing an instrument which will measure accurately the average moisture content of the kind of sand returning from a shake-out and containing sand varying widely in moisture content and existing as anything from single grains to aggregates 1 in or more in diameter.

This is a real difficulty and in order to overcome it, the measuring device is sometimes placed on the output side of the mill. This method has two disadvantages and fails to make the best use of the rapid measuring device. In the first place, the moisture content is measured after the sand has been prepared and if it is not as required, nothing can be done about it except pass the sand back to the returned sand. In the second, water is added as the sand enters the mill, but is measured at the exit and in between there is a quantity of sand rather larger than the actual amount present in the mill. Should the moisture content of the returned sand vary considerably, this method of operation may result in the largest volume of water being added to the sand which requires least, and *vice versa*. Control therefore fails under the conditions which render it most desirable.

It is not a difficult or costly matter to break down aggregates and at the same time reduce local variations in moisture content, and any form of disintegrator will do the job satisfactorily. If therefore a plant to prepare sand in large quantities is being built, a disintegrator and a short length of belt should be included immediately before the mill. The moisture content of the incoming sand can then be measured accurately and the appropriate quantity of water added.

There is no difficulty in making the whole operation automatic and equipment for batch mills is available. In one example, sand passes first to a measuring hopper with a capacity equal to a single batch and the moisture content is obtained by conductivity between a number of spaced electrodes. These are connected to the water supply so that the exact amount of water is added to bring the moisture content up to a predetermined level. A weakness of this procedure is that accuracy of measurement depends on the sand in the hopper having a constant packing density, a condition which is not satisfied if the incoming sand varies. The control functions most satisfactorily when it is not needed, but by attaching the relay system to an accurate moisture meter such as is described in Chapter 1, close control of moisture content can be obtained with either batch or continuous mills without any need for an operator.

The equipment needed to control the water content in sand from both batch and continuous mills has been described by Godding and Bird.[3]

Indirect methods. In view of the influence of water on the physical properties of sand, it is possible to obtain the approximate moisture content from the physical properties, always provided that other factors, and particularly the amount of active and dead clay, are constant.

If a sand is allowed to fall onto a sieve or passes over narrow slots, the amount which passes through the sieve or slots increases as the toughness of the sand decreases, and Gittus[4] obtained a good correlation between shatter index and the 'Loose Flow Number' which is the proportion of a sand which passes through a given sieve onto which it has fallen from a height of 3 ft.

In the Dietert 'mouldability controller',[5] a stream of sand is drawn from the mill and passes over two slots. If the toughness of the sand is low, a large quantity falls through the slots; if it is high a smaller quantity falls through the slots. The information is relayed to the moisture controller and the toughness of the sand increased or decreased by raising or reducing the amount of water added.

Rapid changes in sand toughness are usually due to changes in moisture content and the instrument takes care of these in a satisfactory manner, but performance over a period depends on maintaining a constant amount of both active and dead clay. If, for example, there is a continuous fall in the amount of active clay, the instrument tries to offset this by raising the moisture content, whereas the amount of water needed is less. Eventually a weak sand with too high a moisture content would be obtained, with disastrous results. However, both raising the water content and reducing the active clay content reduce green strength, and the fall in this property can be used to give ample warning of the change. Regular measurement of sand properties is needed to supplement the instrument.

Other indirect methods of controlling moulding properties by control of toughness have been used with varying success. The simplest of these is by fixing an ammeter in the electric circuit of the mill motor and arranging for the discharge door to open when the current reaches a predetermined value. The energy needed to drive a mill increases as the toughness of the sand increases and this procedure, like the Dietert controller, produces sand with a constant toughness. Control depends not on moisture content but on milling time, and success depends on maintaining constant water and clay contents.

Sand control programme

In any sand control programme the most important test is for moisture content and usually some form of test is employed on every batch from a batch mill. In a large majority of cases control is by the operator who judges by 'feel' and actual measurements of moisture content are made in order to check the accuracy of the operator's judgment. Frequency of testing should depend on the success of manual control, but it is advisable to do a check test at least once in each hour, but at irregular intervals.

It is neither necessary nor desirable to make other tests on every batch of sand produced. Except in the case of moisture content, changes are relatively slow and testing at longer intervals is satisfactory. The most useful frequency of testing depends on the kinds of castings being made — if these are uniform in weight and section, changes are not only slow but regular and twice-daily tests are adequate. On the other hand, when the castings vary from time to time, for example if a casting weighing 10 lb is replaced by one requiring the same moulding box but weighing 100 lb, erratic changes in sand properties are likely, and more frequent testing is desirable. It is rarely necessary to test more than once in each hour.

Some tests are needed less frequently than others. Changes in the amount of active and dead clay may be fairly rapid, but changes in permeability are slow unless some deliberate change is made in the mill charge. Nevertheless, since permeability can be measured in a very short time using the test piece which is afterwards to be used for the strength test, it is usual for the two tests to be done together. A mechanical analysis is useful only to determine why there has been a change in permeability and to indicate the appropriate remedial action. In a given sand system, dry strength depends on the amount and kind of clay and on the moisture content, and since there is usually an adequate margin of safety, many foundries omit the test entirely. When, however, the sand contains wood flour, which reduces dry strength, the test is useful. Indeed, when the sand contains both coal dust and wood flour, a fall in dry strength may be the first evidence that the amount of wood flour is increasing.

The sand testing programme should therefore be as follows:

moisture content:	at irregular intervals of not more than 1 h;
green strength shatter index permeability	hourly or twice daily;
dry strength volatile matter	daily;
mechanical analysis:	only when there is an unexplained change in permeability.

Whenever it is necessary to take action because sand properties are out of line, the frequency of testing should be increased until results have remained steady for a day or so.

Occasionally it may be necessary to increase the frequency of testing at selected times. For example, difficulties in mould making may arise at a certain time on each day and persist for a time so short that, with the normal interval between tests, the sand which is the cause of these difficulties is not tested. In one such case, bad lifts were obtained some 2 h after work commenced and when each batch was tested at the appropriate time it became clear that there was a considerable increase in the amount of dead clay present. It was then found that the man clearing up after production ceased deposited sand coated with dead clay onto the shake-out grid and this arrived at the mixer about 2 h after production started.

RECORDING AND INTERPRETATION OF RESULTS

The results should be recorded on a chart which also shows the levels at which action should be taken. This method has the advantage that trends can be seen more clearly than when only a list of figures is available and remedial action can be taken at an earlier stage. A pronounced peak shows that something abnormal has occurred, such as the increase in dead clay recorded above.

There is no difficulty in deciding the reason for a change in permeability. The rate of change in permeability with change in moisture content is slow, so long as the sand is not obviously too wet or too dry, and changes in permeability are almost entirely due to an increase or decrease in the amount of fine material present — either coal dust or the coke derived from it. This, in turn, may be due to the additions being too high or too low, or to an increase or decrease in the amount of fines removed by a dust extractor at the shake-out. For example, replacing a light casting by a heavier one may increase the amount of dry sand passing through the grid with corresponding increase in the amount of fines removed. Control of the amount of dust so removed is difficult and should permeability change to such an extent that the quality of the castings is affected, it may be necessary to change the grading of the base sand.

Control of the moulding properties of the sand is less straightforward. The control tests used are those for moisture content, green strength and shatter index, and the test results depend on the amount of water and active and dead clay present, but of these only the moisture content is measured and there is no practicable method of measuring the amounts of the two kinds of clay in a foundry system sand which contains both. Changes in composition must be deduced from the changes in properties.

There are several complicating factors. Even when there are no other changes in composition, the moisture content has a marked effect on sand properties. Moreover, the amount of water required to produce a mouldable mixture in a sand containing sand, active clay and water only, is proportional to the amount of clay present, *i.e.* if a sand with 5% clay requires 2·0% water, a sand containing 10% clay would require 4·0% water. There are also significant differences between sands containing clays of different kinds. In addition, dead clay increases the need for water, although its influence is less than that of the same amount of active clay. Indeed, in this respect dead clay itself may vary, that which has been heated to a temperature high enough to cause appreciable sintering requiring less water than that which has been heated to the somewhat lower temperature at which the clay breaks down without sintering. Fine materials such as superfine coal dust and the even finer coke arising from its breakdown also increase the need for water, but their combined effects are less than those of clay and should they increase to such an extent that the water content should be raised, there would be a drastic fall in permeability.

The job of the sand controller is to deduce from the results of the tests for moisture content, green strength and shatter index, what changes are taking place in the amounts of active and dead clay present and to decide what changes must be made in the additions of new sand and clay in order to keep sand properties within prescribed limits.

To do this, it is necessary to take into account the relation between moisture and active and dead clay on the one hand, and sand properties on the other.

Clay, water and sand properties

The effect of water content on the green strength and shatter index of two sands containing respectively 5 and 7% clay are shown in Figs. 58 and 59.

At low moisture contents the sand has a high green strength and a low shatter index and is too brittle to be used for mould-making. As the moisture content increases, green strength falls, at first rapidly and then more slowly. Shatter index rises to an indefinite maximum and then falls. The most suitable moisture content is a little below that at which shatter index is at a maximum, although a hand moulder, an operator on a jolt–squeeze machine and a high pressure squeezer have different ideas on this point. The hand moulder would regard the sand from the jolt machine as being a little on the dry side and that from the high pressure machine as too dry for use. It can be seen from the shape of the shatter index/moisture curve that an excess of moisture can produce only a small increase in shatter index, but that a deficiency in moisture content will produce a marked reduction. The influence of moisture content on green strength in this region is rather more pronounced that its influence on shatter index, but again the low water content produces the greater change.

As clay content increases, the curves are displaced towards the direction of higher moisture content. The amount of water needed to produce a sand suitable for moulding is, in the absence of dead clay or any other fine grained material, very nearly proportional to the clay content.

Dead clay itself has no bonding properties and the physical properties of a moist silica sand containing dead clay differ little from those of the same sand with the same moisture content. However, if the two sands are squeezed in the hand, the sand containing dead clay seems the drier of the two.

On the other hand, it was shown in Chapter 4 that when dead clay is present in a sand which also contains active clay, it affects the properties of the sand in three ways: raising the amount of water needed, increasing

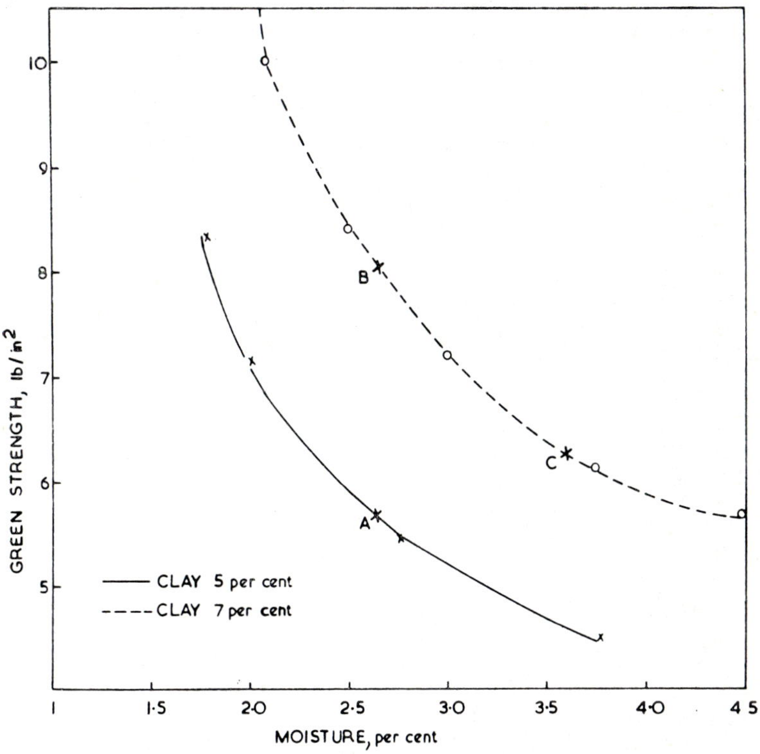

FIG. 58. Effect of moisture content on green strength.

the green strength and reducing the shatter index. These changes are shown in Figs. 60 and 61.

Dead clay is not added to the sand, but is formed when that part of a mould which is near to the poured casting is raised to a high temperature. With any one clay, the amount formed depends mainly on the weight of metal and the time between pouring and shaking out, and the proportion in the sand at the shake-out depends on the ratio of weight of metal/weight of sand and on the time. The proportion in the prepared sand depends also on the amount of new sand added, since this reduces the proportion of dead clay by dilution. In this respect, any sand entering the system as a result of the breakdown of cores with organic binders is quite as effective as new sand, and when the castings are heavily cored it may be unnecessary to add new sand.

Breakdown of clay does not commence until the temperature reaches 400–500°C, depending on the kind of clay and in this range of temperature is slow. The rate of breakdown increases as temperature rises. On the other hand, the temperature of the sand cannot exceed 100°C until all moisture has been removed and the rate of loss of moisture is far greater

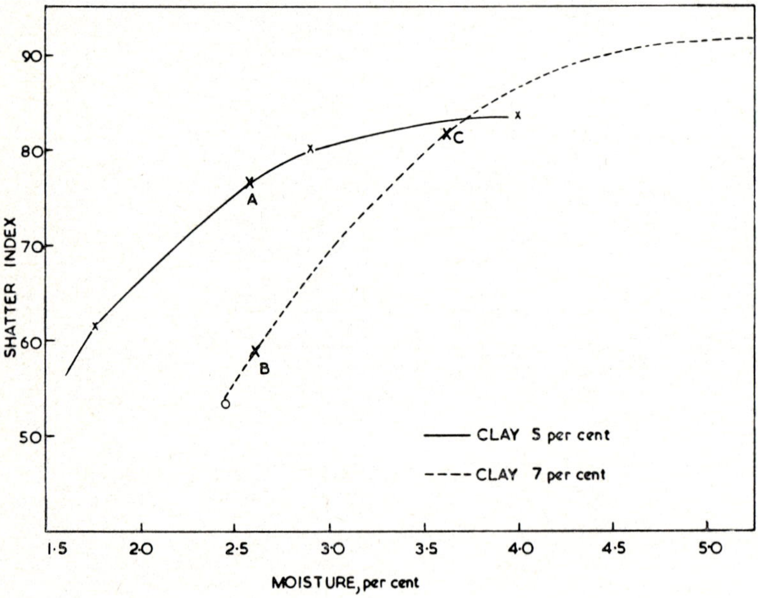

FIG. 59. Effect of moisture content on shatter index.

than the rate of breakdown of clay. It is also more susceptible to changes in conditions, for instance a change in metal/sand ratio or the time between pouring and shaking out.

On the whole, therefore, changes in the proportion of dead clay present in a prepared moulding sand are slow, but there are conditions under which a rapid increase is possible. It was shown in Chapter 4 that the sand adhering to heavy castings is coated with dead clay only, and that should these castings be set aside in one place to cool, a considerable amount of sand may be left behind when the castings are finally removed. The most

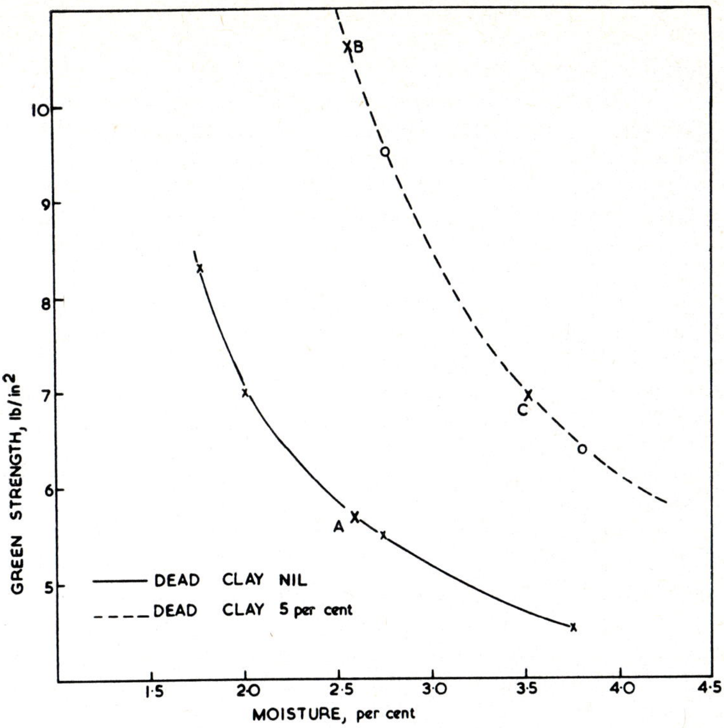

FIG. 60. Effect of 5% dead clay on green strength.

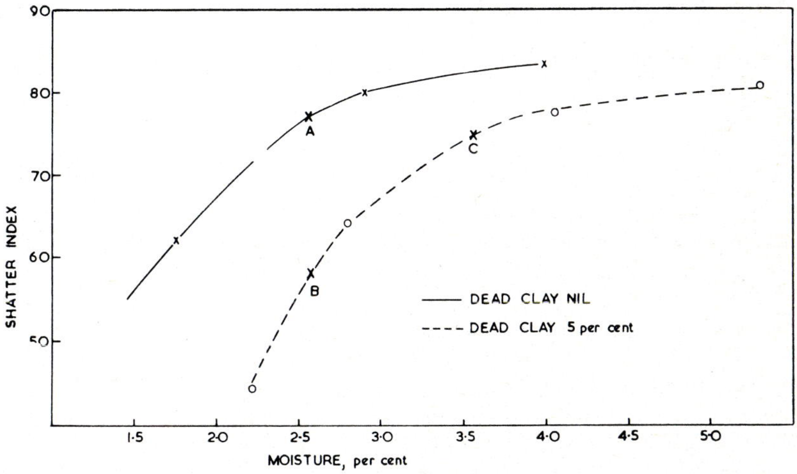

FIG. 61. Effect of 5% dead clay on shatter index.

convenient way to get rid of this sand is to dump it on the shake-out grid, but it has been shown that this is a false economy. The most serious effect, however, is that little mixing may occur as the sand passes round to the mill and a few batches of sand with an unusually large proportion of dead clay are prepared. The whole system becomes unstable and remains so until the sand affected has gone round the system several times. The sand containing dead clay should be removed from the foundry and discarded.

The sand which sometimes adheres to light castings contains only a little dead clay and to discard it would be unnecessarily wasteful, but in practice there is no difficulty in recognising the sand coated with dead clay. It is usually dull white in colour and the patches of dead clay can be seen if the sand is examined under a low power magnification. If the sand is shaken with a large excess of water little clay appears, but if it is first milled or lightly rubbed in a mortar, much of the dead clay is detached from the sand grains and appears on shaking with water.

Methods of control

Of the three variables which have to be controlled, moisture content is the most important since it varies more rapidly than the others. A hold-up at the shake-out, which increases the time during which heat is transferred from metal to mould, has a small effect on the amount of dead clay formed but a major influence on the amount of water removed by evaporation. Knocking out a few damaged moulds does nothing of any consequence to the amount of dead clay, but leads to a batch of sand with a much higher than usual moisture content. In consequence, even when only one kind of casting is being made, there may be appreciable variations in moisture content.

All methods of sand control, therefore, depend primarily on controlling the moisture content. Three methods are in use.

1. The mill operator squeezes a sample of sand in his hand and judges the moisture content by the 'feel'.
2. The moisture content of the sand in a batch mill or entering a continuous mill is measured automatically and the appropriate amount of water added.
3. A quantity of sand passes through a Dietert 'mouldability controller' which controls the addition of water.

The control methods are applied to the water addition only and do not in any way control the amounts of active and dead clay. Changes in these amounts must be deduced from the test results, but the effects of such changes on sand properties depend on which of the three methods of moisture control is used.

Interpretation of the test results is not the same in the three sets of conditions.

1. Control by operator

When this method is used, the operator makes no attempt to maintain a constant moisture content in the sand. Should the amount of active or dead clay rise or fall, he attempts to make the appropriate change in moisture content, *i.e.* if the clay content rose from 5 to 7% in Figs. 58 and 59, he would try to raise the moisture content from that indicated at A to that indicated at C.

However, his estimation is not strictly quantitative, but is in the nature of an informed guess and is likely to fail at times. There are therefore three kinds of change possible:

(i) failure on the part of the operator;

(ii) an increase or decrease in active clay;

(iii) an increase in dead clay.

Failure to control moisture content

In all clay-bonded sands, an increase in moisture content leads to a decrease in green strength and a small increase in shatter index. The test results show a pattern similar to that in Fig. 62.

Too low a moisture content leads to an increase in green strength and a fall in shatter index. It can be seen from Figs. 58 and 59 that too low a moisture content has a greater effect numerically on sand properties than too high a moisture content. It does not follow that it is likely to cause more trouble in the foundry. The most important effect of too low a moisture content is that mould-making becomes more difficult and boxes may have to be knocked out; too high a moisture content may have the more expensive effect of producing scrap castings.

Either of two different kinds of change may occur. Moisture content may rise — or fall — gradually with a few halts on the way, as illustrated in Fig. 62. A change of this kind is always possible, since even the best of operators cannot distinguish small variations in moisture content, and in the absence of a standard for comparison, one batch is usually compared with the previous one.

On the other hand, there may be occasional sharp peaks in the moisture content and corresponding changes in green strength and shatter index. These are due to inattention or, occasionally, to a loss of 'touch' on the part of the operator.

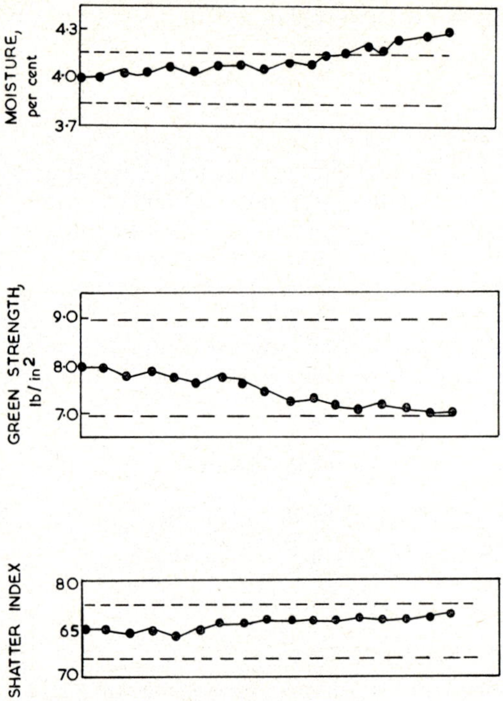

FIG. 62. Effect of excessive water.

Increase or decrease in active clay

Should there be an increase in the amount of active clay, the operator immediately raises the moisture content. At the same time, there is an increase in both green strength and shatter index and the system moves from the point A to the point C in Figs. 58 and 59, although in practice the changes are less than those shown in these figures, which were obtained by raising the clay content by 2%. The pattern of results is shown in Fig. 63.

A decrease in clay content is shown by a fall in moisture content, green strength and shatter index.

Increase in dead clay

Again the operator raises the moisture content, and the system moves from A to C in Figs. 60 and 61, but now there is an increase in green strength and a fall in shatter index. The pattern of results is shown in Fig. 64.

Conversely, a decrease in dead clay is accompanied by a decrease in moisture content and green strength, and an increase in shatter index.

When the test results show that the amount of dead clay is increasing, the amount of new sand added and the amount of used sand discarded would be increased.

The decision as to what action to take when the amount of dead clay decreases appreciably is harder to make. Dead clay itself has a harmful effect on clay-bonded sand, but keeping the amount present to a very low figure by using mainly new sand would be expensive, and the limits on sand properties are set at the level which permits the maximum inclusion of dead clay, without leading to difficulties in the moulding department.

Should there be a continuous decrease in the amount of dead clay, it can be due either to the replacement of a heavy casting by a lighter one or the introduction of a heavily cored casting and an increase in the amount of new sand derived from the breakdown of cores.

In the first case, the remedy is simple — the amount of new sand added should be reduced, but this procedure is not possible in the second case. Quite frequently the amount of sand entering the system from the cores exceeds the total amount of new sand needed and nothing can be done

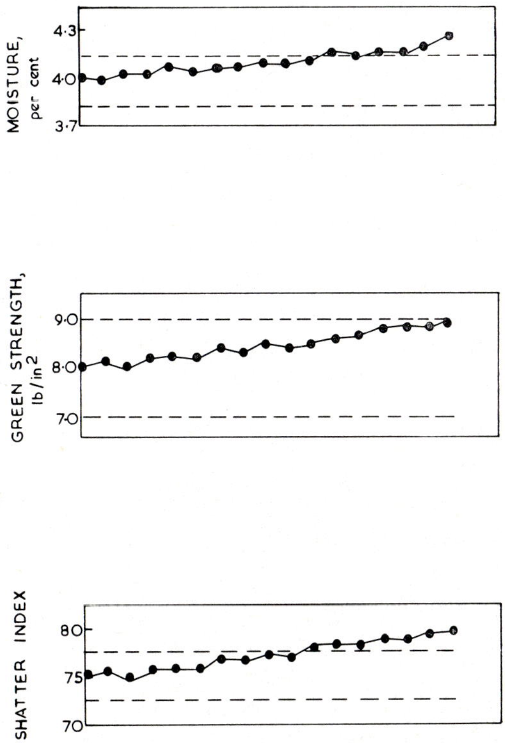

FIG. 63. Effect of an increase in clay content.

about this. There is a fall in moisture content and green strength and an increase in shatter index.

So far as mould-making is concerned, no serious difficulty should arise. Indeed, it may be possible to reduce the active clay still further and operate at still lower moisture content, green strength and shatter index. However, if the castings are such that internal porosity — usually found only after machining — is a possibility, a minimum mould hardness is required to suppress the defect and it becomes increasingly difficult to

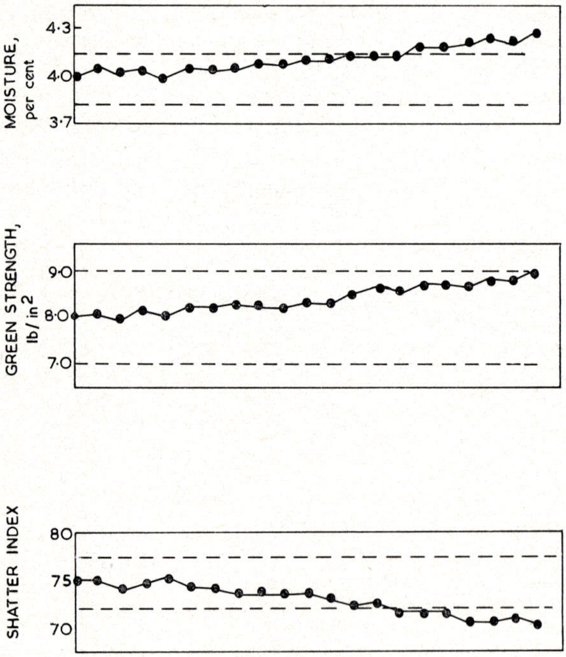

FIG. 64. Effect of an increase in dead clay.

maintain mould hardness as green strength falls. The one change in composition which will do more than any other to offset this change in properties, is a further reduction in moisture content, which will offset the increase in shatter index and fall in green strength due to the decrease in dead clay. The operator should therefore be told to adjust the moisture content so that the sand feels drier when squeezed in the hand.

Simultaneous changes in more than one variable

It is not difficult to obtain from Figs. 58–61 the effect on test results of a change in both active and dead clay at the same time. A change of this kind occurs more frequently than a change in one variable only, since it may be due not only to the additions of sand and clay being slightly off, but also to a change in the kind of casting being made.

Should there be an increase in both active and dead clay at the same time, the operator raises the moisture content by the amount appropriate to the two changes combined. Both active and dead clay raise green strength and when both rise at the same time, the increase in green strength becomes more marked. Active clay raises and dead clay reduces shatter index and the two changes tend to cancel each other out. The final result is an increase in moisture content and green strength, with a small or nil change in shatter index.

Should active and dead clay fall at the same time, the reverse changes occur — a fall in moisture content and green strength with little change in shatter index.

On the other hand, when there has been an increase in active clay and a decrease in dead clay, the kind of change which occurs when a heavy casting is replaced by a lighter one, the changes in moisture content and green strength tend to cancel each other out, but the increase in shatter index becomes more pronounced. Conversely, a decrease in active clay with an increase in dead clay leads to a marked fall in shatter index with little change in moisture content and green strength.

Limitations of control by operator

The one uncertain factor in the foregoing calculations is the reliability of the operator. Most recommendations concerning sand testing recognise this in an indirect way and contain clauses designed to reduce the risk of error to a minimum. The most important is that samples for testing should be taken at irregular intervals, the aim being to eliminate any idea that once a sample has been taken, attention can be relaxed until the next is due. A further safeguard is to test three or four consecutive batches occasionally. Unless there has been an unusual occurrence which produces abnormal sand conditions, such as a breakdown at the shake-out or the stripping mechanism of a machine, and such times can easily be avoided, the amounts of active and dead clay do not change from batch to batch and the method gives some idea of the reliability of the operator.

Nevertheless, there is always a possibility of failure. Even when sand composition is otherwise constant, the best of operators cannot distinguish changes of less than 0·1% in moisture content, and a slow rise or fall may occur. Control becomes even more difficult after a change such as a sudden increase in the amount of dead clay since, whatever the change in moisture content, the two sands never feel the same. A happening such as this can impair the operator's ability to control moisture content for some time.

All the means for deciding when to make a change in the additions of sand and clay are dependent upon the ability of the operator to maintain a certain relationship between sand composition and moisture content. When there is any uncertainty on this point the standard of control falls immediately.

In consequence, no change should be made on the strength of tests on one sample only, and at least three consecutive batches should be tested before coming to a decision. This is not a complete safeguard, but does help.

2. Control by moisture controller

The apparatus now available for maintaining a constant level of moisture content eliminates the uncertainty due to the fallibility of operators, but of itself does nothing to control the amounts of active and dead clay present. Moreover, a change in the amount of clay, active or dead, produces changes in sand properties which are at constant moisture content, very different from the corresponding changes produced when the operator attempts to adjust the moisture content in accordance with clay content.

Suppose that of the sands used to prepare the graphs shown in Figs. 58 and 59, the one with a clay content of 5% is suitable for mould-making with a moisture content corresponding with the point A. Should the clay content rise to 7% an operator would increase the moisture content to that corresponding with the point C and there would be an increase in shatter index and green strength. With a moisture controller in operation, moisture content remains constant and the move is to the point B and not C. There is therefore a large increase in green strength and a decrease in shatter index.

In the first case, there would be no difficulty in making and stripping moulds, but overall hardness would decrease and hardness differences in different parts of the mould would increase. In the second case, the sand would become too brittle for moulding anything more complicated than a flat plate.

This illustrates two points.

1. Whatever method is used for controlling moisture content, changes in clay content must not be allowed to get out of hand.

2. When a constant moisture content is maintained, changes in sand composition produce changes in properties which may be completely different from those obtained in a system controlled by an operator.

Whatever the method of control used, sand composition must not be allowed to get out of line to the above extent and with smaller changes, even the pattern of change in properties may be somewhat different. Suppose that instead of an increase in clay content of 2%, the difference were only 0·2%. The curve relating moisture content with shatter index would be displaced only slightly to the right and the two curves would cross at a moisture content a little below that required to give maximum shatter index, *i.e.* very close to the moisture content at which the sand is suitable for moulding. Any change in shatter index would be less than the error inherent in the testing equipment and could not be observed. The only indication of the increase in clay content would then be an increase in green strength.

Should there be an increase in dead clay, an operator would immediately try to raise the moisture content from A to C in Figs. 60 and 61, but with a moisture controller, the move is from A to B and the increase in green strength and fall in shatter index, which are due to an increase in dead clay, are therefore greater when the controller is used.

It is apparent that when no attempt is made to control the amounts of active and dead clay and either or both of these vary, difficulties in mould making arise more quickly when the moisture is kept constant than when an operator adjusts the moisture content by feel. This does not in any way detract from the usefulness of the moisture controller — all that happens is that smaller changes in sand composition are shown up by the test results and remedial action must be taken sooner. Changes in the amount of new sand or clay added may have to be made at shorter intervals — say each week in place of each month — but this is not a serious drawback.

The supreme advantage of the controller is that control is positive since the accuracy of the instrument is checked precisely each time a sample of sand is tested, whereas there is always some uncertainty in the inspired guesswork of an operator.

When using the controller, charts similar to those shown in Figs. 63 and 64 should be used. There is now no counterpart to Fig. 62, since moisture content is constant.

Again four cases have to be considered — an increase or decrease in active or dead clay.

Increase in active clay

As shown above, this leads to an increase in green strength which is higher than that found when moisture content was raised. There is no appreciable change in shatter index unless the clay content is greatly increased, in which case the index falls. These changes are shown in Fig. 65.

Decrease in active clay

In the operator-controlled system, there was a fall in moisture content, green strength and shatter index. When the higher moisture content is maintained, the fall in green strength becomes more pronounced, but the fall in shatter index is partly offset by the increase in moisture content. It will not be reversed or even completely offset, since in this region the rate of change of shatter index with change in moisture content is slow.

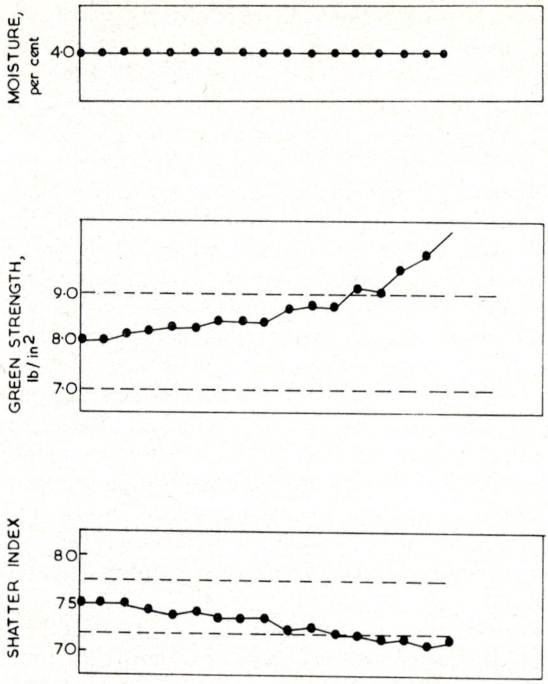

FIG. 65. Effect of an increase in active clay at constant moisture content.

The indication is a fall in green strength with little change in shatter index.

Increase in dead clay

An increase in dead clay raises green strength and reduces shatter index, even when the moisture content is raised. With a constant moisture content these changes become even more pronounced. The changes are shown in Fig. 66.

Decrease in dead clay

A decrease in dead clay produces a reduction in green strength and increase in shatter index.

There is one further point of importance. When either the active or dead clay, or both, fall and the moisture content is not reduced, the sand appears wet to the touch. It has been shown that when the sand is in this condition there is a risk of obtaining blown castings, whatever the level of moisture content. However, the test results indicate the onset of the change well before there is a risk of producing bad castings on this score. The procedure to be followed is to reduce the amount of new sand added, but where

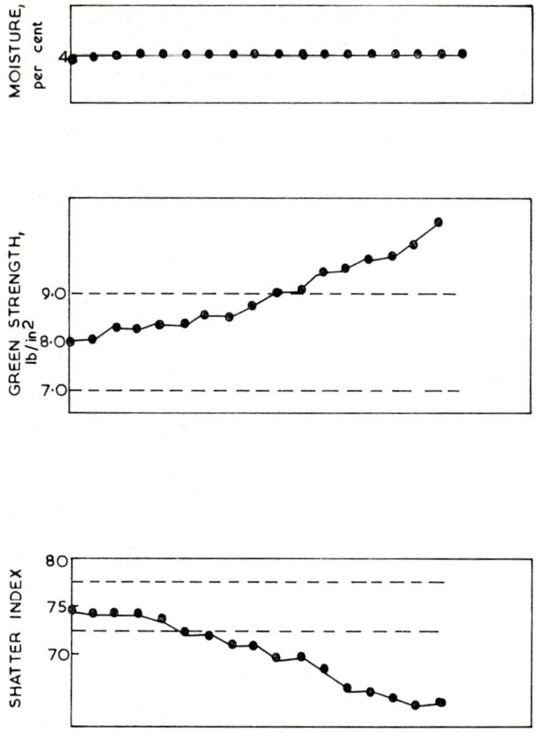

FIG. 66. Effect of an increase in dead clay at constant moisture content.

such sand is derived entirely from the breakdown of cores, to reduce the level decided upon for moisture content.

3. Control by 'mouldability controller'

When the Dietert 'mouldability controller' is used as the basis of control, some sand is withdrawn from the mill and passes along a vibratory conveyor with two slots. The first slot is narrow and the second is wider.

If the sand is very dry, a considerable amount of sand falls through the narrow slot and cuts off a ray of light; this opens a large valve which adds water to the mill. Eventually the moisture reaches a level when no sand passes through the first slot and the large valve is shut off. The size of the slot is such that the sand is a little too dry and the final adjustment is by means of a small valve relayed to the second larger slot.

The amount of sand passing through a slot is controlled mainly by the toughness of the sand.

Gittus[4] compared several such toughness tests and found that all gave an approximate correlation with the shatter index, but the correlation was not exact. Until a similar comparison has been made between Dietert's 'mouldability index' and the shatter index, it will not be possible to say that the controller produces sand with a constant shatter index or the extent of any differences.

This in no way detracts from the usefulness of the controller. Even if the shatter index varies, the variation will be small and changes in sand composition can readily be deduced from changes in other properties, particularly green strength.

Increase in active clay
The controller will immediately raise the moisture content, but not to the same extent as would an operator judging by feel, since the latter would cause a pronounced increase in shatter index. Accordingly, the index changes little, but there is an increase in green strength intermediate between that produced when a water controller, on the one hand, and an operator, on the other hand, is used.

Decrease in active clay
Moisture content will be reduced. The clearest indication is a fall in green strength with a slight fall in moisture content.

Increase in dead clay
The moisture content will immediately be raised, but the increase is likely to be larger than that thought necessary by an operator, in order partly to offset the reduction in shatter index which otherwise follows an increase in dead clay. An increase in moisture content reduces green strength and, at least in part, compensates for the effect of dead clay. There is therefore little change in shatter index, an increase in moisture content and a small increase in green strength.

Decrease in dead clay
The moisture content will be reduced, but the reduction will be rather greater than that selected by an operator. There will be only a small change in shatter index and a marked fall in green strength.

Changes in permeability
The rate of change of permeability with change in moisture content, unless the latter changes to such an extent that mould-making becomes difficult, are small. An increase in clay may cause an apparent increase, because as the toughness increases, it becomes more difficult to pack the sand grains closely, but before the change becomes sufficient to require action, the effects would be serious in the moulding section.

Dead clay also reduces permeability and the permeability of a used foundry sand is always lower than that of the same sand containing new sand, clay and water only. Again, the effect of dead clay on moulding properties is such that action would have to be taken long before there was a significant fall in permeability.

Provided that there has been no change in the grading of the new sand or the amount and grading of the coal dust present, changes in permeability depend on the relation between two factors. There is a continuous production of very fine coke as a result of thermal breakdown of coal dust, and a loss of fine-grained materials of all kinds when the moulds are broken down after casting. The only method for finding out the reason for a change in permeability is by means of a mechanical analysis, with subsequent ignition of the separate fractions to determine how much is sand and how much is combustible matter.

When the moulds are knocked out on the floor, there is a tendency for permeability to fall to a level far below that of the new sand, since the factors which favour removal of fines are absent. In a mechanised foundry, the change may be very different and in one case, the installation of a down-draught dust extractor with a large factor of safety, resulted in the removal of not only fine coke, but also most of the coal dust and many of the finer grains in the natural sand in use. Permeability doubled in a short time and the surface finish of the castings, previously a matter for pride and an important selling point, deteriorated. The offending dust extractor had to be modified.

There is not much which can be done to control changes in permeability when a natural sand is used, but it is only in exceptional circumstances that changes become serious. The influence of mould hardness and mould dimensions on permeability is so great that a wide margin of safety has to be maintained and the changes which take place under most conditions are not significant.

There is no difficulty in controlling the permeability of a synthetic sand provided that two base sands are used — a moderately coarse sand with a small proportion of a finer sand. Should permeability begin to rise or fall it can be restored by raising or reducing the amount of finer sand. By this means there was, on one occasion, no serious difficulty in maintaining a margin of not more than ± 2 points in a sand with a permeability of 60. As an exercise in controlling the properties of sand, this was interesting, but as a contribution to casting quality, it was worthless. Should there have been any risk immediately permeability fell below 58, it would have been necessary to maintain a minimum of well above 70 to take care of changes in pattern and variations in mould hardness.

CONTROL OF NATURAL SAND

When using a natural sand, control is simpler and, at the same time, more difficult than when using a synthetic sand. It is simpler because, in addition to the used sand, only three constituents — new sand, coal dust and water — are added, and there is a fixed relationship between the amount of clay and sand, so that it is not possible to get a build-up of active clay. It is more difficult because some of the means for offsetting undesirable changes in sand properties are lacking. The difficulties in control of permeability have been described above.

The most important test is for moisture content and this should be checked with the same frequency whatever the kind of sand used.

There is little risk of a build-up of dead clay since this is produced only at the expense of active clay, which is added as a fixed proportion of the new sand. Most natural sands contain only a little clay in excess of that needed to produce a mouldable mixture and before sufficient of this clay has decomposed to produce a harmful amount of dead clay, the sand would have become unusable as a result of a deficiency in active clay. To remedy this, more new sand would have to be added, reducing the proportion of dead clay by dilution. The graphs used in the control of properties of a synthetic sand which illustrate the effects of an increase or decrease in active and dead clay are irrelevant and all that it needed is to add sufficient new sand to maintain shatter index and green strength.

Many foundries, although fully mechanised, confine their testing to measurement of moisture content. Since the new sand contains only a small amount of clay in excess of the minimum needed, no harm results from the addition of more than the minimum amount of new sand beyond that of paying for it. Control of coal dust is obtained as a result of inspection of the castings produced.

In such a case, everything depends on the reliability of the sand supply. Here there has been a very great improvement, made possible by concentration into fewer and larger units, and tests made on every delivery from one particular supplier over a period of 4 years show that permeability varied only between 35 and 39 and clay content between 11 and 12%.

It may be cheaper to use rather more new sand than is really needed and to pay a little extra for reliable supplies than to maintain a testing laboratory, but even the most reliable of suppliers may fall down as a result of striking a minor fault in a sand quarry or because of a breakdown in some equipment. The method, therefore, has its disadvantages.

Perhaps the most serious of these, is that should anything go wrong, the information which will enable the fault to be tracked down and put right is lacking. Difficulty in stripping a difficult pattern may be put down to a fault in the machine when it is really due to a gradual change in sand composition or *vice versa.* A tendency for an increase in misruns may be attributed to a fall in sand permeability when it is due to cold metal at the cupola or a slowing up of the metal distribution system.

Regular testing is therefore desirable even when a natural sand is used, but except for moisture content, the frequency of testing can be reduced.

REFERENCES

1. *Grey Iron Founding* (1950). *Anglo-American Council on Productivity.* Paras. 281–286.
2. Zehnder, C. (1959). *British Foundryman*, **52,** p. 278.
3. Godding, R. G. and Bird, D. (1968). *Measurement and Control of Water Content of Foundry Sand. BCIRA J.*, p. 30.
4. Gittus, J. (1951–1953). *BCIRA J.*, **4,** pp. 560–567.
5. Dietert, H. W., Graham, K. L. and Hanna, T. H. (1961). *Controlling Sand Mouldability and Mill Quality. Foundry*, May, p. 238.

CHAPTER 8

Mould-Making and Properties of Moulds

The effect of ramming

When molten metal is poured into a mould, a pressure is exerted on the mould surface which depends on the dimensions of the mould. At a later stage, metals such as grey iron, which expand on solidification, exert a further pressure due to that expansion.

The extent to which pressure from the liquid metal causes the mould to distort, or molten metal to enter the spaces between the sand grains, and the amount of yielding of the mould which is due to the solidification, depends on how closely the sand grains have been packed together by ramming. The various effects of yielding of the mould are discussed in detail in Chapter 10, but it is clear that a more accurate casting must be obtained when the sand grains are more closely packed, *i.e.* when the mould is denser and harder. Of the two parameters, hardness is more convenient than density to use when making tests on moulds since it can be measured, without serious difficulty, at any point on the mould surface without destroying the mould. The average density of a mould does not convey much useful information, since density in different parts of the same mould may vary widely, while trepanning sections is a lengthy operation which destroys the mould without giving results which are sufficiently selective to be of much value.

The hardness, and particularly the pattern of hardness distribution, depends upon the method of ramming and on the physical properties of the sand used. The combination of sand properties which influence the uniformity of mould hardness is frequently referred to as 'flowability', but since this is usually defined as 'that property or properties which enable a sand to be rammed uniformly round a pattern', the definition does not give much assistance in obtaining a quantitative value for the property.

There are two major aspects of flowability with at least one of less importance.

Suppose, for example, that a mould is being made by squeezing sand round a simple pattern as shown in Fig. 67. A loose frame is placed on the moulding box, the whole filled with loose sand and strickled level. Pressure is applied, compressing the sand until no further movement takes place. Suppose the height of the pattern is 2 in, the depth of the moulding box 4 in and of the frame 1 in. If there is no lateral movement of sand from above the pattern then over the pattern, squeezing compresses the sand to two-thirds of its original volume, but at the base of the pattern only to four-fifths. The volume of sand over A is therefore decreased and its density increased to a greater extent than over B, and the mould is hardest at A. On the other hand, if the sand flows like a liquid this hardness difference must disappear. This is one aspect of 'flowability'.

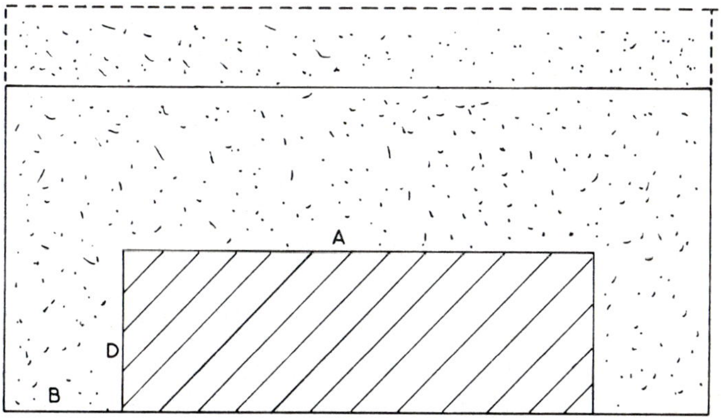

FIG. 67. Squeezing sand round a pattern.

When sand is compressed, individual grains move in relation to each other, and since the total friction between large numbers of sand grains is high in this condition, the applied force decays as the distance from the point of application increases. Mould hardness decreases when under otherwise constant conditions, the depth of sand over the pattern increases and the extent of this decrease can also be described as depending on the 'flowability' of the sand concerned.

A smaller, but nevertheless significant effect of friction, is that friction between sand and pattern is high, and the hardness of the mould against a vertical face of the pattern — at D in Fig. 67 — must be less than would be expected from its location in the mould. In practice, this is invariably the case and unless the method of ramming used is such that sand is actually forced in the direction of the pattern, vertical faces are always the roughest surfaces on a casting.

Many observers have examined the various aspects of flowability and three methods have been used:

(1) measurement of lateral movement during ramming;

(2) measurement of hardness gradients in rammed sand;

(3) observation of the relation between mould hardness distribution and some physical property of the sand.

Lateral movement during ramming

Lissell and Ash[1] placed sand in an otherwise standard test piece holder with a hole 1 in diameter in the base plate. Increasing pressure was applied at the top of the sand column until sand flowed through the aperture.

With dry sand, a pressure of 2400 lb was needed. The addition of oil greatly reduced the pressure and a weak slurry of bentonite and water had a similar but smaller effect. The addition of bentonite and water in the proportions in which they are used in moulding sand prevented extrusion, although the pressure was raised to 26 000 lb, at which point the base of the container was broken.

Lissell and Ash concluded that sand flows mainly in the direction of the applied force and that lateral movement is of minor importance.

Davies[2] was concerned mainly with hardness gradients, but observed that when a loose sand is compressed, there is a slight lateral movement in the initial stages, but that this ceases as soon as the sand attains any cohesion.

Fairfield and McConachie[3] regarded flowability as movement of sand in a direction at right angles to that of the applied force and designed the 'Kennedy flowability test' to measure this property. The test is carried out in the following manner.

A disc 2 in diameter is attached to the standard test piece holder by three pins such that there is a clearance of $\frac{3}{16}$ in right round the lower edge of the container. The sand to be tested is sieved into the holder and placed under the standard rammer in such a way that the rammer head is just inside the top of the tube. The rammer assembly is allowed to fall, when some sand is forced out of the space between the tube and the disc. This sand is weighed and the weight used as an index of flowability.

Fairfield and McConachie found a good relationship between the 'Kennedy' flowability and the smoothness and uniformity of the surface of the test piece after its removal from the container.

Other properties were also measured, including green strength and green deformation, from which toughness can be obtained. Flowability increased as toughness decreased. It was not possible to find any relationship between Kennedy flowability and the flowability test of Dietert and Valtier (see below).

The amount of sand forced out of the aperture between tube and disc is a very small proportion of the sand used. In tests on a series of sands, including oil-bonded and synthetic and natural clay-bonded sands, the highest quantity was 3·1 g and the lowest 0·55 g, whereas the weight of sand taken can hardly have been less than 150 g.

The effect of lateral movement

Of the three observers, Lissell and Ash found that lateral movement did not occur in clay-bonded sand under the conditions of their test, Davies found that a slight lateral movement occurs when compression begins, but ceases when the sand acquires any cohesion, and Fairfield and McConachie

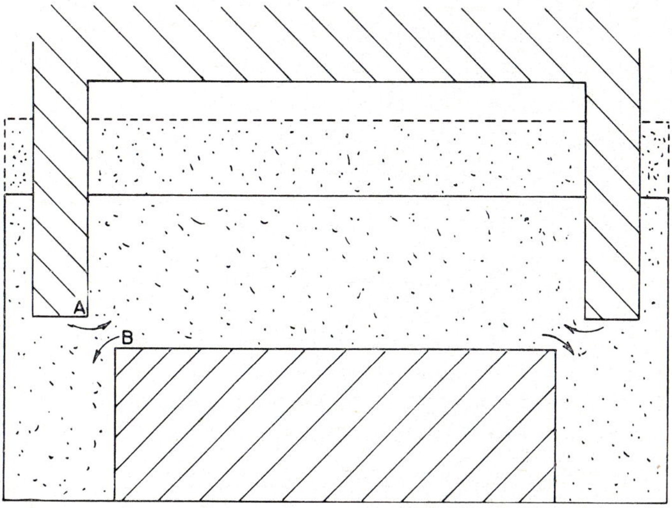

FIG. 68. Lateral movement with contoured head.

measured the amount of sand which moves laterally when the sand is compressed vertically, but differences between the three are apparent rather than real.

The very small weight of sand which was displaced laterally in Fairfield and McConachie's experiment matches up with Davies' observation and would be insufficient to cause sand to flow through the aperture in the experiment of Lissell and Ash. It can be concluded that movement at right angles to the applied force is small.

The movement of such a small amount of sand would have a negligible effect in improving uniformity of packing in a mould such as that illustrated in Fig. 67, but it might be significant when using a contoured squeeze head.

Suppose that a contoured head is used in place of a flat head to produce the mould shown in Fig. 67. When the pressure is first applied, conditions are as illustrated in Fig. 68.

Some lateral movement occurs at A, sand being forced away from beneath the rammer probe. When sand movement begins at the face of the pattern at B, sand is forced away as shown.

Both amounts of sand may be small, but they are both concentrated in the small space which at the conclusion of ramming, is left between the head probe and pattern, and help to increase packing density in that part of the mould which is usually least dense and which otherwise would derive no benefit from the use of a contoured head.

Hardness gradients in rammed sand

Kyle[4] used two methods of measuring flowability which are carried out as follows.

1. A metal box $12\frac{3}{4}$ in high, 2 in wide and 1 in deep, with one $12\frac{3}{4} \times 2$ in face removable, is used to contain the sand. The face is removed and the box filled with sand by sieving. The face is replaced and the sand compacted from one 2 in × 1 in end by squeezing or jolting. The face is removed and hardness at various distances from the bottom end measured, using the standard indentation hardness tester. Flowability is calculated from the area under the hardness/distance curve.

2. Enough sand to form a test piece 2 in long is sieved into a split container. The sand is rammed by means of a standard rammer modified in such a way that the weight falls through a distance of $\frac{1}{4}$ in. Hardness at top and bottom is measured and flowability expressed as:

$$\frac{\text{hardness at bottom}}{\text{hardness at top}} \times 100$$

In Kyle's tests only one blow from the rammer was needed to give a mould hardness of about 60. If the test piece was rammed harder than this, the difference between top and bottom was so small that the test became insensitive.

The two tests placed sands in the same order.

Kyle related the flowability of a number of sands with their moisture contents and also measured permeability. He found that with a fine-grained, natural sand containing 18·4% clay, flowability was at a minimum when moisture content was between 8 and 10%. He also found that permeability was at a maximum when moisture content was 8·6%. Now when the moisture content of a clay-bonded sand is adjusted so as to give maximum permeability, toughness is also very near to the maximum and it can be concluded that Kyle's work shows that flowability decreases as toughness increases.

Davies[2] rammed a sand test piece from one end and measured variations in bulk density, a method which is more precise although more lengthy than the hardness method of Kyle.

Davies' method was to build up a cylindrical test piece container by assembling accurate segments of equal depth, kept in position by an outer cylinder into which they fitted closely. Sand was sieved into the assembled container and squeezed by placing a weight on the top. The segments were removed one at a time and the sand strickled off and weighed, flowability being derived from the bulk density gradient. Davies did not record the strength or toughness of his sand, but made the observation that the higher the bulk density of the sand before compression, the higher the flowability. Now sand which has a high bulk density in the uncompacted condition has a low toughness and it can be concluded that Davies' work also shows that flowability decreases as toughness increases.

Lissell and Ash[1] used the entirely different method of measuring the work done in compressing sand when a compressive load was raised in steps to a maximum of 475 lb on the standard test piece. They concluded that as sand becomes stronger, more work is required to compress it, but whether sand strength is the only, or even the main factor, is doubtful.

The most important observation made was that it is more difficult to compact angular sands than round ones.

Other methods

Many other methods of measuring or illustrating the effects of 'flowability' have been used by a number of observers. Some of these have been empirical tests made on a sand test piece, but the most informative have been those in which hardness variations in a mould of some selected design are related to sand properties.

Dietert and Valtier[5] proposed a method which is interesting mainly because it has persisted for such a long time and the standard testing apparatus can be used with the addition of a dial gauge. The sand is rammed exactly as when preparing the standard test piece, except that four blows are given instead of three. A fifth blow is then given and the further reduction of the length of the test piece measured. Flowability is calculated from the formula:

$$\text{flowability } \% = 100 - 1000r$$

where r is the reduction in length in inches.

There is a clear limitation to the success of this method. If a sand has such a high flowability that it is completely compacted by not more than four blows of the rammer, r is 0 and flowability 100%. However, if flowability is 0, no movement of the sand occurs and the test gives the same result. This criticism is not necessarily valid when comparing sands with a limited range of properties.

Kyle, however, found that the Dietert and Valtier method produced a minimum at a point where his hardness gradient method produced a maximum, and since Kyle's conclusions are in line with those of other workers, his results seem to have greater significance.

Graham[6] endeavoured to develop a test which could be made on a standard test piece and could be used to show whether the hardness distribution on a mould made on a jolt–squeeze machine would remain constant if the number of jolts, the air-line pressure and the squeeze pressure remained constant. He developed the shatter test for this purpose and found that it was quite satisfactory.

The shatter index is an index of toughness and the test is included in Standard Methods of Testing in this country for measuring this particular property. Graham's findings agree with those which can be derived from the work of other observers, including Davies and Kyle.

Gittus[7] carried out a number of tests on a series of sands, including the shatter test, loose bulk density, Dietert flowability and loose flow, the last being the proportion of an uncompacted sand passing a sieve on to which it had fallen from a height of 3 ft. He found a definite relationship between all these tests except the Dietert flowability test, and concluded that the same basic property, allied to flowability, was being measured.

He then prepared two sands by milling a black sand from a foundry with 4·6 and 1·8% water, respectively. The sand with 4·6% water had a high shatter index, a low loose flow number and a low density before compression, and therefore a low 'flowability'. The sand with 1·8% water had a high 'flowability'.

The two sands were then used to prepare a mould for two circular domes 4 in high and 3 in diameter mounted side by side on a plate. Sand was sieved on to the pattern to a height of 1 in above the base plate, the top of the sand layer being flat, not following the contour of the pattern. A thin layer of white sand was added, followed by a second layer of black sand and a thin parting of white sand. This was continued until the box was full. The sand was then squeezed, using a flat squeeze head, and the moulds stripped, dried and sawn through. Photographs of the moulds are shown in Fig. 69 (4·6% moisture) and Fig. 70 (1·8% moisture).

Since the sand layers between the white partings were 1 in thick before ramming, then in one mould the weight of sand between two white layers is constant. Where the layers have been made to move closer together by ramming, there must have been a considerable increase in mould density and therefore mould hardness. Where, after ramming, there were wide differences in the thickness of the black layers, there must be similar differences in mould hardness. When the thickness of the layers is constant, so is mould hardness.

In the above experiment, the controlling factor was clearly bulk density before compression — when this was low, squeezing produced much

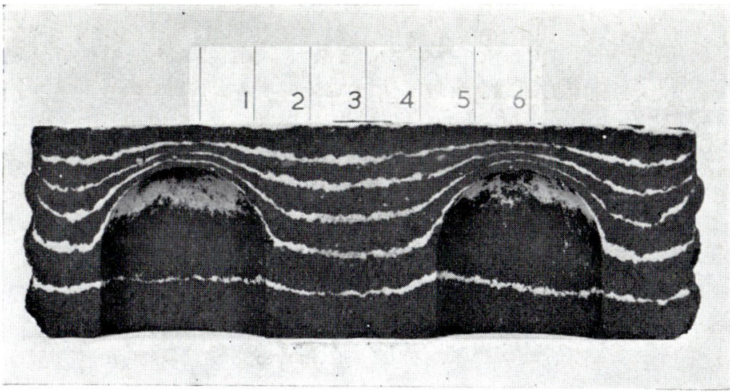

FIG. 69. Sectioned mould with high shatter index.

sand movement, and variable hardness; when it was high there was little sand movement and uniform hardness. Measuring bulk density before compression is a slow operation, not an advantage in a routine test and a toughness test is more convenient.

It is important to avoid pushing the comparisons between the moulds shown in Figs. 69 and 70 too far. The separation between the white layers is an indication of the effect of ramming on mould hardness, but only within one particular mould. It does not allow comparisons to be made between the hardness of the moulds in the two figures. The mould in Fig. 70 is more uniform than that in Fig. 69, but there is no indication whether it is uniformly hard or uniformly soft.

Bosworth[8] and his co-workers observed sand movement in a box containing an irregular pattern. Movement both in the direction of the applied

FIG. 70. Sectioned mould with low shatter index.

force and at right angles to it were observed and the hardness of the stripped mould was measured on both horizontal and vertical faces.

The mould assembly could be regarded as a slice 1 in thick cut from a pattern and moulding box. The section was 12 in high and 9 in wide and the pattern a rectangle 3 in wide and 4 in (a 6 in pattern was also used) high. The front and back of the section could be closed by metal plates.

To carry out a test, the back plate was attached and the assembly laid on its back. Sand in a grid pattern of units 2 in × 1 in × 1 in was then placed in position round the pattern, the same sand being used throughout, except that alternate grids were darkened by the addition of coal dust. The front plate was then fastened on and the assembly placed in an upright position.

The mould was then compacted, three methods being used — jolting, squeezing and jolting with a follow board or contoured head. The assembly was placed on its back and the front plate removed, when the pattern of sand movement could be observed and hardness measured.

The most informative results are those which show that lateral movement is small, but that when using a contoured head, this movement is sufficient to bring about a significant improvement in uniformity of ramming.

Walton[9] studied flowability by ramming sand in the standard test piece container by means of the standard rammer and measuring permeability after each of a number of blows. The number of blows was plotted against permeability on a logarithmic scale, when a straight line was obtained. The slope of this line was used as an index of flowability, a gentle slope indicating high, and a steep slope low flowability. The line becomes horizontal when no further movement occurs on ramming.

Walton made the comment that sands with a high flowability are brittle, which is just one more way of saying that tough sands have low flowability.

Sand properties and 'flowability'

The conclusions which can be drawn from the results observed by the numerous investigators into 'flowability' are clear.

1. When a sand is compacted by ramming, it moves mainly in the direction of the applied force, movement at right angles to this being small and significant only when some form of contoured head is used.

2. As a result of the uni-directional flow of sand, compacted moulds are not uniformly packed and there may be wide differences in hardness between different places on a mould surface.

3. When using any one pattern, differences in mould hardness increase as the toughness of the sand increases.

It follows that measurement of toughness, which is most widely used in conjunction with measurement of green strength as an index of stripping properties (see Chapter 1) has also a value independent of this. The objections to the shatter test, which is the most widely used, and, indeed, the only standard test for toughness, are therefore important at all stages of mould-making. These objections are that the shatter index is not really quantitative and that when using sands with a shatter index exceeding 85, differences are so small that the test is not selective.

The various alternatives — bulk density before compression, loose flow test, product of strength and deformation — are all open to serious objection. The most promising is the impact test,[10] but this has not yet been tested on the scale needed if it is to be adopted as a standard.

Ramming moulds

The oldest method of ramming sand is by hand, using a rod with a flat or tapered head, and this method is still in use even in some foundries making large numbers of identical small castings. The practice has been declining for some years, not because the method is invariably more expensive than producing on a mechanised plant, but because when there is a shortage of labour heavy and dirty jobs are most seriously affected. Moreover, pouring moulds containing cores with organic binders and shaking out the castings all over the foundry floor gives rise to a dust and fume problem which can be solved technically, but only at great cost. Castings required in large numbers are therefore usually made on machines and machine makers frequently make pattern changing as simple and as rapid as possible, in order to extend the field of use.

Mould-making machines do, however, have quite a long history. One simple machine which has long been a museum piece rather than a foundry tool, was used for making moulds for flat castings, rammed by means of heavy rollers suspended from a frame. The moulding box was filled with sand, the frame drawn forward so that the rollers rested on the mould between the box bars. The mould was rammed by pushing the frame and rollers backwards and forwards. This kind of machine disappeared many years ago, but rolling as a method of compaction is still used, although in a different manner.

However, jolting and squeezing and combinations of the two replaced other methods of ramming by machine. Later these were supplemented by the slinger.

Most modern machines also depend on jolting and/or squeezing, but modifications have been introduced which enable moulds to be produced with higher and more uniform hardness than could be obtained earlier.

The ideas behind these modifications are not necessarily new. Squeezing or jolting with a weight on the back of the mould, contoured to 'follow' the pattern, is a very old method to which Bosworth[8] turned when investigating the movement of sand during ramming. A separate board or contoured head is required for each pattern, but changing over can be avoided by using a modern machine with a diaphragm or sectioned head. Even this is not new and in 1920 there was an advertisement in the *Foundry Trade Journal* for a machine with 'a head which automatically adjusts itself to any shape of pattern and capable of producing 50 moulds an hour.' A blow–squeeze is by no means a new method of making moulds, but it never really got going until used in the Wallwork and Disamatic automatic machines.

Not all machines depend on jolting and squeezing. Some moulds for small sand-spun pipes are prepared by placing a fixed volume of sand in a cylindrical moulding box, causing the box to rotate at high speed and thus throwing the sand to the circumference, and finally inserting a mandril which falls off centre and compacts the sand by a rolling action. The slinger, and other machines of a similar kind, cut off a small amount of sand by means of an impeller rotating at high speed and throw it vertically downwards into the moulding box. In the Waformat, a machine recently developed in Germany, moulds are rammed by means of weights which are lifted by compressed air and allowed to fall.

Each type of machine has its own field of use, although there is considerable overlapping. The simplest and most widely used machines are the squeezer for flat castings and the consecutive jolt–squeezer for deeper work. For heavier castings a plain jolter is used, but as jolting leaves the sand at the back of the mould unrammed, final ramming by hand or with a power-assisted rammer is needed. A machine which can be used either for jolting followed by squeezing, or for simultaneous jolting and squeezing, produces more uniform moulds than can be obtained by jolting followed by squeezing. Quite high rates of production can be obtained with some of the machines, but in general, those most recently developed have the highest output and are to some extent automatic in operation. The machines with pneumatic or sectioned squeeze heads are rapid in action, but the fastest are the Wallwork and Disamatic blow–squeeze machines and the Waformat machine which rams by means of weights. A single machine produces so many castings that it can be used as a separate unit with its own melting plant.

Each method of moulding produces its own characteristic pattern of mould hardness distribution and this has a profound influence on the properties of the mould. Moreover, as can be seen from Figs. 69 and 70, the properties of the sand also affect the mould hardness distribution and on both counts, it is necessary to use a sand with the properties appropriate to the method of ramming employed.

Whatever the method of moulding, it is difficult to ram sand into deep, narrow spaces such as may be found when a vertical face of a pattern is close to the moulding box, since downward movement of the sand is restricted by friction between the sand and both the pattern and box side. In these circumstances, even a fairly light pressure with the fingers gives closer packing than can be obtained by even prolonged jolting. Ramming in narrow spaces becomes progressively more difficult as the toughness of the sand increases.

When large numbers of one particular casting have to be made, the boxes chosen should be such that there is ample clearance between pattern and box. An attempt to reduce the amount of sand in circulation by cutting the size of the boxes must lead to difficulty in producing good moulds and is of doubtful value on other grounds, since reducing the weight of sand/weight of metal ratio is likely to lead to an increase in sand temperature with its attendant difficulties.

The clearance necessary depends on pattern height and taper, sand toughness, and pattern finish and with four variables, no precise recommendations can be made, but with a height of 6 in, a clearance of less than 3 in is likely to cause trouble.

If a foundry has to make modest numbers of each of a range of castings differing widely in size and shape, it may be difficult to arrange production in such a way that there is always an adequate clearance between pattern and box. In this case, it may be more satisfactory to use auxiliary hand-ramming to obtain a proper degree of sand compaction.

Jolting

During jolting, sand is compressed partly when its downward velocity is reduced to zero at the bottom of the stroke, but mainly by the impact of the column of sand above. The back of the mould is relatively uncompacted

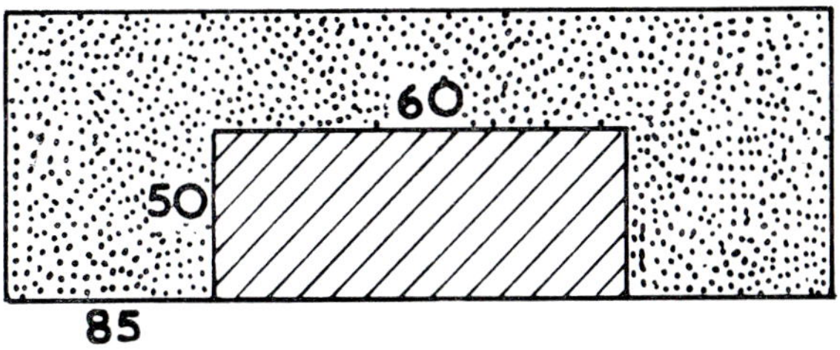

FIG. 71. Typical hardness distribution after jolting.

and generally the deeper the mould, the harder the sand at the mould face. Vertical faces are less densely packed than would be expected from the depth of sand above them, because of friction between sand and pattern. Using a mould hardness tester as a measure of the degree of compaction, distribution is of the kind shown in Fig. 71. The figures are indicative only.

The effect of prolonged jolting

There is an upper limit to the degree of packing of a sand, even against a horizontal pattern face, and when this has been reached, further jolting is likely to cause the mould to crack.

Suppose the mould for the pattern shown in Fig. 72 is being jolted. All the sand in the mould except that actually in contact with the pattern at the commencement, moves downward. After a time, downward movement becomes very small. With further jolting, the sand grains continue to move together, but with a decrease in mould dimensions in all three directions.

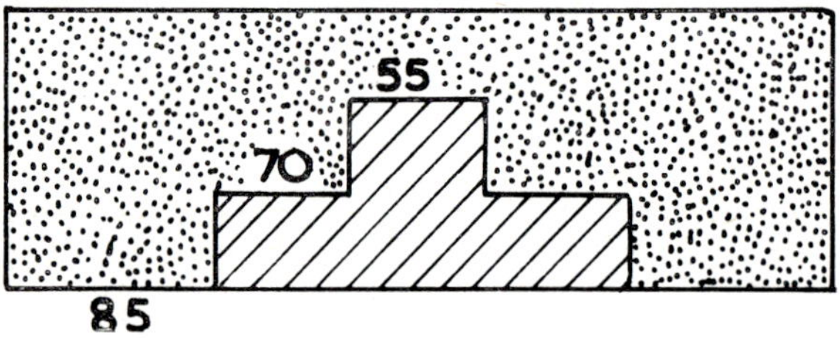

FIG. 72(a). Mould hardness distribution after normal jolting.

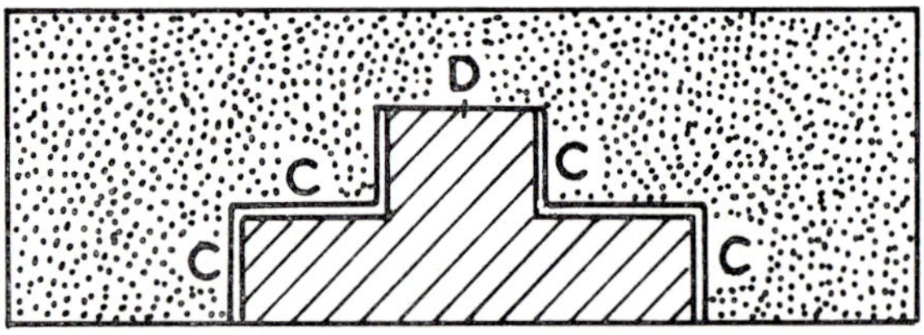

FIG. 72(b). Cracking after over-jolting.

Figure 72(a) shows the mould hardness distribution at the time when downward movement becomes very slow. Further jolting results in closer packing, but the change in dimensions in all directions results in a clearance between mould and pattern at C. The mould is now supported only by the top face of the pattern and if jolting continues cracking occurs at D.

The reason for this is clear. Sand grains cannot be arranged in close packing formation by being compacted from one end. For example, a standard test piece prepared from a silica sand with uniform grains arranged in close-packing formation would weigh a little over 200 g, but even with prolonged ramming it is difficult to obtain a test piece weighing more than 150 g and to get 200 g of sand into the test piece, it would be necessary to use a load sufficient to crush the grains. Quite large voids are present and can readily be seen using a low-powered magnification. In this condition, the sand is quite stable so long as the load is applied only in the original direction, but a very much smaller load at right angles to the first results in further compaction. If a sand compact is squeezed in one direction with a pressure of 100 lb in^{-2} until no further movement occurs, a pressure of 50 lb in^{-2} applied at right angles to the first will cause further compression. For a similar reason, a rammed cube of sand has a higher strength when tested in the direction in which it was rammed than when tested at right angles.

The lateral pressure is due to the vibration which is always considerable when moulds are jolted.

Squeezing

A totally different pattern is obtained if the mould is squeezed. Since lateral movement of the sand is negligible and the overall reduction in sand depth is everywhere the same, the percentage reduction is greatest over the high spots which are, therefore, rammed hardest. A typical hardness distribution is shown in Fig. 73.

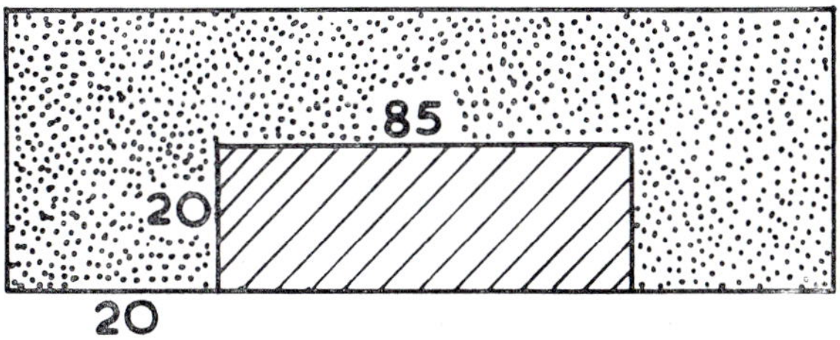

FIG. 73. Mould hardness distribution after squeezing.

If, instead of the mould shown in Fig. 73, that shown in Fig. 72a is made by squeezing, then the hardest part of the mould will be at D in Fig. 72b, which will take almost the whole of the pressure. Hardness on the shoulder will be little greater than that at the base plate.

Owing to friction between the sand grains, pressure falls as the distance from the squeeze head increases. There is therefore a limit to the depth of sand which can be effectively compacted by squeezing.

Jolt–squeeze

With a jolt–squeeze machine, a combination of the two effects is obtained and when using the pattern illustrated in Fig. 71, a hardness distribution of the kind shown in Fig. 74 is obtained.

If, however, a pattern of different shape is used, mould hardness distribution may be entirely different. In particular, if a small, flat area is well above the remainder of the mould, it takes the whole of the squeeze

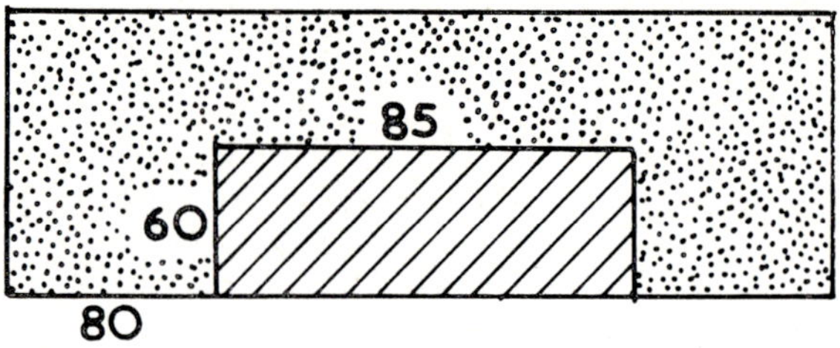

FIG. 74. Typical mould hardness distribution after jolting and squeezing.

pressure and is harder than elsewhere in the mould. Even when, as is usual in jolt–squeeze operations, a very light squeeze is used, the hardness distribution is influenced to a minor degree only.

Figure 75 shows a mould rammed on a jolt–squeeze machine using a squeeze pressure of 10 lb in^{-2}. Jolting produced a hardness of 60–70 at the bottom. The hardness at the top was low after jolting, but the 'light' squeeze raised this to 80.

The two small flat areas at the top were 2 in diameter and had a combined area of rather more than 6 in^2. The mould itself was 18 in^2, so that the total area was more than 300 in^2 and the total pressure at 10 lb in^{-2} was 3000 lb. Since almost the whole of this pressure was taken by the two high spots, the pressure at the top approached 500 lb in^{-2}, which is far higher than that used in high pressure squeezing.

Machine moulders frequently move the sand about whilst a mould is being jolted, pushing sand from above the high spots in order to minimise hardness differences. They are successful to some extent, but in an extreme

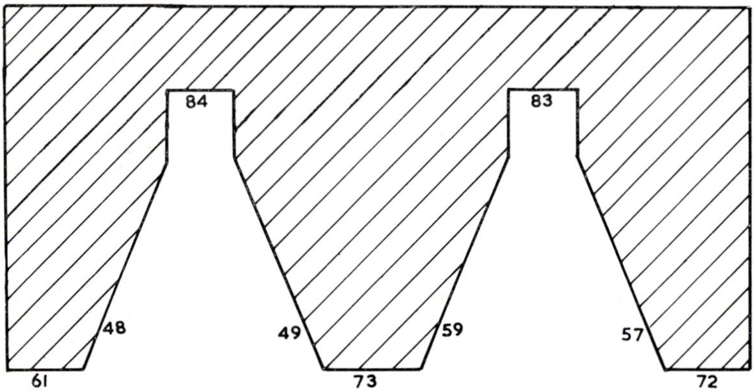

FIG. 75. Actual mould hardness distribution after jolting and squeezing.

case, shown in Fig. 75, their success is limited. The figure shows the actual mould, but the hardness numbers are taken from two different moulds. The first mould was jolted, filled up and squeezed with no effort to equalise hardness, the results obtained being shown at the left. The pattern was then given to an experienced machine moulder with instructions to ram as uniformly as possible throughout without using hand ramming, but moving the sand around in any way he wished during the jolting operation. The hardness distribution of the most uniform of a number of moulds so made is shown on the right. There has been an increase in hardness near the bottom, but no reduction in the excessive hardness at the top.

Simultaneous jolt–squeeze

When two methods of ramming are used together, it is usually possible to obtain a more uniform mould hardness, than when the same methods are used in sequence. For example, if a moulding box is filled with sand and vibrated, some compaction occurs, but if the mould is then squeezed, the final pattern of mould hardness differs little from that obtained when the initial vibration is omitted. If, however, the mould is vibrated while it is being squeezed, a harder and more uniform mould is obtained. The difference is not very marked and as there is a considerable increase in noise, the procedure is hardly worth while.

There is a real increase in hardness when a mould is jolted while the squeeze pressure is on, and since performing the two operations together reduces the time required to ram a mould, the method is being used on an increasing scale by foundries. Even with very difficult patterns, it is possible to obtain high and uniform hardness by using a preliminary free jolt followed by jolting under pressure, always provided that sufficient energy is used.

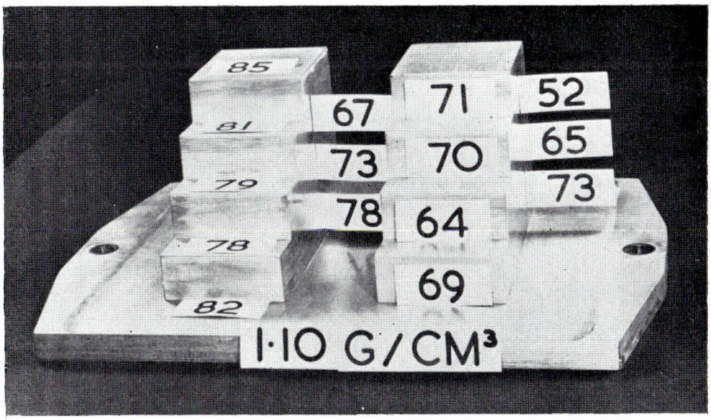

FIG. 76. Plain jolt 30 s at 120 lb in^{-2}. (After K. E. L. Nicholas.)

FIG. 77. Simultaneous jolt–squeeze 30 s at 120 lb in^{-2}. (After K. E. L. Nicholas.)

The effect of simultaneous jolting and squeezing was studied by Nicholas.[11] The pattern used consisted of two step blocks separated by a cod of sand 3 in wide, with a clearance of $2\frac{1}{2}$ in between pattern and moulding box. The overall bulk density of the mould was measured and hardness tests made on all vertical and horizontal faces.

The overall bulk density of the mould increased as the air-line pressure was raised to 120 lb in^{-2} and the time of jolting was raised to a maximum of 30 s. The highest mould density and most uniform mould hardness

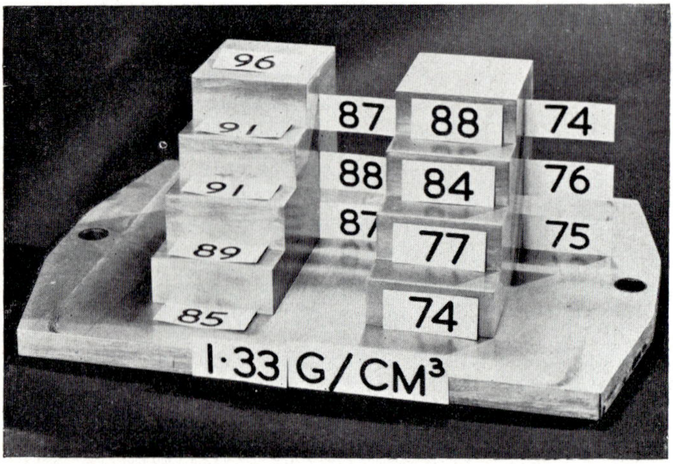

FIG. 78. Prejolt 30 s, simultaneous jolt–squeeze 30 s at 90 lb in^{-2}. (After K. E. L. Nicholas.)

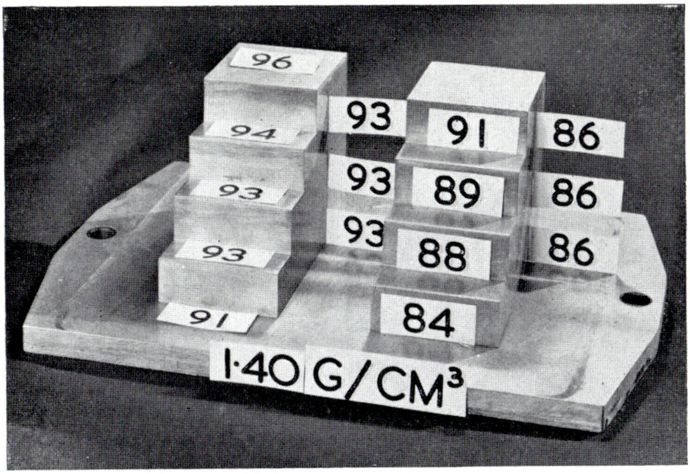

FIG. 79. Prejolt 30 s, simultaneous jolt–squeeze 30 s at 120 lb in^{-2}. (After K. E. L. Nicholas.)

were obtained by jolting alone for 30 s followed by a simultaneous jolt–squeeze for the same time with an air-line pressure of 120 lb in^{-2}.

Some results obtained by Nicholas are shown in Figs. 76–79.

Since speed of production is one of the features of modern moulding machines, a procedure which requires jolting for 60 s is unacceptable, but designers of castings rarely think up anything as difficult to ram as the mould used by Nicholas. For most moulds, all that is needed is a simultaneous jolt–squeeze of a few seconds with the more usual air-line pressure of 100 lb in^{-2}.

Squeezing with sectioned squeeze head

The use of a contoured squeeze head has the disadvantage that with every change of pattern, a different head has to be used. This does not seem to be a serious limitation when a very large number of identical castings are required, but nevertheless machines have been designed such that the head automatically adjusts itself to the pattern contour.

In one of these, the head is made up of a number of independent feet, 4 in × 4½ in, each actuated by hydraulic squeeze pressure, which can be set at any required level up to 200 lb in^{-2}. In practice not much is to be gained by increasing pressure beyond 100 lb in^{-2}.

The behaviour of a machine of this kind was also studied by Nicholas[12] using a pattern which incorporated many of the features of that used when investigating the performance of jolt–squeeze machines.

Nicholas used a range of moulding sand and found that very uniform

FIG. 80. Mould hardness of sand compacted at 200 lb in^{-2}. (After K. E. L. Nicholas.)

hardness was obtained with all of them, although it does not follow that all are equally suitable.

Mould hardness distributions obtained with a squeeze pressure of 200 lb in^{-2} on a synthetic sand containing 7% Western bentonite are shown in Fig. 80.

Squeezing with a diaphragm head

In this machine, the squeeze head is a diaphragm behind which air, at any required pressure, can be introduced. The action of the machine is shown schematically in Fig. 81.

Various methods, designed especially to control the rate of deformation of the sand have been introduced.

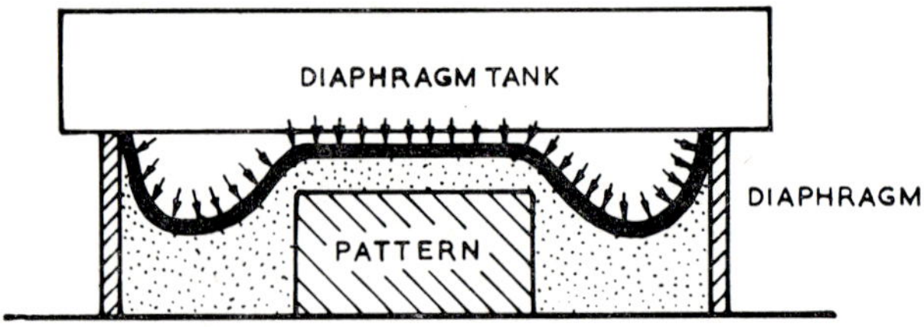

FIG. 81. Squeezing with diaphragm head.

This is the only method of ramming by machine which pushes sand towards vertical faces and, in theory, it should therefore be the best method of compacting moulds with a very deep pattern draw which incorporate such a face. However, sufficiently uniform hardness can also be obtained by means of the simultaneous jolt–squeeze and divided head machines and the selection of the most suitable depends more on the cost of obtaining the machine and keeping it in operation.

With a diaphragm head, pressure is exerted on all vertical faces including the sides of the moulding box, which must be made sufficiently rigid to withstand the high pressure without distortion. Indeed, with all high pressure squeeze machines there is considerable lateral thrust on the box which may distort, in which case the box tends to return to its original shape when pressure is removed and may cause the mould to crack. Usually boxes without bars are used for two reasons. In the first place, bars restrict the initial flow of sand round the pattern and tend to reduce uniformity of packing, and in the second, removing the bars permits quick and relatively silent punch-out of the mould and casting, in place of removal on a vibratory shake-out which is a slow and noisy operation when the moulds are very hard.

Blow–squeeze machines

When sand is blown into a box, considerable packing occurs and a subsequent squeeze produces only slight movement when obtaining the final compaction. It is therefore possible to obtain the uniform packing illustrated in Fig. 70 when using the kind of sand which would, in the absence of the preliminary blow, give the variable packing illustrated in Fig. 69.

This is not a particularly new method, but it had no great success until the automatic Wallwork and Disamatic machines were developed. In both machines, the sand is blown into the mould cavity and then squeezed at high pressure, producing a mould of high and uniform hardness. Box and pattern are withdrawn, leaving a block of sand which carries half of the mould impression on one side and the other half on the opposite side. The mould is ejected on to a band and pushed along one space by a ram. The second and subsequent moulds are similarly ejected and the assembly pushed along the band. By the time a dozen or so moulds are present, the exact number depending on the mould dimensions and the smoothness and temperature of the band, the force required to move the line of moulds is sufficient to keep the moulds nearest the ram closed and pouring commences. When the castings are solid, the poured moulds are carried on and dumped into a rotary screen, where the sand is separated from the castings and broken up. The process is almost completely automatic and the operation of the foundry is greatly simplified by the absence of moulding boxes.

Up to the present, the moulds produced have been mainly small, and in this range of work are harder and not less uniform than those produced by any other means.

Limits of compacting sand under high pressure

If the squeeze pressure on sand is continually increased, a point is reached when the sand grains are in contact and can no longer move closer together. When this point has been reached, a further increase in pressure deforms the sand grains themselves. This deformation is elastic and when the pressure is removed the grains regain their original shape and the dimensions of the rammed sand increase, leading to inaccuracies in mould dimensions and sometimes to cracking of the mould. This is known as 'spring back'. The maximum pressure which can usefully be employed is variously quoted as 200 or 300 lb in^{-2} or some figure in that region. It is influenced by the characteristics of the bond.

Jolting with weights

Heavy jolters have been in use for a long time for ramming moulds of medium weight. Occasionally a weighted board was placed on the top of the box, but usually compaction of the back of the mould was obtained by means of a hand or pneumatic rammer.

In some modern machines, independent weights which rest on the sand during jolting, eliminate the need for subsequent ramming. By suspending the weights from a frame and adjusting the length of the suspender, the action of the weights can be cut off when any required degree of compaction has been obtained. In this way it is possible to control the hardness of moulds with a deep draw and a near vertical side, for example, a mould for a bath, to within a few points at any desired level.

Ramming with a falling weight

Provided that the final thickness does not greatly exceed 4 in, one of the quickest and most effective methods of compacting sand is by allowing a weight to fall on it. This is the method used for preparing the standard test piece, and a 14 lb weight falling on the sand three times from a height of 2 in gives a degree of compaction exceeded only by moulds made by very high pressure. By using a large number of independent weights suitably arranged in rows, very uniform hardness is obtained.

This method is used in the Waformat machine, the weights being raised pneumatically to a height of 8 cm and allowed to fall. By using a large number of small weights, each covering only a small area of the pattern, uniformly hard moulds can be obtained whilst maintaining individual impact at a low level, and plaster patterns can be used if needed.

Ramming by explosion

From time to time, suggestions for ramming sand by exploding a charge above it are made. There do not appear to be any accounts of the use of this method on a production line, but at least one patent[13] has been taken out.

The method should be interesting since it illustrates one point concerning ramming which has not been fully explored. There is some evidence that a mould is rammed more uniformly and harder if the load is applied very rapidly. For example, when soft moulds are being produced in a squeeze machine, a trick sometimes used is to release the pressure and then apply it suddenly, producing the effect of a blow. This may not improve the subsequent performance of the machine, but it does produce a harder mould. On the diaphragm machine, mould hardness is improved if the pressure is applied rapidly.

Whatever its other effects might be, an explosion would certainly give a rapid rate of application of the load.

The sandslinger

The sandslinger and other machines of the same kind, ram sand in a different way and produce a different pattern of mould hardness. An impeller cuts off small amounts of sand and projects them downwards at high speed and in rapid succession.

The mould is hardest where the sand is thrown directly against the pattern, which is more rigid than rammed sand and produces more rapid deceleration, and it is sometimes found desirable to reduce this hardness by placing a little loose sand on the pattern. This sand is rammed by the impact of the first sand thrown by the machine.

Away from horizontal faces, compaction is not affected by varying sand depths, as it is when jolting and squeezing, unless the machine is badly used. Should movement of the head be too slow, a large volume of sand is delivered to one place, forming a high spot from which some sand rolls to a lower level, where it can be consolidated only by the stream of sand from the impeller. As the hard backing of the pattern is not present, a soft area is formed.

A further hardness variation which is more difficult to control is frequently found on vertical faces. The hardness differences are illustrated in Fig. 82.

If the sand projected by the machine is seen in strong sunlight, two streams appear, the main stream proceeding vertically downwards and a smaller secondary stream away from the first at a small angle. Where the secondary stream is directed towards the pattern, the mould is harder than where it is directed away from the pattern.

The hardness differences increase as the impeller wears, but are present to some extent soon after a new one has been fitted. They are most pronounced when the sand is very brittle, and the differences shown in Fig. 82

were obtained when using a coarse sand bonded with a ball clay. Sand of this kind has a very narrow moisture range and even the small variations in moisture content which occur when sand control is good, lead to wide variations in shatter index.

When using a slinger, very tough sands should be avoided or soft moulds will be obtained. A sand with a shatter index of about 60 is usually satisfactory, although some foundries use even weaker sands. In this connection, the effect of the slinger head on the toughness of sand which has been very

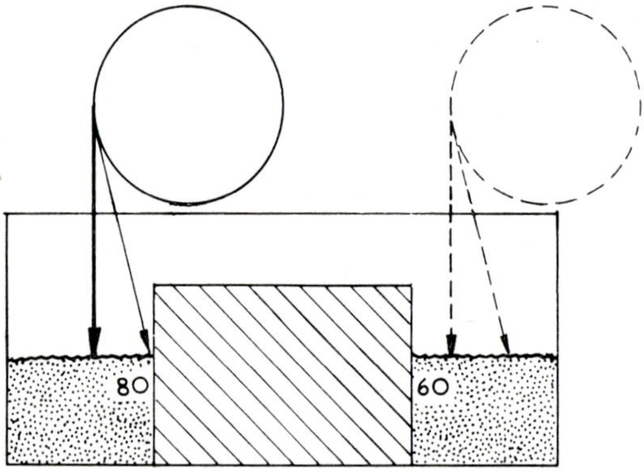

FIG. 82. Hardness differences with a slinger.

lightly milled is important (see Chapter 7). Such a sand may have a shatter index of 60 before it reaches the slinger head, but if the clay content is high and the low shatter index is due to absence of milling, an increase to 80 or more may be developed as the sand passes through the slinger head. The machine cannot produce a reasonably hard mould from a sand such as this.

The most suitable fields for use of the slinger are in making moulds of medium or large size, when only a few of each kind are wanted and boxes of different size are needed, and for very large moulds for which a very large and expensive jolter requiring special foundations would be necessary. The slinger is always likely to lose out to a jolter for the production of, for example, bath moulds which are required in large numbers in standard boxes, since a pair of jolters can produce 60 moulds in 1 h, but it has no rival for filling medium and large boxes of assorted sizes. When first available, the slinger was used for the quantity production of castings such as baths and small pipes, but it is being used on a decreasing scale for such work and on an increasing scale for jobbing work.

From time to time, modifications have been made to the machine in order to improve its performance or widen its field of use. The size of the

required for making the mould by hand and more than that needed on a machine. Accordingly, the operation is skimped; the mould is stabbed all over with a wire in the shortest possible time, with the result that some holes pass right through the mould and become inoperative as soon as the metal reaches them, and others terminate so far from the mould face that their effect is negligible.

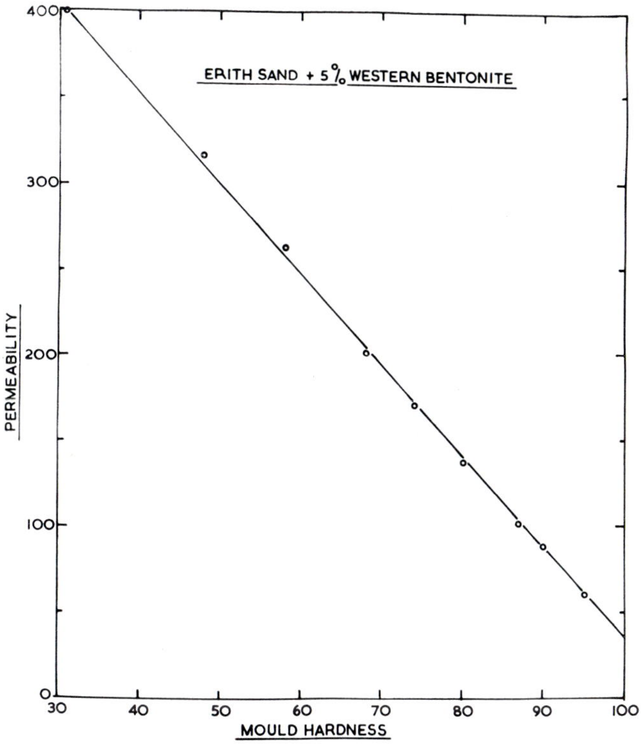

FIG. 84. Mould hardness and permeability — synthetic sand.

Belief in the efficacy of this kind of venting is supported by the quite obvious fact that gas passing through the mould burns at the entrance to the vents. This is to be expected, since passage through the back of the mould is so much easier at the vents, although the change at the face of the mould is so small. It is possible that venting may save an occasional casting when the factor of safety is small, but the method cannot be relied upon since unavoidable variations in other factors, particularly mould hardness and moisture content have an overriding influence. If the factor of safety is so small that an increase of a few per cent in permeability has an effect, a sand with a higher permeability should be used. On the whole, it is likely that venting costs more in damage to patterns than it saves in misrun or blow castings, and no harmful results have been observed in foundries where the practice has been abandoned.

2. Placing large holes from the back of the mould, particularly into cods of sand which will later be almost surrounded by metal is, on the other hand, beneficial. Suppose the bottom half of a mould is of the shape shown in Fig. 85; in the absence of a vent, gas given off at A has to travel

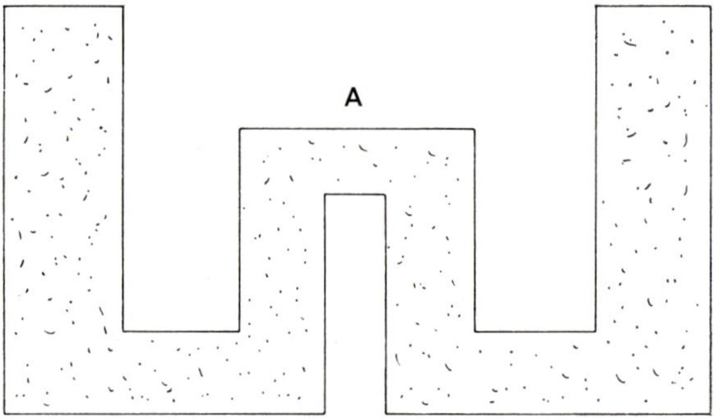

FIG. 85. Venting a cod.

through sand right to the back of the mould, but with a vent as shown, the distance is greatly reduced. To be effective the vent must be large; poking a few holes with a wire is not satisfactory.

It is not sufficient to ease the passage of gas just to the back of the mould. If this is all that is done, and there is no easy means of escape to the atmosphere, pressure will build up rapidly and the vent will be useless. Indeed, when a mould is such that the bottom half needs venting, then the most important place to vent is the floor. Placing the mould on a thin layer of sand and venting this, shows a degree of optimism not always justified by results, and where the boxes have always to be laid out on the same strip of concrete, it is safer to cut grooves in the floor. When large moulds are bedded in the floor, a most important preliminary is the preparation of a very open base into which gases can pass through mould vents and then escape through other vents to the atmosphere.

When weights covering an appreciable proportion of the mould surface have to be used, they should be grooved on the under side.

3. If a hole $\frac{1}{8}$ in diameter is made right through a standard test piece instead of stopping $\frac{1}{4}$ in from one end, permeability is so high that it cannot be measured on the standard apparatus. If a similar hole is made

through a mould from the highest point in the cavity, the escape of any gas in the mould cavity itself will be made easy.

Suppose that a core is placed in the mould shown in Fig. 75, such that metal thickness is about $\frac{1}{4}$ in. When the casting is poured, the air in the mould expands rapidly, exerting a pressure which slows up the rate of entry of metal into the mould. Since the first metal to enter has to pass over most of the mould and core, losing heat the whole time, a misrun or small blows around the core at the highest point may occur, and anything which restricts the rate of flow increases this possibility. Placing a through vent or 'whistler' prevents the build-up of pressure which otherwise restricts the rate of entry of metal.

When the mould is full, metal enters the vent and it becomes inoperative and of no assistance when gas is given off by the mould or core covered by metal.

Risers on large moulds have a similar effect, but they are not needed for the same purpose. Indeed, some moulders attempt to plug risers with some combustible matter in order to build pressure inside the mould cavity. This is said to prevent the top surface scabbing or 'being pulled down', but as the pressure inside even a completely closed mould without a riser is very small and the plug used is extremely open, any effect on scabbing, if it exists, is not due to pressure inside the mould. The most useful function of the plug is to keep out dirt.

Green strength and shatter index

As mould hardness increases, green strength also increases. Shatter index rises at first and then becomes almost constant. Now increasing green strength with a constant shatter index must be accompanied by a fall in deformation, and sand therefore becomes more brittle and more difficult to strip as mould hardness increases.

In addition, when a pattern is rapped or a mould jolted in order to obtain a small clearance between mould and pattern and so assist stripping, the volume of the sand between the outside of the pattern and the box side must be reduced. It becomes increasingly difficult to obtain this reduction as the original mould density is increased.

On both counts, stripping becomes more difficult as mould hardness increases.

There is rarely any serious difficulty in removing patterns from moulds made by machines other than those which produce extremely hard mould, but stripping of the hard moulds made by high pressure squeezing is possible only when sand composition is suitably adjusted.

Sand for high pressure moulding

There is no real need for using sand with an exceptionally high permeability. An increase in hardness beyond 90 as measured by the ball indentor, is accompanied by only a small increase in mould density and decrease in pore size, and therefore a small decrease in permeability.

It is, however, necessary to use sand with a strength and shatter index outside the range needed when using other methods of moulding.

In the first place, clay content must be high. Nicholas[15] showed that the maximum hardness obtained by squeeze pressures up to 400 lb in^{-2} depends on the amount of clay present. Figure 86 is reproduced from the paper by Nicholas.

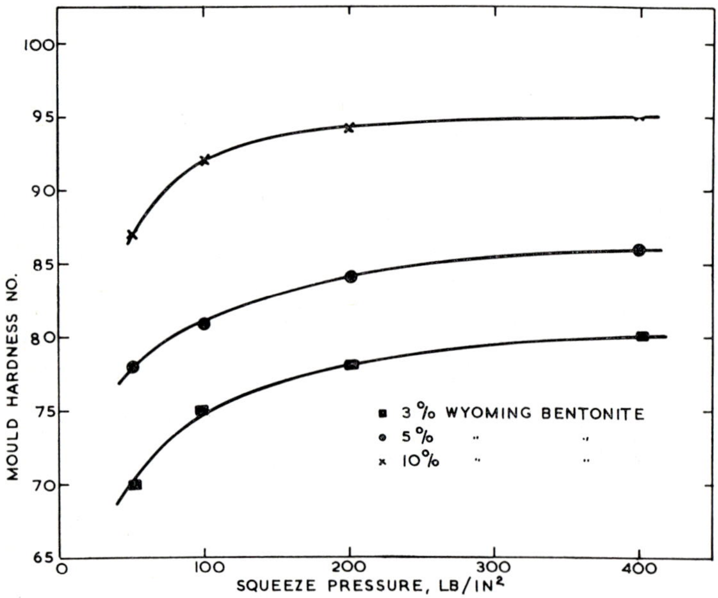

FIG. 86. Squeeze pressure, clay content and mould hardness. (After K. E. L. Nicholas.)

At the same time, it can be shown that mould stability is greatest when the water/clay ratio is low and, since reducing the moisture content increases brittleness, this factor also makes stripping more difficult. To remedy this, an organic binder such as dextrin is added, raising the shatter index and reducing brittleness.

As a result, the sand used with very high squeeze pressures may have a green strength of 20 lb in^{-2} and a shatter index of 100, at which point the test becomes non-discriminatory. An alternative to the shatter test, suitable for use with very tough sands is urgently needed.

With very high squeeze pressures and either a squeeze head which adapts itself to the shape of the pattern, or a blow–squeeze machine, the use of such tough sands does not lead to differences in mould hardness. Indeed, there is some evidence that the addition of dextrin increases lateral flow, which is otherwise small and occurs only while the sand is non-coherent, and may give moulds of greater uniformity than can be obtained in its absence.

Dry strength

Dry strength increases with increasing mould density, and the resistance of a mould to erosion improves when mould hardness increases.

However, at the same time, difficulty in removing and breaking up the sand from the box also increases, and users of hard moulds are turning more and more to the use of boxes without bars, in which case the mould may have to be weighted. With boxes without bars, sand and casting can be punched out, an operation much faster and less noisy than using an oscillating or vibratory shake-out.

Expansion

The linear expansion of a dense silica rock composed of a large number of grains with random orientation is rather less than 1·4% over the temperature range 0–573°C. At the higher temperature there is a change in the crystal structure of the quartz and with further heating there is a slight contraction.

A rammed moulding sand also consists of a large number of silica grains with random orientation and its expansion should follow the same pattern, provided the sand grains are rammed until no further movement takes place.

The relations between sand composition, mould hardness and expansion are discussed in Chapter 9, where it is shown that:

1. Expansion depends on the composition of the granular fraction of a sand.

2. Very hard moulds with a pure silica base have an expansion equal to that of a dense silica rock.

3. With sands of any one composition, the expansion of a hard mould is about 0·1% greater than that of a soft mould. When the base sand is of fairly pure silica the range of expansion to be expected in rammed moulds is from 1·2–1·3%.

How effective is high pressure in squeezing?

Most modern squeeze machines are equipped with contoured, sectioned, or flexible squeeze heads and are capable of applying pressures in excess of 200 lb in^{-2}. It is, however, doubtful whether the ability to apply pressures of this kind is an important feature of their usefulness.

The real benefit to be obtained is two-fold.

1. The machines are rapid and, to a considerable extent, automatic in action and the demand for labour and space is low.
2. The method by which the squeeze is applied is such that a uniform mould hardness is obtained.

Very high squeeze pressures increase the risk of scabbing and unless a sand with a very high clay content is used, it is probable that the actual sand grains will undergo deformation. This deformation is elastic and removal of the pressure is accompanied by a recovery, which results in a change in mould dimensions and loss of accuracy. There have also been reports[14] that with very hard moulds, there has been metal penetration which is not easily explained, such penetration increasing as squeeze pressure is raised.

In consequence, it is doubtful whether there is any benefit to be obtained by increasing squeeze pressure to more than 100 lb in^{-2}.

REFERENCES

1. LISSELL, E. O. and ASH, E. J. (1942). *Trans. Amer. Foundrymen's Assoc.*, **50,** pp. 637–653.
2. DAVIES, W. (1946). *J. Iron Steel Inst.*, **CLIV**, pp. 24–26. See also *Foundry Sand Control*, **CLVI**, p. 210, United Steel Co., Ltd, Sheffield.
3. FAIRFIELD, H. H. and MCCONACHIE, J. (1953). *Amer. Foundryman*, **23,** April, pp. 127–132.
4. KYLE, P. (1940). *Trans. Amer. Foundrymen's Assoc.*, **48,** pp. 175–189.
5. DIETERT, H. W. and VALTIER, F. (1934). *Trans. Amer. Foundrymen's Assoc.*, **42,** pp. 199–210.
6. GRAHAM, E. L. (1948). *Foundry Trade J.*, **84,** April 15, pp. 365–367.
7. GITTUS, J. (1951–1953). *BCIRA J. Res. Devel.*, **4,** pp. 560–567.
8. BOSWORTH, T. J. (1960). *British Foundryman*, **53,** March, pp. 137–146.
9. WALTON, C. F. (1954). *Gray Iron News*, Oct., pp. 4–10.
10. OUTHWAITE, J. G. (1967). *Modern Castings*, **51,** April, p. 67.
11. NICHOLAS, K. E. L. and ROBERTS, W. R. (1964). *BCIRA J.*, **12,** No. 2, March, pp. 212–228.
12. NICHOLAS, K. E. L. Unpublished work at BCIRA.
13. British patent no. 1 036 617.
14. BARLOW, T. E. (1965). *Amer. Foundryman*, **13,** July, pp. 177–186.
15. NICHOLAS, K. E. L. and ROBERTS, W. R. (1963). *BCIRA J. Res. Devel.*, **11**(1), Jan., pp. 61–75.

CHAPTER 9

The Effect of Heat on Clay-bonded Sand

When molten metal is poured into a green sand mould, all the constituents of the sand — sand grains, clay, water and organic materials — are affected. The extent of the changes and the order in which they occur, control the behaviour of the mould during pouring and the difference between the properties of the original sand and those of the recovered sand.

(*a*) *Water*

The water evaporates and the temperature of the mould face cannot exceed 100°C until all the free water has evaporated. As a result of the loss of water, the face of the mould becomes harder, stronger and more brittle.

Some of the evaporated water, carried through the mould with hot air from the mould cavity, condenses in a layer behind the mould face. As a result, the mould consists of a thin, hard and brittle layer, behind which is a thin, weak and plastic layer, backed up by the unchanged sand.

(*b*) *Sand grains*

The sand grains expand. When the grains consist of quartz only and are rammed so that there is actual grain-to-grain contact, in which case additional ramming cannot produce further consolidation, the expansion is the same as that obtained by Sosman when testing quartz rock and shown in Chapter 3. There is a high but fairly regular expansion until the temperature reaches 573°C, when the α quartz changes to β quartz with a further expansion. On raising the temperature still further, β quartz contracts slightly. These changes are reversible and it is not possible to obtain a sand which does not expand on heating simply by raising the temperature to above 573°C followed by cooling.

When the grains are not closely packed, part of the expansion can be taken up inside the compact and the expansion as a whole is lower, although the individual grains expand just as much.

If quartz is maintained at a high temperature for a long time, it changes to tridymite and then to crystobalite, both of which have a specific gravity of 2·35 whereas that of quartz is 2·65, so that there is an expansion of more than 10%. On cooling, the quartz can be obtained only by allowing to stand for a long period at the temperature just below that at which the high temperature modification is formed. Quartz does not re-form when the grains cool to room temperature in a short time, and the change is, for practical purposes, irreversible. Quartz is usually described as the stable modification at temperatures up to 873°C, tridymite between 873 and 1400°C, and crystobalite from 1400°C to the melting point. Small amounts of the high temperature modification have been observed on the surface of steel castings, particularly when the metal has penetrated between the grains, but it is rarely possible to find any in the used sand.

Silicate-bonded sands are an exception. The presence of a fusible silicate gives an enormous acceleration to the change and appreciable quantities of crystobalite, but no tridymite, may be formed on the surface of moulds and cores adjacent to heavy castings. In one case, the sand from a small core in a heavy casting had a specific gravity of 2·35, showing complete conversion and that adhering to the surface of a casting 3 in thick and weighing 3 tons, had a specific gravity of 2·5, showing 50% conversion.

Nevertheless, the change is not so rapid that any appreciable conversion occurs in the time required to fill even a large mould, and the incidence of defects due to sand expansion is not raised. Traces of crystobalite can often be found in the used sand and there may be a small amount in the dust in the foundry atmosphere, although far less than the amount of quartz.

(*c*) *Clay*

When a clay is heated to a sufficiently high temperature, it loses its combined water and its characteristic structure. As a result, plasticity cannot be regained by adding water and, moreover, when new clay is added, the dead clay has an effect on the properties of the mixed sand, as described in Chapter 4.

This change in the composition of a clay does not cause disintegration and a clay-bonded sand test piece, heated so that decomposition is complete, does not lose strength unless the temperature is sufficiently high for at least part of the clay to melt, in which case strength falls rapidly. Indeed, if a sand test piece is held at an intermediate temperature for some time, sintering occurs and the combined effect of this and of decomposition of the clay is a pronounced increase in strength.

Decomposition of a clay is not instantaneous and sintering is a slow process, while loss of strength due to fusion occurs immediately the appropriate temperature is reached. When a sand is heated rapidly, the softening point may be reached before there has been sufficient time for sintering to proceed to a marked extent, and no intermediate increase in strength is possible.

The strength of a clay-bonded sand at a high temperature therefore depends not only on the kind of clay and the temperature, but also on the time required to reach that temperature.

(d) Organic materials

The effect of prolonged heating on any of the organic materials added to clay-bonded sand is to drive off the volatile matter leaving, provided air is excluded, free carbon and ash. If there is free access of air, free carbon is removed, leaving only ash.

Heating to an intermediate stage may produce different results. Coal dust softens when the temperature reaches the point at which the most fusible constituents melt. Decomposition begins at an even lower temperature, although it is then slow, and the amount of liquid formed, or whether any forms at all, depends on both the temperature and the rate of heating. Pitch is quite liquid at 100°C, but thereafter several changes occur, some compounds polymerising to form others of higher viscosity, whilst some simply decompose. Eventually all volatile compounds disappear, leaving a small amount of fixed carbon or, with access of air, a negligible amount of ash.

Both coal and pitch swell whilst they are decomposing and any carbon left is weak and does not retain the structure of the original particles. This also has an effect on strength.

Wood behaves in an entirely different manner. Volatile matter is removed, but in the absence of air, the residual charcoal retains the shape and size of the original wood particles and is only a little less rigid. The loss of volatile matter is not accompanied by a fall in strength.

Most of the fibrous materials added to sand behave in a manner much like that of wood, although the rate of decomposition and the rigidity of the residual carbon may differ.

The two kinds of organic material may therefore have very different effects on the properties of a sand at high temperature. Either may or may not influence properties at room temperature to some degree, but at high temperatures a liquid replaces the relatively hard particles of coal or pitch and some loss in strength is probable. On the other hand, when wood is heated in the absence of air, the charcoal formed retains the original particle size and shape and, to a considerable extent, the rigidity. Any change in strength should therefore be small.

TESTING SAND AT HIGH TEMPERATURES

When drawing up a programme for high temperature testing, two points have to be considered:

(1) the nature of the tests;
(2) the rate of heating.

Most testing of sand at high temperatures has been carried out in order to determine whether the rammed sand is likely to scab. It has been assumed that the type of scab under consideration is due to expansion of the thin, dried and heated layer on the mould surface, since such a layer is found separating the scab from the body of the casting and no scab forms if quartz sand is replaced by a granular material such as olivine or zircon with a low coefficient of expansion. However, if the sand can deform without cracking to an extent greater than the expansion, no scab will form unless, as a result of being deformed, the sand layer has been weakened to such an extent that it cannot support its own weight. The important parameters are expansion, deformation and strength.

The rate of heating provides a more difficult problem which must be solved if the results are to indicate what is likely to happen to a mould surface. When a standard test piece is placed in a hot furnace, the time required for the centre to reach a temperature not more than 50°C below that of the surface is about 20 min. In order to reduce this time, much of the earlier tests at high temperature were made on test pieces with a cross section of 1 in^2, but even so, a soaking time of 12 min was needed. The surface of the test piece reached the furnace temperature quite rapidly and was maintained at that temperature for a time sufficient to allow the clay to sinter. When an organic material such as pitch or coal dust was included, this was decomposed or, with access of air, burnt out by the time the test could be made. However, no method of reducing the heating time was known and tests under the above conditions were made by Dietert,[1] Hudson[2] and, during the period 1943–1955, by a committee of the American Foundrymen's Society.[3] All these tests showed that heating produced a very great increase in strength and that the increase was smaller when a combustible material was present.

Even the reduced time of 12 min was long enough to suggest that sintering of the clay and burning out of the organic materials might have a significant effect on the test results, and that more rapid heating might be desirable. It is difficult to measure the rate of increase in temperature of the surface of a mould heated by radiation whilst a mould is being filled, but there is indirect evidence which gives a reasonably accurate indication. Petterson,[4] investigating the formation of scabs on the top face of a mould, separated the two halves of the mould, introduced some molten metal and observed the time which elapsed before the top face began to spall. He reported that if no spalling occurred during the first 50 s after the commencement of pouring, it did not occur at all. Since quartz begins to contract when the temperature passes 573°C, it is reasonable to suppose that once this temperature has been exceeded there is no further risk that the mould will fail. The heating equipment to be used for high temperature testing should therefore aim at reaching a temperature of 573°C in not more than 50 s. Where the sand mould is in contact with molten metal, the rate of rise in temperature is much faster, but there appears to be little chance of heating a sand test piece at this rate.

Heating the test piece

The problem of heating the test piece at a speed approaching that called for above was solved by Rew and Godding,[5,6] using a dielectric heater.

In this kind of heater, a non-conductor, which is to be heated, is placed between two parallel plates acting as a condenser and supplied with power from a source alternating with a high frequency. The rate of heating which can be obtained depends on the potential difference between the places, the loss factor of the non-conductor and the frequency. The potential difference must be below that at which a flash-over occurs, nothing can be done about the loss factor of silica which is low, and the only variable is the frequency. Using valves which gave a frequency of 100 Mc s^{-1}, the highest which could then be obtained at a reasonable price, Rew and Godding were able to heat a standard test piece to 1000°C in less than 3 min, rather more slowly than was aimed at, but not too far different. There was no temperature gradient except that the surface lost a little heat by radiation.

Strength and deformation

Measurement of the deformation of a test piece composed of granular materials cannot always be derived from a stress/strain curve obtained when a steadily increasing load is imposed. When the test piece is brittle, the curve is a straight line up to the point where the maximum stress is reached, after which the test piece fails suddenly. When, however, the test piece is plastic, a similar straight line, but showing a somewhat higher deformation, is obtained, until maximum stress is reached, but thereafter the test piece collapses more slowly. It can be shown that deformation continues with a gradual loss of strength, but without collapse, and it is this deformation which is significant. A sand mould is used only once and provided the thin, dried layer is strong enough to hold together, the actual loss in strength is of no consequence. When the machine is such that the maximum load is maintained on the weakened test piece, collapse is so rapid that deformation cannot be measured.

It is possible to obtain an approximation to the stress/strain curve required by placing the test piece under a press, imposing a known deformation and measuring the strength of the deformed test piece. This has been done with dried cold test pieces containing wood flour and several fibrous materials, since the influence of these materials is to a considerable extent independent of temperature, but the method fails completely when the sand contains pitch or coal which have an effect dependent upon whether they have or have not reached the temperature at which fusion begins.

The method cannot be used with hot test pieces as it is slow and cooling of the surface would influence the results.

This problem was also solved by Rew and Godding.[5,6] In their testing apparatus, the load was applied by means of a constant speed motor to a knife edge, pressing on the centre of a very stiff calibrated spring steel beam. The beam was supported on two knife edges, attached to a block which rested on the test piece. When the load was applied, the centre of the beam moved down and the outer knife edges also moved down in proportion to the deformation of the test piece. The load was determined from the bending of the spring. After maximum stress had been reached, a plastic test piece became weaker and began to collapse rapidly under the

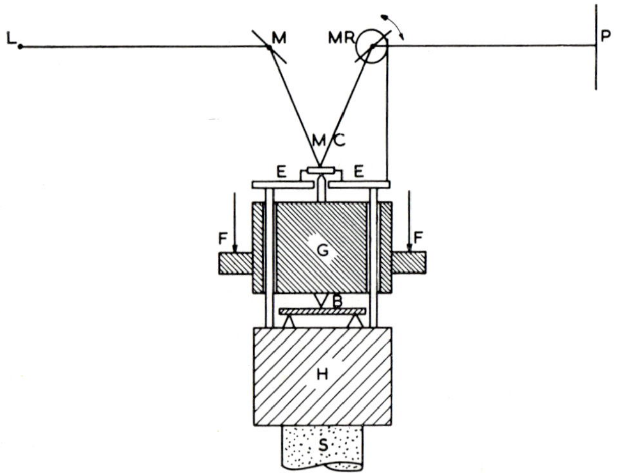

FIG. 87. Apparatus for obtaining stress/strain curves.

stress. Accordingly the outer knife edges move down rapidly, away from the centre knife edge which was moving at a constant speed and the load relaxed. Stress and strain are recorded automatically. The apparatus is shown in Fig. 87.

The test piece, S, is placed on a rigid support beneath the block, H, carrying the knife edges and the calibrated beam, B. The load is applied at the two positions F on the block, G, and its knife edge by means of a screwjack driven by a synchronous motor.

A mirror carriage, E, is attached to the test head, H, by means of two rods passing through the block, G. This mirror carriage rotates when there is any relative movement between the block, G, attached to the central knife edge and the block, H, attached to the outside knife edge. The amount of tilt is proportional to the degree of bending of the spring, *i.e.* to the load on the test piece. The weight of the superstructure is taken up by springs round the supporting rods and there is no resultant load on the test piece.

A second mirror, MR, is attached to the block, H, by means of a rack and pinion and this mirror tilts in proportion to the movement of the block, *i.e.* to the deformation of the test piece. The directions of rotation of the two mirrors are at right angles and when light from the source, L, is deflected by a fixed mirror on to the two rotating mirrors in turn and then on to a photographic plate, P, a stress/strain diagram which shows the extent of deformation and the weakening of a plastic test piece is drawn automatically on the plate. The source of light is a lamp which flashes at 5 s intervals and the rate of change of load and deformation can be observed.

Heating the specimen and measuring temperature

The sand specimen is isolated from the supporting block and test head by hollow glass insulators and is separated from these insulators by silica plates.

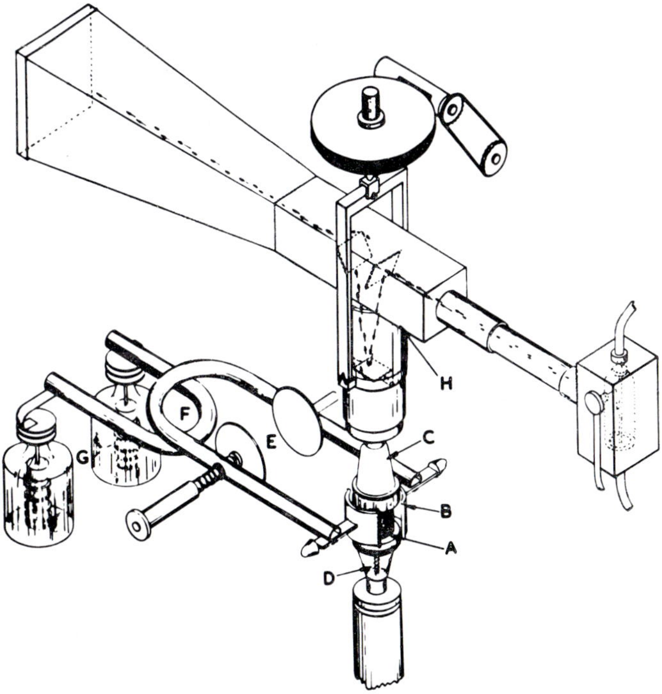

FIG. 88. Apparatus for testing at high temperatures.

The rate of heating is to 1000°C in less than 3 min and it is necessary to measure the temperature continuously in order to determine exactly when the required temperature has been reached. To do this, a hole is drilled along the vertical axis of the test piece, reaching from the bottom to the centre, and a thermocouple pushed upwards through a hole in the lower silica plate. A small hole in such a position has no effect on the breaking of the test piece, and the thermocouple, being positioned along the neutral axis, is not directly heated, but records the temperature in the centre of the test piece.

The two plates forming the condenser are placed round the test piece and connected to the dielectric heater. These plates are curved so as to enclose as much as possible of the curved surface of the test piece and to minimise the loss of heat by radiation.

The heater is switched on and as soon as the required temperature is reached, the thermocouple is withdrawn and the test completed. The complete apparatus is shown in Fig. 88.

Measurement of expansion

The springs supporting the loading device are replaced by others such that there is a residual load of 2 lb on the test piece. The screwjack is wound back so that there is no other load on the test piece and the mirror, MC, remains stationary. As the test piece expands, the rack and pinion linkage causes the mirror, MR, to rotate proportionally.

When measuring expansion, the automatic is replaced by a handoperater flasher. The test piece is heated and a light flash given for every 20°C, or other suitable interval, rise in temperature. The photographic plate then shows a row of dots, the distance between any two neighbouring dots showing the expansion due to a rise of 20°C at the particular temperature concerned.

This method of measuring expansion has one disadvantage. There is nothing to show the order in which the dots were made and the highest dot shows the maximum expansion. It is not possible to see whether the test piece first expanded and then contracted as, indeed, happens with quartz. This limitation could be overcome by causing the photographic plate to rotate, but such a modification would greatly increase the difficulty of construction and to little purpose. Since silica ceases to expand when the temperature passes 573°C, and it has been shown by Petterson and others that the risk of scab formation also ceases, most tests were carried out at 600°C in order to ensure that the maximum expansion had been reached.

It was shown that the expansion of the test piece was the same whether it was heated rapidly in the dielectric heater or slowly in a heated furnace, and by using the latter, the contraction above 573°C could be obtained. However, to obtain the complete picture of expansion, a large number of tests would be needed, each requiring a heating and soaking time of 20 min and the small advantage in the way of improved detail would not have justified the very large reduction in speed.

THE PROPERTIES OF SAND AT HIGH TEMPERATURES

Hot strength

The first objective was to determine whether there was any kind of agreement between the results obtained using the dielectric heater and those obtained using a hot furnace and allowing the test piece to stand in the hot zone until a reasonably uniform temperature was obtained.

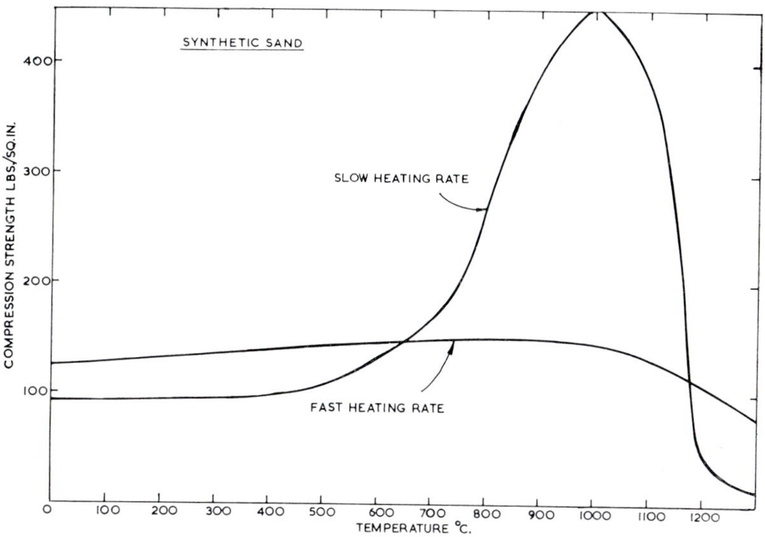

FIG. 89. Hot strength of sand with different rates of heating.

The strength of a silica sand containing 5% bentonite was measured over a range of temperature from the cold up to 1200°C. Some test pieces were heated in the dielectric heater, the time required being up to a maximum of 3 min, depending on temperature, while others were placed in a hot furnace and allowed to stand for 20 min while the temperature became uniform. The results are shown graphically in Fig. 89.

The strength of the sand which had been rapidly heated changed only a little until the temperature reached 1000°C, when strength fell.

On the other hand, the test piece which had reached the appropriate temperature only after soaking for 20 min had very different properties.

When the temperature reached 600°C, strength began to rise and at 1000°C was about four times as great as at room temperature.

The reasons for the difference are clear. At 600°C the clay begins to sinter and strength to rise. As temperature increases, the rate of sintering, and with it the rise in strength, also increases, but at 1000°C the most fusible constituents of the clay begin to melt and strength falls rapidly. When the sand temperature is raised to 1000°C in less than 3 min, there is no time for appreciable sintering and strength remains almost constant, again until the temperature exceeds 1000°C, when fusion commences and strength falls.

On the surface of a mould, temperature rises even more rapidly than in a dielectric heater. Most castings made in green sand moulds are filled in less than 50 s and the rise in temperature becomes very rapid as soon as metal comes into actual contact with the mould wall and even the small rise in strength which can be seen in the test piece heated in the dielectric heater cannot happen. It is, of course, likely that if the casting is allowed to stand for some time in the mould, then at some point well behind the mould face, the temperature will reach a level between 900 and 1000°C and not increase further, in which case there will be an increase in strength, but such a change can have no influence on the formation of scabs or any other casting defect.

The effect of coal dust and wood flour

The effect of increase in temperature on the strength of two sands containing, respectively, 5% coal dust and 2% wood flour is shown in Fig. 90.

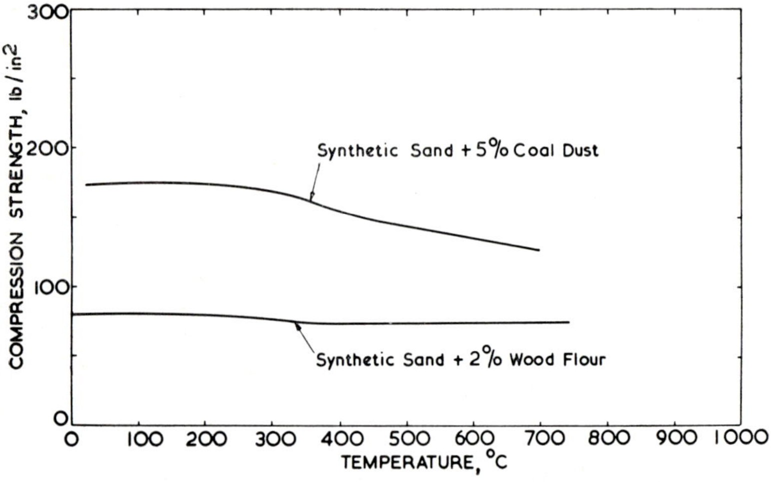

FIG. 90. Hot strength of sands containing wood and coal.

The time needed to reach the appropriate temperature is short and should the organic materials decompose, a reducing gas is given off during the whole of the time needed to carry out a test. Although some volatile matter is lost, none of the fixed carbon can burn and the residue consists of partly decomposed coal or wood with some coke or wood charcoal. The same conditions occur in a mould wall.

When the temperature reaches 400°C the coal begins to soften and strength falls. Any liquid formed also begins to decompose and further changes in strength depend on whether the amount of liquid being formed is more or less than the amount decomposed. With the rate of heating obtained in the test illustrated in Fig. 90, the amount of liquid formed increased — and its viscosity probably decreased — right up to the final temperature of 700°C.

There is one uncertainty. If the rate of heating is reduced, any liquid formed may decompose as fast as it is formed and any decrease in strength can only be due to replacement of coal particles by finer and less coherent coke particles. If, on the other hand, the rate of heating is raised sufficiently, most of the fusible compounds would liquefy before appreciable decomposition occurred and there would be a greater fall in strength.

The effect of wood flour is completely different. The dry strength at room temperature is reduced, but thereafter, the only change which occurs is the replacement of wood particles by equally rigid charcoal particles and there is no change in strength as the temperature rises.

Expansion

The expansion of a fairly pure quartz sand to which had been added 5% bentonite was measured, using a series of test pieces rammed to different hardness. Four of the curves obtained are shown in Fig. 91.

The dotted line shows the expansion obtained by Sosman for a dense crystalline silica rock. The expansion of the very hard sand test piece is almost identical, any variations being due to the difficulty in producing a flash at the exact temperature intended when the rate of increase exceeds 5°C s^{-1}. When a sand is rammed to this hardness, the clay has no influence on expansion.

The expansion decreases as mould density, or mould hardness, decreases.

The standard test piece is rather harder than hard foundry moulds except, perhaps, those made by very high pressure squeezing, whilst test piece no. 3 is softer than most foundry moulds. The results show that the expansion of a hard mould prepared from a clay-bonded silica sand is 1·3% and that of a very soft mould 1·2%.

The expansion of a red sand

The corresponding curves for a red sand containing rather more than 10% clay are shown in Fig. 92.

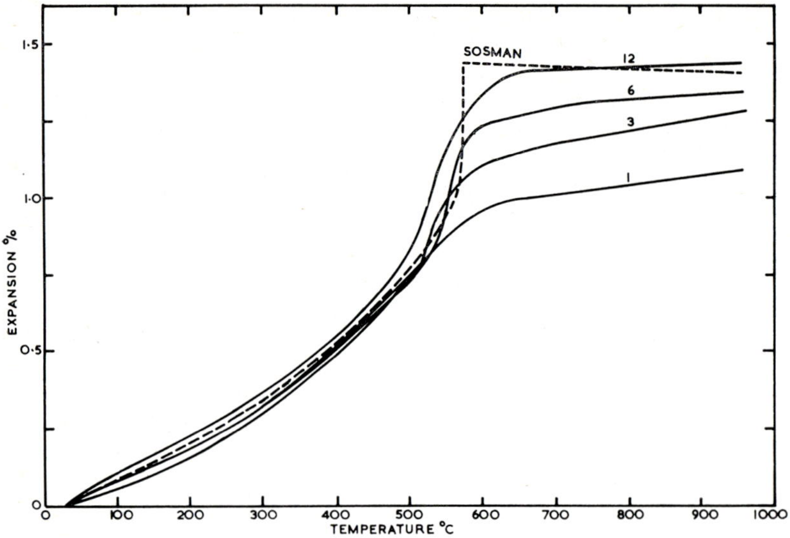

FIG. 91. Expansion of quartz sand with 5% bentonite at different hardnesses. Number 12: rammed until no further compression could be obtained; mould hardness 100. Number 6: the standard test piece; mould hardness 90. Number 3: a rather soft test piece; mould hardness 30. Number 1: the softest test piece which would hang together; mould hardness 0.

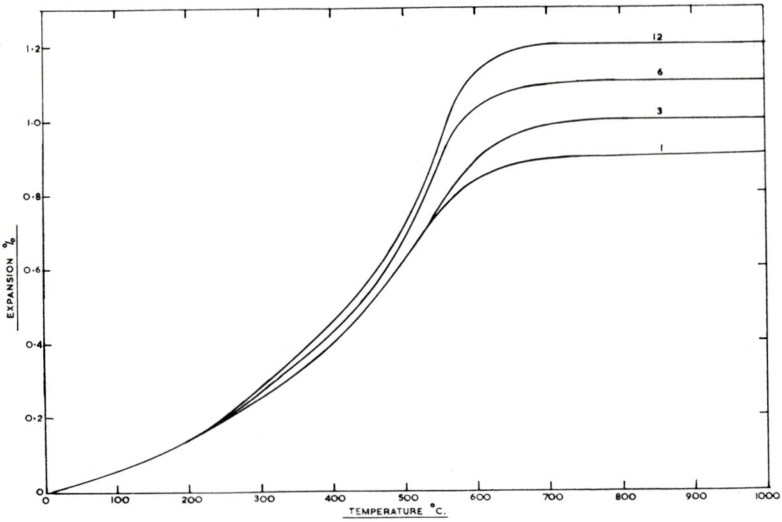

FIG. 92. Expansion of red sand at different hardnesses.

The expansion of a hard red sand mould is 1·1% and that of a soft mould 1·0%.

The lower expansion of the red sand is due to the presence of felspar which has a coefficient of expansion much lower than that of quartz. The size distribution of the felspar grains is almost identical to that of the quartz grains and the expansion of a mixture of the two can be calculated approximately from the proportions and expansions of the two constituents.

The effect of clay

The expansion of the hardest synthetic sand test piece is the same as that of a clay-free dense silica rock. This is to be expected. In the undried sand which is being rammed, the quartz grains are coated more or less uniformly with a layer of the much softer moist clay, and hard ramming forces the sand through this layer until the grains are in actual contact. The clay then merely occupies a part — and only a small part — of the voids in the rammed sand and has no effect on the expansion of the whole. Indeed, when this stage has been reached, any further increase in ramming intensity has no effect on the clay, but causes the quartz grains to deform elastically, producing a further reduction in overall dimensions of the rammed sand. When the compacting force is removed, the grains recover, causing an increase in the size of the compact. This occurs when squeezing moulds using very high pressures, in which case 'spring-back' may cause difficulty.

When the moulds are less densely packed, a layer of clay separates the grains. The moist clay contracts on drying and expands little when heated, thus reducing overall expansion. The major part of the reduction, particularly in the very soft moulds, numbered 12 in the graphs, is due to the ease with which the sand grains can expand into relatively large voids.

Effect of organic materials

The expansion of the synthetic sand used to prepare curve no. 6 shown in Fig. 91, but with the addition of:

(a) 6% coal dust (B),
(b) 1½% pitch (D),
(c) 2% wood flour (C),

is shown in Fig. 93. It has been claimed that these materials reduce scabbing because they reduce sand expansion. Standard test pieces were used.

The expansion of the sand alone is shown by curve (A). Coincident curves (B) and (D) were obtained for the sands containing 6% coal dust and 1½% pitch. These curves show an expansion slightly higher than that of the original sand, but the difference is less than possible inaccuracies inherent in the instrument and all that can be said is that adding coal or pitch to a sand does not reduce expansion.

The reduction resulting from the addition of wood flour is real and cannot be attributed to inaccuracies in measurement. Even so, it seems probable that the reason for the reduction is indirect, since all that happens to wood flour when it is heated in a reducing atmosphere is a gradual change from wood to charcoal of similar shape, size and rigidity, and such a change is not likely to reduce expansion. For the fragments of wood to cause an overall reduction, it would be necessary for these fragments,

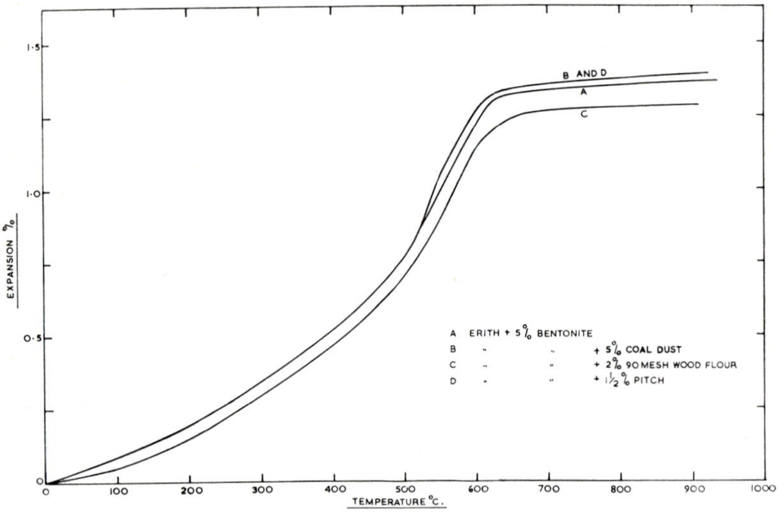

FIG. 93. Effect of organic materials on expansion.

which are of similar size to that of the sand grains, to contract appreciably. Commercial grades of wood flour contain all sizes below that named in the specification and a large proportion is so fine that its volume changes can take place in the voids without having a marked effect on overall expansion.

However, it was shown in Chapter 6 that when a sand contains a large proportion of wood flour, it expands after it has been compressed. This behaviour is just apparent when the amount present is about 2%, and there are fewer grains in the test piece than would have been present if the after-expansion had not taken place. Expansion is correspondingly lower. When the amount of wood flour was reduced to 1%, expansion was the same as that of the original sand which is shown in curve (A).

It is clear that the expansion of a sand cannot be reduced by adding a small amount of this or that organic material. It depends on the composition of the sand grains and on the mould hardness, and when these are fixed by other considerations, as is usually the case, nothing can be done to reduce expansion.

Deformation

Two kinds of material have been used for increasing deformation:

1. A substance which partly or completely melts to form a viscous liquid in the surface layers of a mould during pouring.
2. A fibrous material.

The synthetic sand used in the expansion test was heated in the dielectric heater and the complete stress/strain curve obtained as described earlier.

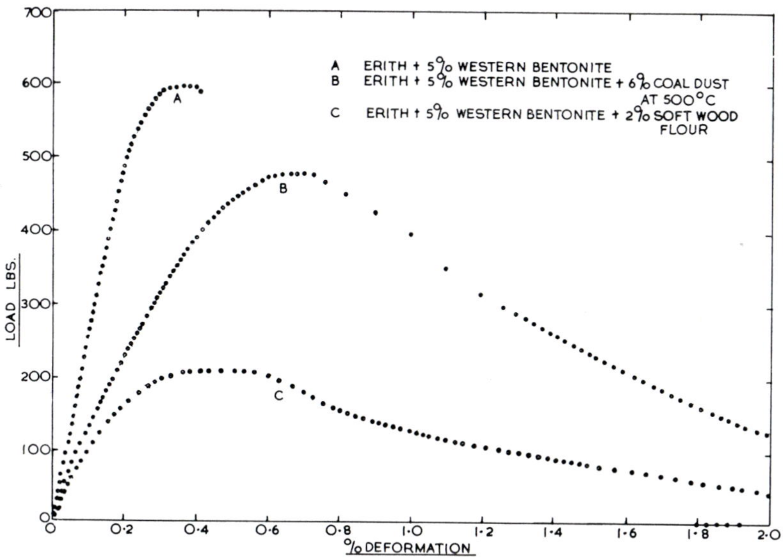

FIG. 94. Strength and deformation — effect of coal dust and wood flour.

The tests were then repeated on two sands containing, in one case, 6% of a fine coal dust and in the other, 2% of wood flour. The stress/deformation curves are shown in Fig. 94.

The tests were made at 500°C. If it were possible to increase the rate of heating in the dielectric heater until it was equal to that of the surface of a mould heated by radiation, the most suitable temperature would be just above 573°C. However, the extent of changes which take place in heated coal dust depend on time as well as temperature and would have proceeded farther in the more slowly heated test piece. Accordingly a somewhat lower testing temperature was used when carrying out the test illustrated in Fig. 94. Tests have been made at various temperatures up to 600°C, but only slight differences, which in no way influence the interpretation put on the results, were obtained.

As an increasing load was applied to the heated sand containing no organic additions, there was an increasing deformation and the stress/strain curve obtained was a straight line. When the maximum stress was reached, there was a further slight deformation, after which the test piece shattered completely. Maximum deformation was 0·4%.

In the presence of coal dust, a similar straight line, but with a higher deformation for unit stress, was obtained when the load was first applied. Before the maximum stress was reached, the rate of deformation increased and the line began to show a distinct curvature. When the maximum stress was reached, there was a further slight deformation, after which the test piece failed, but the type of fracture was different. Instead of shattering, the test piece broke into a few pieces which were prevented from falling away by the electrodes, and after a period of rapid deformation a second series of dots appeared, representing the slowly decreasing resistance of the broken test piece. The test piece failed at the first point when the spacing between the dots increased.

When the sand contained 2% wood flour, the rate of deformation began to increase as the load increased, and the initial straight portion of the curve was short. After maximum stress had been reached, the test piece continued to deform with a gradual loss of strength, but without fracture, and had not failed when deformation was several times greater than that at maximum stress. There is one further important point: almost identical stress/strain curves were obtained whatever the testing temperature in the range room temperature — 600°C.

Each of the fibrous materials tested had an effect similar to that of wood flour, increasing deformation to failure, whatever the temperature.

The deformation of the sand containing 5% coal dust is appreciably lower than the expansion. Increasing the amount of coal dust raises deformation slightly, but the maximum amount which can be used is limited by the rapid evolution of gas from the heated coal and only rarely do foundries use as much as 10%. Doubling the amount of coal raises deformation by less than 50%, which is still lower than the expansion, and coal dust cannot be depended upon to offset the effects of expansion in causing a mould face to buckle in all cases, although it is successful in a majority of cases. Where coal dust does not provide the answer, wood flour is always satisfactory. Its success in some cases may be in part due to the raising of deformation at all temperatures. When a mould with a large surface area is filled in a time as short as 5 s, it is possible that a fairly coarse coal dust may not liquefy in time to produce even its limited effect, whereas wood flour is equally effective at all temperatures.

THE PERMEABILITY OF HEATED MOULDS

When a cold gas enters a heated sand test piece, its temperature rises and it expands. The volume of the gas passing through the test piece is therefore greater than the volume of the same gas in the bell of the permeability apparatus at room temperature. On this account, the apparent permeability of a hot test piece attached to a normal tester must seem to be lower than that of an otherwise identical cold test piece.

This effect was first described by Saunders[7] who used a pressure testing device for indicating permeability. Later, Hudson[8] measured the permeability of a standard test piece rammed inside a tube and heated to various temperatures. He found that the apparent fall in permeability was appreciably greater than calculated from the increase in volume which took place as the air was heated. However, Lorig, in a discussion on the paper by Saunders, had previously pointed out that the viscosity of a gas increases as temperature rises and the apparent discrepancy in Hudson's results was due to this factor.

Hudson used test pieces of 'sands' other than silica sands, choosing grains with a low coefficient of expansion. He found that the results obtained were not significantly different from those obtained when using either synthetic sands based on quartz or natural sands, and concluded that the expansion of the sand grains has only a negligible effect on permeability at high temperatures.

The apparent fall in permeability when the test is made on a heated test piece is due to the increase in both the volume and viscosity of the air as it passes through the hot sand. When attempting to apply these changes to moulds heated by molten metal, it must be remembered that only a thin layer of moulding sand is heated during the time required to fill a mould, and once the mould gas has passed through this layer, it is rapidly cooled and its volume and viscosity decrease.

REFERENCES

1. Dietert, H. W. and Woodliff, R. E. (1939). *Trans. Amer. Foundrymen's Soc.*, **47,** p. 309.
2. Hudson, F. (1935–36). *Proc. Inst. British Foundrymen*, pp. 155–172.
3. *Trans. Amer. Foundrymen's Soc.*, see AFS Committee on Hot Strength reports, 1943–1955.
4. Petterson, H. (1955). *Proc. Inst. British Foundrymen*, **48,** A.201–A.216; see especially A.203.
5. Godding, R. G. and Rew, R. (1953–1955). *BCIRA J. Res. Devel.*, **5,** pp. 278–295.
6. Rew, R. and Godding, R. G. (1955). *J. Sci. Instruments*, **32,** March, p. 93.
7. Saunders, W. M. (1930, 1931). *Trans. Amer. Foundrymen's Assoc.*, **38,** pp. 259–272; **39,** pp. 440–441.
8. Hudson, F. (1935). *Foundry Trade J.*, Dec. 5, pp. 411–416.

CHAPTER 10

Behaviour of Moulds During and After Pouring

DEFECTS DUE TO MOULDS

To be satisfactory, a casting must be dimensionally accurate, have an acceptable surface and be free from scabs, inclusions, blowholes and internal porosity. Failure to satisfy any one of the above conditions may be due to a failure of the mould, but is not invariably so.

Defects due to mould failure can be divided into four groups:

1. Swelling and internal porosity due to the inability of the mould to withstand the pressure exerted by the metal, either while the metal is liquid or, in the case of metals which expand on solidification, shortly after solidification commences. The latter type of failure is usually known as mould dilation or mould wall movement, to distinguish it from simple swelling.

It is convenient to include metal penetration and other surface roughness in this group, since some of the factors which control these also influence swelling.

2. Mould erosion during pouring.
3. Misruns, blows and solid scabs.
4. Expansion scabs.

Of the above defects, some, for example scabs, are clearly due to failure of the mould; others may be due to causes independent of the mould. Gasholes may be derived from the mould or cores, or may be due to metal composition, or to a reaction between mould and metal. Inclusions may be derived from the mould or may be introduced with the metal. Most defects are influenced by the arrangement of the ingates and the speed of pouring.

It is important that all possible causes of defect should be examined before action is taken. Raising the permeability of a sand system, for example, is a lengthy operation and to do this only to find that the gasholes are due to high manganese and sulphur in the metal is not helpful.

(1) SWELLING, INTERNAL POROSITY AND METAL PENETRATION

Swelling

Swelling occurs when the pressure on the mould face is sufficient to compact the rammed sand still further and therefore increases as mould hardness decreases.

This pressure has two components, static pressure proportional to height of the metal in the pouring bush above the appropriate point in the mould, and dynamic or velocity pressure due to the sudden arrest when a fast moving stream of metal strikes a mould wall.

Static pressure is highest when the castings are of considerable depth, but its effects may be pronounced even in light castings set horizontally for pouring. Prior to 1940, stove plate castings were sometimes made by hand moulders and by machine from identical patterns in the same foundry, and observations made over a period of weeks, in one case, showed that the average weight of the castings made by the hand moulders was 15% higher than that of the castings made by the machine moulders. There were no obvious defects, but the thickness of the hand moulded castings was greater, the difference being greatest at the centre of the casting and least at the edges. The differences were due mainly to the effects of static pressure on moulds differing in hardness. The hand moulders produced softer moulds for two reasons:

1. Ramming large volumes of sand by hand is hard work and moulds were rammed only just hard enough to produce a casting which had no obvious defects.

2. It is difficult to organise pouring as efficiently when the moulds are spread out on the floor as when they are on a mechanised plant, and some of the metal is a little cold. Misrunning is less likely when the mould is soft enough to yield and give an oversize casting.

Swelling due to dynamic pressure is almost entirely confined to light castings, since in heavy castings, the velocity of the metal falls when the ingate is passed. The defect may occur at irregular intervals and may be confined to one of a number of castings in the same mould. Swelling may be due to a method of pouring sometimes adopted when the pourer thinks that the metal is so cold that a misrun is likely. The ladle, held so that metal impinges directly on the down sprue, is raised, as soon as pouring has commenced, to a height of 18 in or so above the mould. This gives a high dynamic pressure and a swollen casting is obtained. Cold metal is not as a rule included in the causes of swollen castings, but it is frequently an indirect cause.

Roughness and metal penetration. The smoothness — or lack of it — of a casting depends mainly on the size of the sand grains and provided that the sand has been properly aerated so that no large voids are present, surface finish improves as mould permeability decreases.

When, however, the pressure exerted by the metal exceeds a certain critical value which depends on the diameter of the pores in the sand, metal may be forced in between the sand grains. When the castings are large so that the pressure is high and the temperature of the mould is maintained for a long period, metal may penetrate to a considerable depth, since the sand behind the mould face is usually less densely packed and penetration is easier. With a given sand packed in a given way, penetration commences when the depth of metal exceeds a certain limit, which depends on the surface tension of the metal. Hoar and Atterton[1] measured the depth of steel required for penetration to commence with a series of moulds of different composition. They also showed that the presence of a small amount of iron oxide reduced surface tension and enabled penetration to commence at a much lower pressure, but the formation of iron oxide on molten metal containing 3% carbon and 2% silicon is unlikely, especially when the mould atmosphere is reducing. Even so, moulds for large castings and any cores needed are usually coated with a layer of very fine inert carbonaceous material, held in place by clay and an organic binder. Decomposition of the organic binder helps to maintain a reducing atmosphere, although less strongly reducing than that obtained when the sand contains coal dust. Application of this coating or blacking not only prevents metal penetration, but gives a very smooth finish without reducing the permeability of the mould behind the face.

The smoothest finish is obtained when the carbonaceous material is graphite which is composed of thin, flat plates, but the cost is high and finely ground coke is frequently used instead.

Metal penetration is most likely to occur in pockets and reentrant angles where the sand is usually less densely packed.

As a rule, blacking is used only on moulds which are to be dried and is applied as a suspension in water. Occasionally, hand moulders dust unbonded blacking on to the surface of green sand moulds and force the blacking on to the mould surface by replacing the pattern in position and applying a moderate pressure, a process known as 'printing back'. This procedure is used mainly when producing ornamental castings.

Metal penetration increases fettling costs, since the excess metal usually has to be removed by chipping and grinding. Moreover, a mixture of metal and sand has to be removed and in the process the sand grains are broken down to dust, which may escape into the atmosphere. Since quartz is the most toxic of the normal constituents of atmospheric dust in foundries, precautions must be taken to see that this dust is not inhaled.

The static pressure in moulds for small castings is not, as a rule, sufficient to cause metal penetration, but the defect can occur on such castings as a result of dynamic pressure. This may occur even on moulds made by high pressure squeezing, in which case there is no swelling. Usually, with small castings made in hard-rammed moulds, metal penetrates to a depth of one or two sand grains only, since there is a steep temperature gradient in the mould and the advancing front of the penetrating metal solidifies rapidly. At the same time, the defect may occur over a considerable part of the total area of the casting, since the causative pressure is independent of the difference in height between casting and pouring basin, and in such a case it may be cheaper to make another casting than remove the penetrating metal.

Penetration with gross swelling may occur when moulds are unusually soft. In this case the penetration may take the form of a long fin instead of passing between individual sand grains. This can happen only if the mould is so soft that there is some risk of a run-out.

Mould dilation and unsoundness

When a mould is filled with molten iron, the metal contracts until solidification commences. In a grey or nodular iron casting there follows an expansion as graphite separates. Provided that the method of running and feeding is such that liquid metal can make up for the liquid shrinkage, a grey iron casting should be free from the shrinkage troubles which occur with metals such as steel, which contract on solidification. Nevertheless, internal porosity has been a major source of trouble in foundries producing grey iron castings.

Recognition of the cause was gradual. One of the old moulder's axioms, and usually these were factually correct although the accompanying explanation appeared strange, was that a casting made in a dried mould is sounder than a duplicate made in a green mould. No conclusions were drawn from this until Longden[2] pointed out the relation between mould rigidity and casting soundness.

Even then, the relevant conclusions did not receive wide recognition until Womochel and Sigerfoos[3] published their paper on the subject. These observers made two castings, one a near-spherical bowl with a long straight neck, and the other a plain cylinder with the length about twice the diameter. Moulds were made in baked oil sand, dried clay-bonded sand and green clay-bonded sand with various additions, including coal dust.

The castings obtained were sectioned, the bowl casting by a cut along the axis of the bowl and through the neck, and the cylinder along the axis. Sections through the bowl castings made in baked oil sand and green sand moulds are shown in Fig. 95.

The casting made in baked oil sand is completely sound, but that made in green sand has a large diffuse cavity.

Since the moulds were originally the same size, the same quantity of metal was needed for each, but the amount in the neck of the casting made in green sand is clearly the lower, and the amount in the bowl must therefore be the higher. The diameter of the bowl is also greater. The neck can be regarded as the runner system and the bowl as the casting, and it follows that unsound castings are larger and heavier than sound castings, always provided that the unsoundness is due to mould dilation. Weighing

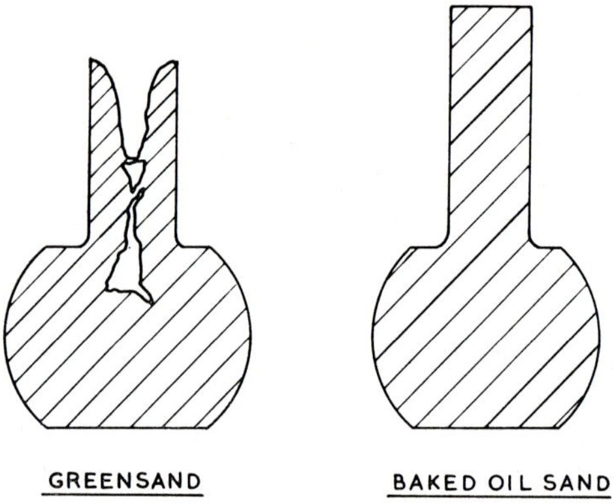

FIG. 95. Castings made in green sand and in baked oil sand.

and measuring can therefore be used to distinguish between sound and unsound castings.

When the cylindrical casting was sectioned, it was found that when the casting was made in green sand, porosity invariably extended into the bottom half. The conditions favoured feeding of the bottom half of the cylinder to a greater extent than it is usually practicable to feed castings, and the test shows that when a casting is unsound as a result of mould dilation, the defect cannot be eliminated by feeding.

The relation between mould hardness, casting weight and casting soundness was shown in greater detail by Greenhill[4] when dealing with unsoundness in a brakedrum weighing about 170 lb. The proportion of castings showing porosity after machining was high. A section through the wall of a defective drum is shown in Fig. 96.

There is considerable swelling in the region where the porosity occurs.

By weighing the castings, it was possible to determine which were likely to be unsound, without first performing an expensive machining operation.

Examination of the hardness distribution in the cod of sand shaping

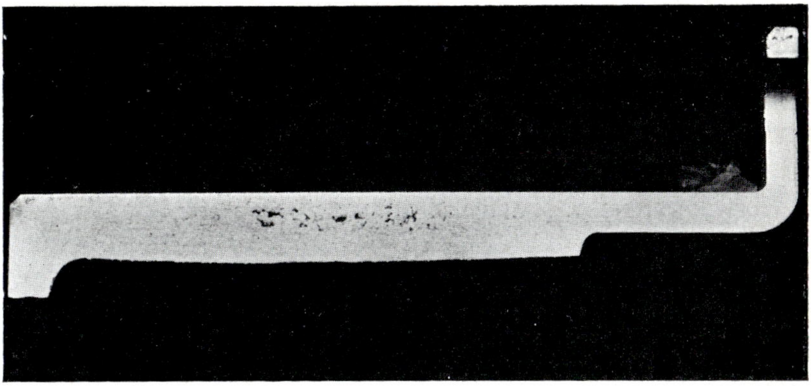

FIG. 96. Porosity in brakedrum. (After J. M. Greenhill.)

the interior of the casting showed that in the region where swelling and porosity occurred, the mould was softer than elsewhere. Hardness distribution is shown in Fig. 97.

The method of ramming was modified in such a way that the minimum hardness was raised to 70 and no castings were found to be porous after machining.

Each casting in a batch made before mould hardness was increased was weighed and a note made as to whether the casting was sound or not. A similar batch was examined after the hardness had been changed and a third some two months later. The weight distribution and incidence of porosity are shown in Fig. 98.

In the first period, rather more than half of the castings weighed 169 lb or more, and most of these were rejected. All castings weighing 168 lb or less were sound.

In the second batch, all castings weighed less than 168 lb and were sound.

In the third batch, control of ramming had been relaxed to some extent

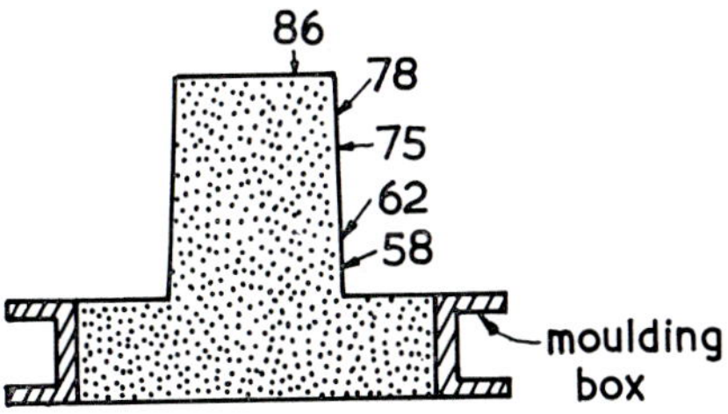

FIG. 97. Hardness distribution in mould. (After J. M. Greenhill.)

and several castings weighed more than 168 lb. About half of these were porous.

The importance of Greenhill's work is that it shows clearly that the deciding factor, as far as soundness is concerned, is the hardness of the softest part of the mould and not the average mould hardness. A quite small soft area will lead to porosity in the immediate neighbourhood and if this happens to be in a position such that it will be revealed by machining or by some test for soundness, the casting may be rejected.

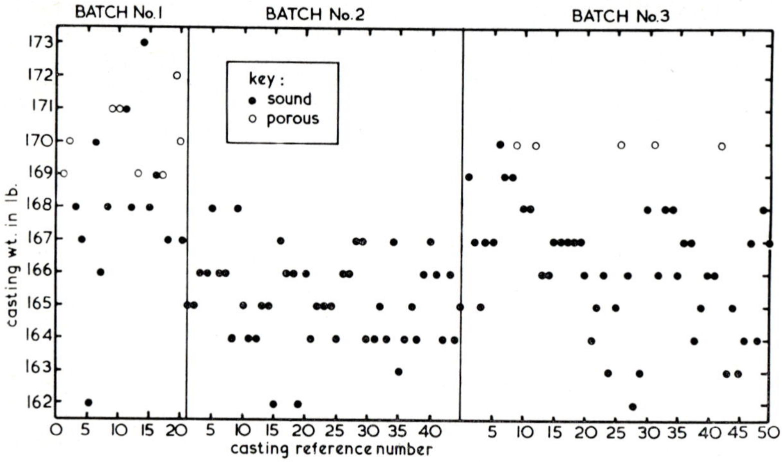

FIG. 98. Weight distribution and porosity. (After J. M. Greenhill.)

In the above example, all castings were sound when the minimum mould hardness was 70, but, although this figure will be found to be satisfactory in a majority of cases, it is not universally applicable. The brakedrums which were acceptable varied in weight and it is possible that a small trace of porosity was present in the heavier castings, but was not revealed on machining. If the tests made had been more stringent, even a small amount of porosity might have resulted in rejection. Suppose, for example, the castings had been fly wheels and the basis of acceptance had been a tensile test on a test piece cut from that part of the rim where any porosity would be concentrated, then a trace of porosity which could not be detected by the naked eye might lead to rejection of a whole batch. In such a case, an appreciably higher mould hardness would be needed.

Effect of sand composition

Clay content and green strength. It is not possible to obtain a very hard mould when using a sand with a low green strength. Mould dilation depends on how far the mould is deformed by the pressure exerted by the solidifying and expanding casting, and the mould hardness tester is so useful because it also measures the deformation which results from the imposition of a fixed load, although in this case the load is applied by means of a spring. A certain minimum deformation to rupture is needed to permit separating mould and pattern and if strength is low, deformation under a given stress is correspondingly high.

Moreover, increasing the shatter index of a sand while maintaining a constant ramming effort on a machine has two effects on mould hardness distribution:

(1) it reduces overall hardness;

(2) it increases hardness differences between different parts of a mould by reducing still further the hardness of the softest sections, unless the moulding machine has a head which is contoured by one of the methods described in Chapter 8.

Using conventional squeeze or jolt–squeeze machines, it is necessary to compromise between a weak sand which produces overall soft moulds, and a very strong sand which produces moulds varying in hardness. Under most conditions, sands in the range green strength 8 lb in^{-2} with shatter index of 70, to green strength 12 lb in^{-2} with shatter index of 80, can be used to give reasonably hard moulds.

Moisture content. An increase in moisture content above that needed to produce a sand suitable for mould-making increases shatter index slightly and reduces strength appreciably, and on this account, tends to increase hardness differences and reduces overall hardness.

Too high a moisture content therefore increases the risk of obtaining unsound castings.

The moisture content needed depends on the nature of the sand. Different kinds of sand need different proportions of moisture in order to develop their moulding properties and it is an amount in excess of this which causes unsoundness. A system sand based on a red sand may need 5% water before it can be used for mould-making, but does not produce castings less sound than those obtained when using a synthetic sand of similar green strength which requires only 3% water. On the other hand, if the moisture content of the synthetic sand is raised to 5%, the risk of obtaining unsound castings is increased.

The general rule that if a sand feels wet to the touch it will produce unsatisfactory moulds and unsound castings, applies to a number of defects.

Coal dust and pitch. In their original experiments, Womochel and Sigerfoos[3] observed that when 'sea coal' was included in green sand, sounder castings were obtained. This has been confirmed and extended to include pitches by a number of observers.[5,6]

The mechanism by which the improvement is obtained is uncertain. Most coal dusts soften at 450°C and pitches at 100°C, and it is difficult to see how the addition of such a material to a clay-bonded sand can increase the resistance to deformation. In addition, the coal dusts used differed in swelling index and the pitches in hardness and softening point, but these differences had little influence on their effect.

The amount of coal dust or pitch included in a moulding sand is controlled by considerations of surface finish and resistance to scabbing and although the improved soundness of castings is an acceptable by-product, increasing coal or pitch cannot be used as a means of improving soundness.

Skin-drying, mould facing and soundness. Drying a mould increases its rigidity markedly, but when only a thin layer of sand on the mould face is dried and is backed up by a weak and easily deformed green sand, the support is greatly reduced. When a mould is skin-dried, a thin layer of high moisture content forms behind the dried layer, but no quantitative evidence is available to show whether this weak layer has any significant effect in reducing the resistance of the mould as a whole. It cannot increase it.

The minimum thickness of the dried layer needed depends on the size of the casting and on the hardness of the undried sand. A thicker layer is needed for a large mould and when the supporting sand is weak. It is doubtful whether skin-drying to the depth of $\frac{1}{4}$–$\frac{1}{2}$ in, which is frequently carried out on moulds of medium size, helps with casting soundness, although it may eliminate many surface defects. On the other hand, moulds for quite large castings are not always completely dried, but the dried layer is usually several inches thick and provided the mould is not allowed to stand in the cold for a long period, no harm results. This can hardly be described as skin-drying.

Moulds made in silicate bonded or air-set sands are extremely resistant to deformation and produce sound castings. In order to keep costs down, only the face of the mould may be made with the hardenable sand and the box filled up with green sand. The minimum thickness of the facing material, which must be used if sound castings are to be produced, depends on the shape and size of the casting, but unsoundness has been observed in castings about 3 ft square, using a facing of silicate bonded sand nearly 1 in thick.

Unsoundness has also been observed when the clay-bonded sand was completely dry but had inadequate support. The core for a cylindrical casting weighing several tons was made with clay-bonded sand on a small core barrel covered with several layers of straw rope. At some period after pouring, the rope was burnt out and the unsupported dry sand yielded to the pressure of the solidifying and expanding metal and the casting was porous over nearly half of its length. By reducing the thickness of rope and increasing the size of the barrel, the support for the dry sand was improved and sound castings were obtained.

(2) MOULD EROSION AND SAND INCLUSIONS

Sand inclusions occasionally present the foundry with an awkward problem, since it is not always easy to determine the source of the inclusion. The loss of a few grains of sand from points on the surface of the mould or runner system may not be obvious, but the same sand grains collected at one point on the mould surface, particularly at a thin metal section — and this is the most usual place for them to be caught — may result in rejection of the casting.

When there is an obvious surface defect, for example a scab or a wash, the cause is plain to see, but in the absence of any such defect a careful search may be necessary.

Frequently the sand is derived from some place in the runner system and it is useful to select a few runners at random from the pile and to shot-blast and examine them for defects. All the metal entering the mould passes through the runner system and the sand here is usually subject to the most severe conditions, but frequently this source of sand inclusions is overlooked.

Several tests have been devised for comparing the resistance of different kinds of sand to erosion. Usually these involve measuring the amount of sand removed under standardised conditions of abrasion. However, the usefulness of such tests is limited, since the sand behind the surface is much less resistant than the face of a mould and the tests do not compare resistance to the initiation of sand removal.

The most frequent causes of sand inclusions are:

low dry strength;
air-drying of the mould surface;
loose sand in mould or runner system;
mis-use of strainer cores.

Low dry strength

This is sometimes put forward as the main cause of sand erosion and sand inclusions, and is justified to some extent in that under identical conditions sand is removed more readily from a mould with a low dry strength than from one with a high dry strength. Other factors are, however, of importance and a high dry strength does not guarantee freedom from sand erosion, nor does a low dry strength make its occurrence certain.

The effects of low dry strength are most pronounced on medium and heavy castings. Pouring time is necessarily longer and any failure of the surface layer, even when quite small, exposes the less resistant sand behind the mould face to attack, which may persist until pouring has ended.

When pouring time is short, castings have been made successfully in sands bonded with low dry strength clays even when conditions have been severe and the almost complete neglect of these clays by users of synthetic sands has been due to factors other than low dry strength. For example, moulds for baths were made for many years by means of a sandslinger using a synthetic sand bonded with Dorset ball clay, which has a very low dry strength. Moisture content was kept as low as possible without having bad lifts, and dry strength was about 20 lb in^{-2}, but sand washing did not occur although the metal fell for 2 ft from the pouring bush on to unprotected sand.

The use of cores bonded with water only, in moulds for rainwater pipes 6 ft long and 3–5 in diameter, a practice used by several foundries, was even more interesting. Occasionally a sand containing 1% or so of clay was used, but frequently the cores were made with Southport sand and water only, when the dry strength was 0. A thin layer of sand was strickled on to a core barrel and metal poured directly on to the top of the core, but no washing occurred, although if the core was allowed to stand until it was dry, the sand fell off the barrel.

These two examples do, however, beg the question of the influence of dry strength. Pouring time was short and for the pipes four ingates were used and if pouring time was raised by using only one of these, the 'casting' was just a mixture of sand and metal.

In both cases, green sand was more resistant than dry sand to the wash of metal, not because it was stronger, but because it was less brittle. When pouring time is longer, the sand dries out before pouring is complete and dry strength becomes a factor in the resistance of the mould to erosion.

Air-drying

When the surface of a mould dries to a depth of a few sand grains only, it is more friable than that of either a green or a fully dried mould. The reason for this is obscure, but air-drying is at times a serious source of trouble on this account. Erosion by molten metal is not the only means of removing sand grains from a mould surface and the core-setter finds it difficult to place cores in position without dislodging grains from an air-dried surface, particularly near the core prints. As he would have little difficulty in placing the same cores in a completely green mould, it is clear that brittleness, and not dry strength, is the controlling factor.

Surface drying may occur in the interval between removing the mould from the pattern and closing. When the sand is cold, the rate of evaporation is slow, since the vapour pressure of water is low, and any evaporated water helps to provide a moist atmosphere over the mould surface. Moreover, migration of water from the interior of the mould helps to maintain moisture content at the mould face and these factors greatly extent the time needed for an air-dried layer to be formed. When the sand is hot, the rate of evaporation is greatly increased, since the vapour pressure of water in mm of mercury rises from 17 at 20°C to 149 at 60°C and 355 at 80°C. Moreover, the hot saturated air near the mould face rises and may be replaced continuously by cool air with a much lower moisture content.

An additional factor is that when the sand temperature is appreciably higher than pattern temperature, it is difficult to obtain a good strip and it becomes necessary to heat the pattern, which still further raises sand temperature. Frequently, patterns are heated almost continuously, usually by built-in heaters, but these are not always satisfactory, for example when large patterns are used on a jolter, since failure of the heater may cause a lengthy hold-up. Intermittent external heating is then used, but in order to increase the time interval between heatings, the temperature is raised until it is appreciably higher than that of the sand. The mould face is then at a temperature above that of the sand supply and the rate of air-drying is increased. Under such conditions, there is appreciable sand washing immediately following heating of the pattern, after which the incidence of the defect decreases.

Effect of sand composition

The nature of the sand in use affects the rate of air-drying and its effects in three ways.

1. Sands bonded with a clay with a narrow moisture range air-dry quickly because they need to lose only a small proportion of water.
2. Sands with high permeability air-dry more rapidly than sands with low permeability.
3. Coarse, uniform sands are more friable than fine sands or sands with a wide grain size distribution.

On all counts, synthetic sand air dries more rapidly than natural sand and is more friable when dry. When really hot sand is used, edges may dry out in the short interval between stripping and core-setting and it becomes difficult to lay cores without rubbing some sand away, especially near the core prints.

Sand temperature and sand properties

The increased friability found when hot sand is used is due to air-drying of the surface of the mould and not to any difference between the properties of hot and cold, moist sand. So long as drying is prevented, sand temperature has little effect on sand properties and sand as hot as 80°C has been used on plants with automatic mould-closing without any failures due to sand washing, simply because the atmosphere inside the closed mould is saturated and no surface drying can occur. Cold cores must not be placed in a warm green mould since water would condense on them and this might cause blows.

Loose sand in mould

Loose sand in the mould cavity does not invariably result in an obvious sand inclusion in the casting and some observers have spread a small quantity of loose sand over the bottom of the mould without finding an obvious inclusion. Such a dispensation cannot be relied upon and precautions should be taken to see that any such sand is removed before pouring. In particular, the small aggregates which may be dislodged when cores are being placed in moulds which have air-dried, can be a nuisance. They are invariably moved by the incoming metal and frequently maintain their shape, so that there is an obvious inclusion somewhere on the mould surface.

Poor finish in runner system

When making moulds for medium and large castings by hand, a sand of known properties — even though the sand is not tested, a given proportion of a known sand is added and the moisture content is judged by feel — is used to face the mould and a backing sand, which is often variable in properties and more so in moisture content, is used to fill the box. The runner through which all the metal has to pass is frequently cut by hand through this backing sand. The finish is poor and sand unreliable and many sand inclusions are derived from such runner systems.

Wherever possible, runners should be moulded and in all cases the sand used should be up to the standard of the facing sand.

Strainer cores

A thin core containing a number of holes $\frac{1}{8}$ in or so in diameter is sometimes placed at the top of the down sprue or elsewhere near the beginning of the runner system. The description of the core implies that its main function is to strain or filter off any sand aggregates, slag or other foreign matter which may be poured from the ladle, but provided that the height of metal in the pouring bush is maintained, any such impurities should be retained there. The most critical times are at the commencement of pouring when the bush is empty and from then until the runner is full and some metal is in the casting, a period during which the rate of metal flow is at a maximum and the pourer has most difficulty in manipulating the ladle so as to maintain a full bush. The strainer core acts as a mild regulator of the rate of flow of metal during the critical period.

In foundries using oil–cereal cores, the strainer core is frequently made with one of the mixtures used in the core shop, but when metal temperature is high and pouring time anything but very short, the binder may be almost completely destroyed before pouring has ended. The metal stream then washes away the loose sand which enters the casting, when the strainer core produces the very defect it is designed to suppress. Whenever strainer cores bonded with an organic binder are used, runners should be regularly inspected to see that the 'fingers' formed where the metal passes the core are intact. Even undue surface roughness indicates that some sand has been removed.

The risk of premature breakdown is reduced if the proportion of oil in the mixture is increased. At the same time, the amount of cereal should be reduced, a change which does not cause any difficulty in stripping a flat disc. It is, however, safer to use one of the ceramic strainer cores which are commercially available and it is often worthwhile to modify the shape of the mould so as to make possible the use of one of the standard shapes.

When a separate pouring bush is used, the strainer can be replaced by a shim of thin tinned steel, a procedure which enables the pourer to fill the bush before metal enters the runner itself. When pouring temperature is very high a plain shim is satisfactory, but with lower pouring temperatures entry into the mould may be prevented. If this occurs, the shim should be pierced to form a single small hole. Only very thin metal should be used, in order that it shall disappear completely by the time the bush is full; otherwise a very slow stream of metal may enter the runner system.

Mixed inclusions

Inclusions in castings are sometimes mixtures of sand with a slag-like material. In such cases there is usually no evidence as to whether slag has a particularly erosive effect on the mould surface or merely collects dispersed sand grains which would otherwise pass unnoticed, but unless the sand preponderates, removal of the slag usually results in disappearance of the defect completely.

Cupola slag is rarely found in iron castings and its presence is usually due to carelessness, since its melting point is high and it can easily be skimmed off. Sources more difficult to deal with are the ladle lining, from which a slag rich in iron silicate is sometimes obtained, slag from the treatment of the metal with sodium carbonate, or with some inoculants, a slag due to the use of metal with high manganese and sulphur contents, and a slag which is formed when a fusible botting clay is used.

Identification of slaggy inclusions in iron castings has been described by Morgan.[7]

(3) MISRUNS, BLOWS AND SOLID SCABS

When molten metal is poured into a green mould, the air in the mould cavity expands to several times its volume at room temperature, and at the same time gas, mainly water vapour, is given off by the heated mould surface.

In order to enable the mould to be filled, air and other gases must escape from the mould, the controlling factors being the volume of gas and the permeability of the mould.

At a later stage, and especially after the mould is full, gas still being evolved must pass through the mould wall and not through the metal, where it may be trapped beneath a core or under a thin skin of solid metal, usually on the upper surface. Should evolution of gas continue for some time, blowholes are not formed so long as the whole of the metal remains liquid, but individual sand grains or quite large aggregates may be removed from the mould surface.

There are thus three distinct types of defect due to the presence of gas evolved so quickly that it cannot escape through the mould walls.

1. The metal is prevented from filling the mould completely — a misrun.

2. Gas is trapped beneath a core or the upper surface of the casting — a blow.

3. Sand is forced off the mould wall by gas pressure behind the mould face — a scab or, in very mild cases, a rough patch. In both cases a sand inclusion will be found somewhere in the casting.

Misrun

When metal enters a mould, expansion of the air present raises the internal pressure. Air flows through the mould wall. As is implied by the method used for measuring permeability and described in Chapter 1, the rate of flow of air is proportional to the difference between pressure on the inside and on the outside of the mould wall, and the inside pressure builds up until the rate of flow is sufficient to prevent any further increase. This pressure restricts the rate of flow of metal into the mould. At the same time, heat is being lost to the mould wall, and should the front of the advancing stream solidify before the cavity is full, a misrun casting is obtained. The front edge is rounded and, particularly when coal dust or pitch are present in the sand, shiny.

In less severe cases, the mould may be filled almost completely, but two converging streams of metal may be at the point of solidification when they meet and a defect occurs which may appear as a mere crack, but also exhibits the tell-tale bright surface.

Misrunning occurs most frequently in thin castings poured on the flat.

When a small amount of iron is poured on to a large flat sand surface it does not spread out to cover the whole of that surface. Surface tension at the metal–sand–air interface is high and if, for example, 10 lb of iron is poured into a mould for a casting $\frac{1}{8}$ in thick and weighing 20 lb when complete, the result is not a casting $\frac{1}{16}$ in thick, but half a casting $\frac{1}{8}$ in thick. The molten metal flowing into the mould is therefore in contact with the mould both above and below. These conditions favour loss of heat to the mould and little or no mixing takes place, so that most of the heat loss is from the front of the advancing stream.

The defect may occur on thin projections on larger castings, but is of less frequent occurrence. Whilst the mould is being filled, considerable mixing of metal takes place and loss of heat is not concentrated on a small amount of metal until the thin section is reached. Unless the area of the thin section is large, misrunning is unlikely.

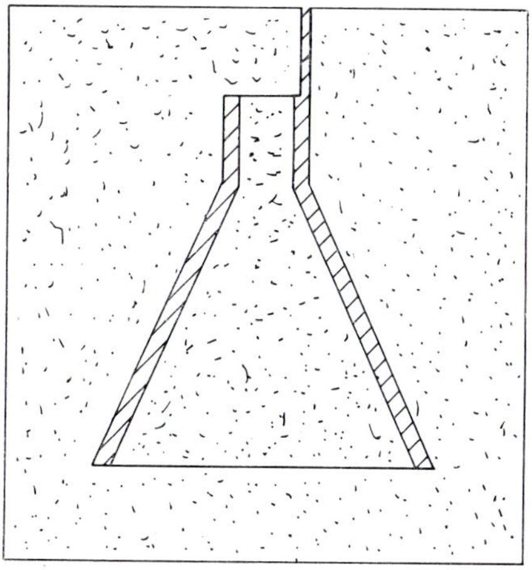

FIG. 99. Misrun on deep casting; effect of a 'whistler'.

The rate of removal of air from the mould can be raised and the restrictive pressure reduced by increasing mould permeability. This is a lengthy job and before it is undertaken, it is well to ascertain that the metal temperature and ingate area are adequate. For most flat plates with mould thickness of 2 in and a mould hardness not exceeding 70, a sand permeability of 25 or more is adequate.

Suppose the mould shown in Fig. 99 is being filled from near the bottom. Metal rises in the space between core and mould and loses heat to both. Little mixing of metal occurs, whatever the thickness of the metal section, and the loss of heat is most severe at the head of the metal stream. It was shown in Chapter 8 that in a mould such as that illustrated in Fig. 99, hardness is usually much higher over the narrow part of the pattern at the top and although mould thickness is least here, permeability is quite low. Conditions are therefore similar to those found when pouring a plate on the flat and a misrun is possible. At other times, a group of small blows may be found at the top edge, but again their presence is due to the fall in temperature at the front of the metal stream. In a case such as this, the most effective method of releasing the air from the top of the mould is to pierce with a hole or 'whistler' of about $\frac{1}{8}$ in diameter.

Blows

When a mould is being poured, gas is given off from uncovered parts of the mould heated by radiation and by all parts of the mould in contact with molten metal. Gas from the uncovered part of the mould exerts a pressure which restricts the rate of entry of metal, but gas given off from any part of the mould which is in actual contact with the molten metal must either escape through the mould wall, or pass through the metal itself. In the latter case, it may be trapped underneath a solid skin which may have formed on the surface of the casting. The controlling factors are the rate of evolution of gas and the permeability of the mould. Gas is also given off by any cores present. This is discussed in Chapter 11.

The main source of gas in a green sand mould is the free water, which evaporates rapidly as soon as the temperature reaches 100°C. Gas is also given off by coal dust and by the clay (combined water), but in both cases, the rate of evolution is negligible at temperatures below 400°C and increases comparatively slowly at higher temperatures. It was shown in Chapter 7 that there is no direct relationship between the moisture content and permeability of a sand and the tendency for moulds to blow, but that any sand with a moisture content appreciably higher than that needed to develop the moulding properties of the sand is likely to produce blowholes. However, although this excess moisture is the main cause of this type of defect, permeability does have a considerable influence and a red sand with a permeability of 20 is more likely to cause blowholes than another red sand with a permeability of 60, even when the moisture content of each is at the appropriate limit.

In practice, blows from moulds are comparatively rare except under a few easily recognisable conditions. The major constituent of the gas is steam and when this passes through the mould it condenses as soon as it reaches an area in the mould where the temperature is less than 100°C, thus reducing the volume and pressure. The effective thickness of the mould, so far as the passage of steam is concerned, is the distance between the face and the point at which the temperature is below 100°C, which, except with large castings, is usually quite small. Since permeability is inversely proportional to mould thickness, the effective permeability of the mould is high.

When blows do occur, they are in most cases due to high moisture content or inadequate mould permeability. Blows are usually found underneath the top surface of the casting or beneath a core, and are frequently completely covered by a thin shell of metal, although they sometimes break through to the surface.

Occasionally, the blow takes the form of a shallow depression on the upper surface of the casting, partly or wholly covered by a skin similar to an eggshell in thickness. The surface of the depression is quite smooth. This type of blow usually described as a 'flutter', is almost invariably due to a high moisture content in the top surface of the mould.

Most blows are fairly large and smoothly rounded in shape and although they are usually found at the top of a casting, the gas producing the defect is more frequently derived from the bottom of the mould and rises through the molten metal until it is trapped beneath a solid skin at the top. There is usually no serious difficulty in allowing gas to escape from the top of the mould, but if the moulding box is placed on a metal plate or concrete floor, both of which have no permeability, the gas has to travel down, then across the box until clear of the casting and finally up to the top of the box. The distance travelled is several times greater than that by the direct path from the top of the mould, and permeability is lowered in proportion. Using boxes provided with a number of holes helps, but the most effective method is to ensure an easy passage through the bottom of the mould as described in Chapter 8.

When moulds are large and the means of escape of gas from beneath the mould are inadequate, pressure can build up for a considerable time and after the mould has been filled there is pronounced disturbance, resulting from the penetration of this gas into the mould cavity. This may cause a scab as well as a blow and the final casting may have three obvious defects: blow, scab and sand inclusions.

Occasionally small deep holes — typical pinholes — may also be due to mould gases. These are found on thin castings of large surface made in moulds in which the sand is brittle. Raising the permeability of the sand has no influence on the incidence of the defect and it has been seen on castings made in moulds where sand permeability exceeded 300. Reducing the moisture content of the sand does not help and occasionally makes matters worse by increasing the brittleness of the sand. The mechanism by which the defect is formed appears to be that the brittle sand cracks when in contact with molten metal. The crack formed is not wide enough to allow entry of metal, but when it is formed, a large area of moist sand is exposed to intense radiation and there is a rapid evolution of gas which leaves a deep and narrow hole in the solidifying casting. This type of defect is particularly serious when the castings are to be finished by a process involving heating the casting, such as vitreous or stove enamelling, since the air in the hole expands and contracts as the temperature changes and there is either a hole, a protuberance or a depression in the finished article. The defect can be eliminated by reducing mould hardness, but in cases where such a change leads to swelling, the nature of the sand should be changed. Cracking of the mould is more pronounced when the sand is coarse and uniform and the moisture content below that most suitable for moulding. In addition, when such a sand contains a clay, such as a kaolinite, which has a narrow moisture range, small variations in moisture content, which are unavoidable in practice, are likely to produce some sand which is a little drier than is desirable. A finer sand with a wider range of grain size should be used, even if this change reduces permeability by half, together with a clay with a wider moisture range.

Gas holes not due to mould condition

Holes resembling blowholes in that they are smooth and rounded, may be due to some factor wholly or partly independent of mould conditions and it is important to recognise that these cannot be eliminated by changes in the sand composition or the method of ramming.

1. *Holes due to high manganese and sulphur in metal*

These have been examined by a number of observers and it has been shown that the incidence of the defect depends on the product of the manganese and sulphur contents and on the pouring temperature. Tonks[8] gave graphs showing the relation between manganese and sulphur contents on the one hand, and the minimum pouring temperature needed to avoid the defect on the other, for one particular casting.

A typical example is shown in Fig. 100, also due to Tonks.

FIG. 100. Defect due to high manganese and sulphur. (After W. G. Tonks.)

The holes may be as large as typical blowholes, but can usually be identified by the presence of a small slag inclusion.

2. *Holes due to high hydrogen content of the metal*
Hydrogen is much less soluble in solid than in molten metal and if the molten metal is saturated with hydrogen, much of the gas comes out of solution on solidification. The gas holes so formed are usually small and, although usually more numerous towards the top of the casting, they may be found anywhere.

FIG. 101. Gas holes due to use of damp ladle. (After W. G. Tonks.)

Figure 101, also due to Tonks, shows pinholes of this kind in a machined casting. These were produced by using a ladle which was not completely dry.

3. *Holes due to mould–metal reactions*
The most frequent cause is a reaction between aluminium and titanium in the metal and moisture in the sand which produces hydrogen. This is superimposed on the hydrogen already present.

The defect was examined by Dawson and Smith[9] who showed that it mainly occurs immediately below the surface of the casting and although an occasional hole may appear after shot-blasting, the majority are revealed only after machining. The holes usually contain a complete envelope of graphite, which can sometimes be detached whole. Typical examples are shown in Fig. 102, also due to Dawson and Smith.

Other elements have a minor influence, but in the foundry aluminium, which may enter with the scrap and is present in many inoculants, is the most frequent cause.

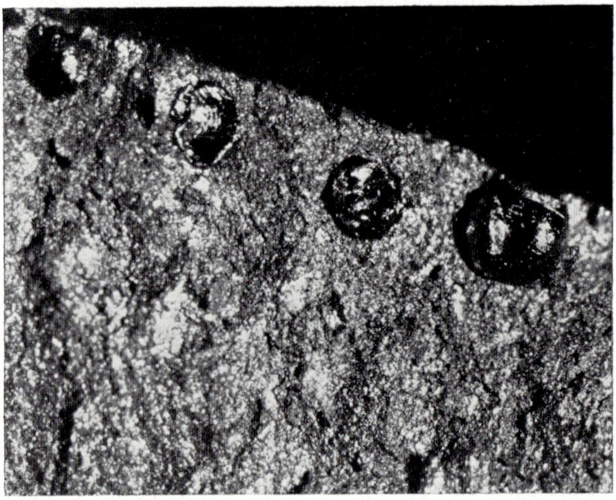

FIG. 102. Pinholes showing graphite film. (After J. V. Dawson.)

Dawson and Smith showed that the composition of the moulding sand has some effect in that the defect becomes more pronounced when the moulding sand contains only very small amounts of coal dust or pitch. In one particular case, raising the amount of coal dust from the unusually low 1 % to the more normal 5 % eliminated the defect. However, the amount of coal dust or pitch depends on other factors and an excess can lead to defects other than pinholes. Raising the amount should not therefore be regarded as the first line of attack on pinholes; metal composition should be examined first.

Other sources

Pinholes, again in this case, mainly due to hydrogen, may result from the use of certain corebinders containing nitrogen. On being strongly heated, these give off ammonia, which dissociates into hydrogen and nitrogen.

The hydrogen dissolves in the molten metal and comes out of solution on solidification.

Nitrogen, itself, can produce unsoundness, but the defect appears as a fissure rather than a round hole and is not likely to be mistaken for a blowhole. This defect is most frequently found in irons with a low carbon equivalent produced in the cupola by melting a charge which consists mainly of steel. The occurrence of this particular defect was studied by Mountford[10] who showed that it could be eliminated by adding to the metal small amounts of titanium. The appearance of the defects is shown in Fig. 103, due to Mountford.

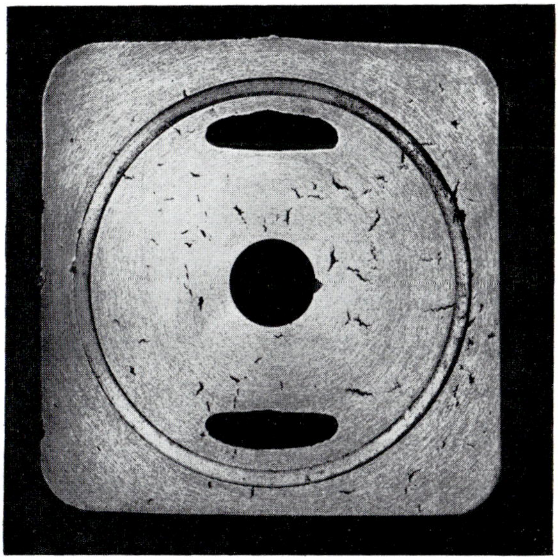

FIG. 103. Defects due to nitrogen. (After F. A. Mountford.)

Solid scabs

A scab is simply a protuberance on the surface of a casting resulting from removal of part of the mould which has then been replaced by metal. There are two kinds of scab, but these have been provided with a wide variety of labels — dumb, blind, hollow, solid, rammer, expansion — with no universally accepted agreement as to what the labels mean.

The two types of scab can usually be distinguished, in that in one kind, which is due to expansion of the surface of the mould, there is a layer of sand between the scab and the parent metal, whereas in the other no such layer exists. Under some conditions, the thin layer of sand which would otherwise separate the scab from the parent metal has been washed away by a moving stream of metal, in which case it becomes difficult to decide whether the original cause of the defect was or was not expansion.

Prolonged erosion of the mould wall by a stream of molten metal may result in the removal of an appreciable amount of sand and it becomes difficult to distinguish such excessive sand-washing from a solid scab. Indeed, the two may be identical in their appearance.

More frequently, the defect is found on the bottom of fairly large castings made in green sand moulds and the usual causes are high moisture content and failure to make adequate provision for the escape of gas from behind the mould face. In either case, pressure increases until it exceeds by a considerable margin the pressure inside the mould cavity and sand is finally forced off the mould face.

The factors which favour the formation of solid scabs other than those due to simple erosion are therefore the same as those which tend to produce blowholes. The defect becomes more pronounced as moisture content rises and mould permeability falls, and is most likely to occur on the bottom face of castings, since it is more difficult to arrange for the ready escape of gas from beneath the mould. As with blowholes, there may be some delay before pressure behind the mould face builds up sufficiently to form a scab and the two defects frequently occur together. Both are due to a high rate of gas evolution and inadequate means for it to escape, and the same methods of elimination are used for both.

(4) EXPANSION SCABS

When molten metal is poured into a green mould, the uncovered surface of the mould is heated by radiation. At the same time the air present in the mould cavity is heated and expands and most escapes through the mould wall, since the volume not filled with metal continually decreases.

The hot air becomes saturated and the mould surface dries out, but the moisture removed from the face condenses when the air reaches cool sand. As the temperature of the sand behind the mould face rises, moisture evaporates from the damp layer so formed and condenses still further back. During the period after casting, a moist layer passes back from the mould face, becoming broader and less intense as it recedes. Its progress has been followed by a number of observers.

However, dry sand has a low heat conductivity and the surface layer may reach a high temperature whilst the wet layer persists immediately behind it. This heated layer expands and since the wet layer is weak and does not inhibit lateral movement of the dry layer, the expansion results in one or a limited number of buckles. When molten metal reaches the top of the mould, it enters a crack at the apex of the buckle, forming a projection which is separated from the body of the casting by a layer of sand, except at the points of entry. The metal or scab can be removed without too much difficulty, leaving a depression at the apex of which the narrow line of the junction can be seen. The sides of the depression are as smooth as the remainder of the casting, but the face of the 'scab', which had been in contact with that part of the mould from which the surface had been removed, is much rougher. The mechanism of scab formation is illustrated in Fig. 104.

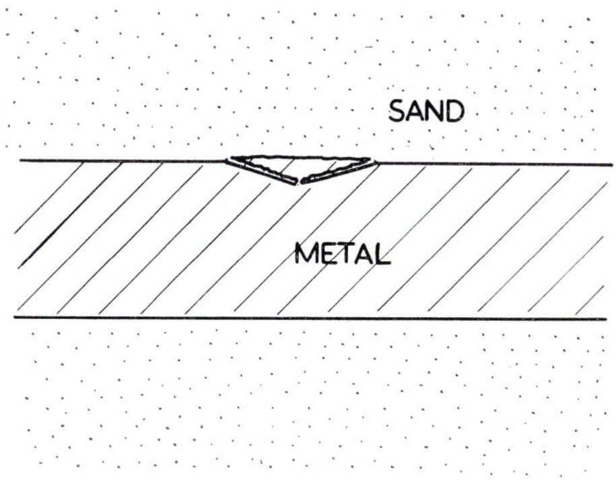

FIG. 104. Formation of a scab.

The presence of a thin layer of sand, separating scab from casting, is characteristic of the scab due to expansion. The sand layer is quite thin. If a mould with a flat top is partly filled with molten metal and allowed to stand for a short time, the surface layer which would buckle and form a scab becomes completely separated and falls on to the molten metal. A hole through the top of the mould permits visual observation and if the top is immediately removed, the detached layer can be examined. The separated layer on the hot metal is shown in Fig. 105. The casting being made was a small die block and the time between pouring and separation about 20 s.

The thickness of the separated layer is about 1 mm, but the depth of penetration into the mould cavity, which depends on width and not on its thickness, is usually very much more. It is therefore impossible to grind out the scab unless it is very small.

A scab of this kind occurs only during the period when the mould is being filled. Once the mould is full, pressure from the molten metal will prevent buckling and the worst which is likely to happen is formation of a rough patch, unless some pressure is exerted from behind the mould face. Such pressure usually detaches the buckled layer completely and, although the effect of expansion may have contributed to formation of the defect, the scab is solid. Moreover, elimination of the defect is best tackled by

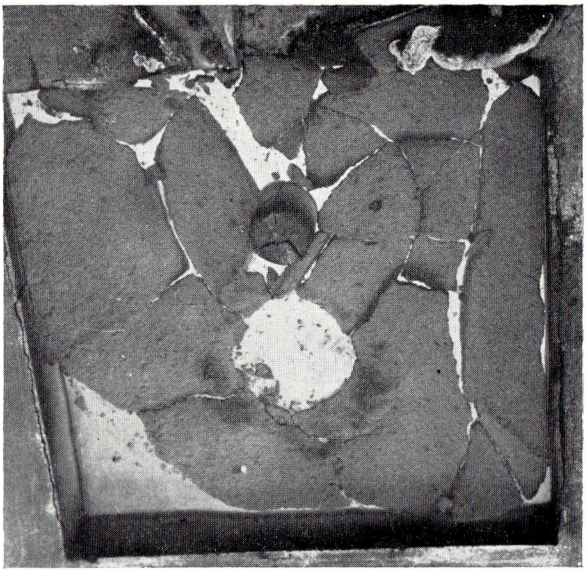

FIG. 105. Separated layer on die block.

removing the cause of the pressure, which is precisely the method used for preventing solid scabs.

In the absence of pressure from inside the mould wall, a scab will not form once the mould is full and an increase in the rate of pouring is sometimes the best method of eliminating the defect, particularly on a thick casting with a flat top.

Alternative form of scab

Another form of scab has been described by Boenisch and Patterson.[11] This kind of scab is also due to expansion, but is formed when the expanding layer of sand is free to move as at point A in Fig. 106.

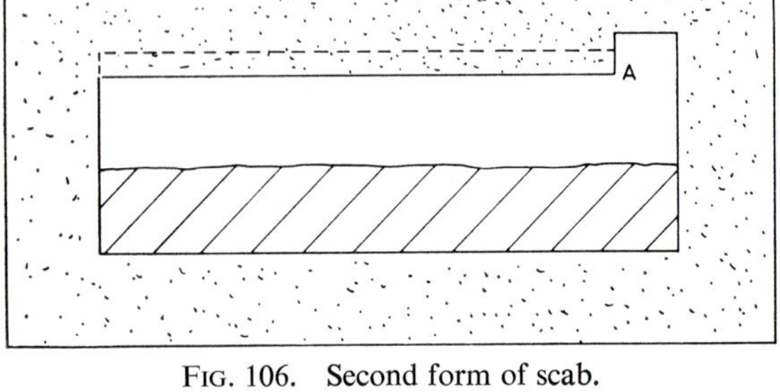

FIG. 106. Second form of scab.

This type of scab is rare and most of the scabs met with in the foundry are buckles.

Prevention of scabbing

There are three methods of approach when considering the elimination of scabs.

1. Reduce expansion of the thin dry layer of sand which fails and forms the scab.
2. Increase the strength of the wet layer between the dry layer and the remainder of the mould.
3. Increase the hot deformation of the dry layer to such an extent that it exceeds expansion.

Usually, elimination of scabbing is not a particularly difficult matter, but the nature of the sand used may limit the usefulness of one particular method; for example, it is difficult to do anything to raise the wet strength when using a natural sand. Under very difficult conditions, it may be necessary to use a combination of two methods and, moreover, some of the methods used under less difficult conditions also have a combined effect.

When making an addition, any side effects on the properties of the sand must be taken into account. Some additions, including those most widely used, reduce permeability, whilst others raise the shatter index, which may lead to uneven hardness distribution and unsound castings unless the method of ramming is suitable for dealing with tough sands.

Reducing expansion

Replacing quartz by a granular material with a lower coefficient of expansion is the only method of reducing the expansion to a marked extent, although other methods have a small effect. The expansion of materials such as olivine, zircon and chamotte is so low that any expansion can be taken up by the deformation without a scab forming.

Two sands with the same grading were prepared from quartz and olivine by milling each for 5 min with 5% calcium montmorillonite and 3% water and used to prepare moulds for a casting 1 ft square and 2 in thick. Mould hardness was 70. The castings were poured slowly, the time for filling being 60 s, conditions which favour scabbing.

The expansion and deformation of the two sands are shown in Fig. 107 and the upper surfaces of the two castings are shown in Fig. 108(a) and (b).

The casting made in olivine sand is completely free from scabs, but the defect is pronounced on that made in silica sand.

Quartz has a very high expansion in the temperature range up to 573°C, but at that temperature changes to the β form and thereafter contracts very slightly. Olivine has a much lower expansion in the low temperature range, but continues to expand more or less uniformly as the temperature is raised further. As a result, the two materials have similar expansion by the time the temperature reaches 1100°C. Nevertheless, no scab forms because at that temperature the clay softens and the rammed sand becomes plastic. A similar mould prepared from olivine sand was filled to a depth of $1\frac{1}{2}$ in with molten steel at a temperature of 1600°C and allowed to stand until the metal was cold, but no cracking of the upper surface of the mould occurred.

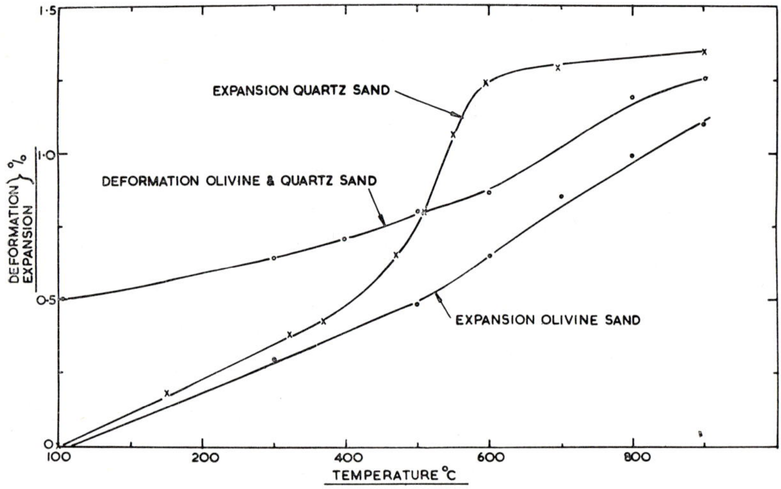

FIG. 107. Expansion and deformation, olivine and quartz sand.

Granular materials suitable for producing sands with a low expansion are too costly to be used widely by iron founders, but they are used on a small scale, particularly by steel foundries, as facings on parts of moulds where metal penetration and scabbing are likely to occur. Chamotte is used by makers of large steel castings.

Some natural sands, particularly the red Bunter sands, contain about 20% felspar, which has a grading very similar to that of the quartz grains. Since the expansion of felspar is low, these sands have an expansion lower than that of synthetic sands based on relatively pure quartz, and the risk of scabbing is lower. Indeed, in the very early days of synthetic sands, their use in some jobbing foundries, previously accustomed to the use of red sands, led to a marked increase in scabs and some time elapsed before synthetic sands lived down the definition of ‘the sand which scabs’. All that had occurred was that precautions, which had hitherto been found

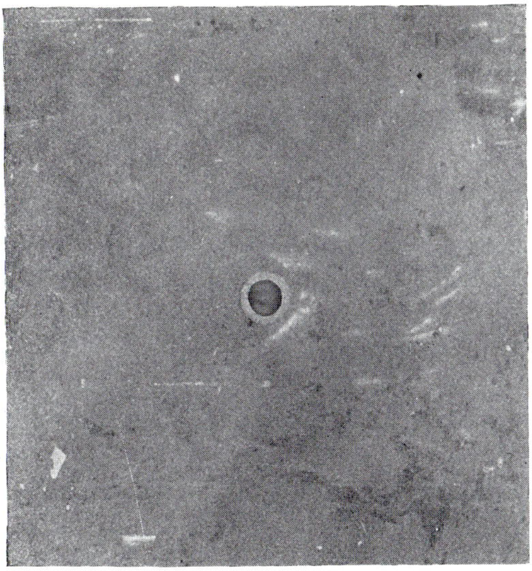

(a)

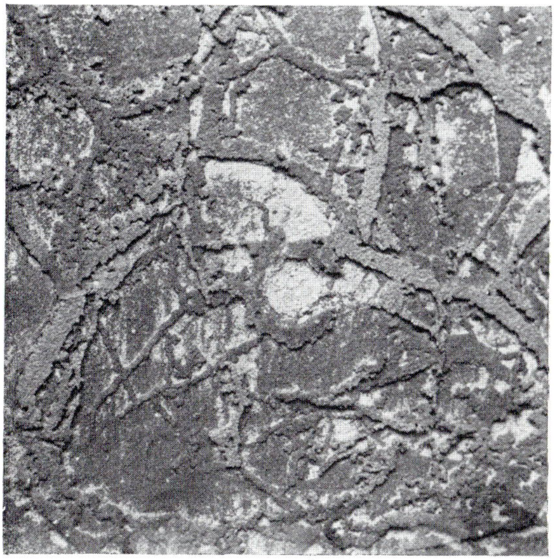

(b)

FIG. 108. Effect of type of sand on scabbing; (a) olivine sand, (b) quartz sand.

adequate when using red sand, were not effective when that sand was replaced by a synthetic sand with a much higher expansion.

The effect of clay

It was shown in Chapter 9 that the expansion of a very hard clay-bonded sand based on quartz grains is the same as that of quartz rock, the clay itself making no appreciable difference. Under some conditions, clay may have an effect which is decisive when scabbing is minimal.

By milling a sand for a long time, for example 10 min, in a laboratory size Simpson mixer, which is a most effective tool, quite high green strength and shatter index may be obtained. If milling time is then reduced and the original strength and shatter index regained by raising the amount of clay used, the sand has a lower tendency to scab.

The difference is in part due to the lower expansion and in part to lower sensitivity to changes in moisture content. Differences in expansion disappear when the sand is rammed very hard, but are significant in moulds other than those made by very high pressure squeezing. In all cases, 'wet strength' increases with increase in clay content.

Prolonged milling thus reduces the amount of clay needed to maintain moulding properties, but may increase the tendency to scab. The effect is sometimes observed in foundries in which the sand plant has changed. If, however, the limit described in Chapter 7, of two horse power for each hourly ton of sand to be treated is not greatly exceeded, reducing milling time and increasing clay content is not the best way of avoiding scabs, since the cost of the clay is likely to be high and control of sand properties difficult.

Increasing the strength of the wet layer

The importance of the strength of the wet layer on control of scabbing was first described by Levelink.[12] At about the same time Patterson and Boenisch[13] built apparatus to measure 'wet strength' under conditions which simulated those found when the face of a green sand mould is heated by radiation or conduction from molten metal. A standard test piece was heated from one end and as a result, the face dried and a wet layer formed immediately behind the dried face. The apparatus was so arranged that the tensile strength of this layer could be measured.

It was found that the highest tensile strength was obtained when the clay used was a montmorillonite in which all the exchangeable cations were sodium. The same result was obtained whether the sodium montmorillonite occurred naturally or was formed from a calcium montmorillonite by treatment with sodium carbonate. Boenisch described such a montmorillonite as 'fully activated'. If an excess of soda was used, wet strength fell and the clay was described as 'over activated'.

Later Boenisch designed test equipment to measure the compressive stress produced in the dried surface layer by the restriction on expansion. He then used the formula (compressive stress/wet tensile strength) to give an index of the tendency of a sand to scab. This formula can apply only to the buckle type of scab, since under the rare conditions shown in Fig. 106, the dry layer is free to expand and the compressive stress is negligible.

The equipment used and results obtained with a number of sands have been reported by Boenisch and Patterson.[14]

The greater resistance of synthetic sands bonded with a sodium montmorillonite to scabbing is recognised, particularly in the steel casting industry, although in this country most steel founders use Wyoming bentonite in which some of the exchangeable cations are calcium in preference to the completely 'activated' montmorillonite. Scabbing is not a serious problem and can usually be completely suppressed by the addition of starch, which has an influence on both the wet strength and the hot deformation.

While the wet strength conception does explain why sands bonded with a sodium montmorillonite have comparatively little tendency to scab, it is of no help to users of natural sands — and although there is a continuing tendency for synthetic to replace natural sands, present usage is split fairly evenly between the two. Whether the nature of the few exchangeable cations present has any influence on the properties of clays such as the kaolinites has not been investigated, but it is likely to be small and is irrelevant. There are slight changes with use in the acidity or alkalinity of foundry sands and some changes take place in the nature of the exchangeable cations. These are small and of little significance when using a bentonite with a base exchange capacity of 100 units, but would be decisive with a clay having a capacity of 10 units. It is not practicable to ensure that a large proportion of the cations are sodium when the total present is 10 units without having at some time an excess of soda.

Some other method of overcoming the defect, independent of the nature of the clay, is needed. Besides being applicable in foundries using natural sands, such a method enables bonding clays other than sodium montmorillonites to be used in synthetic sands.

Increasing hot deformation

If the deformation that the heated mould face can undergo without cracking exceeds the expansion, the mould surface will not crack when heated and no scab will occur. A smaller increase in deformation reduces the tendency to scabbing. With the nomenclature of Boenisch, raising deformation until it is higher than the expansion reduces the compressive stress in the dried layer almost to 0, and there is no tendency of the sand to scab.

It was shown in Chapter 9 that the addition of any substance which melts or partly melts to form a viscous liquid increases hot deformation.

Coal dust does just this and, together with pitch, which behaves in a similar manner, is used far more than all other materials combined to prevent scabbing.

Attempts have been made in the past to eliminate coal dust from the foundry on the grounds that it was a source of much of the dust and dirt present in the atmosphere. Although these attempts were reported only occasionally in the literature,[15] they occurred with considerable frequency at one time. Invariably, elimination of coal dust led to a disastrous increase in the incidence of scabbing and the addition of coal dust had to be resumed.

Coal dust does not increase the wet strength and Boenisch and Patterson concluded from their formula that coal dust did not help to eliminate scabs. This is not in accordance with the facts. A possible explanation of this discrepancy is that when measuring the compressive stress, in their test the sand layer was not heated to a temperature high enough to cause partial melting of the coal dust; otherwise there would have been an appreciable fall in the compressive stress and in the tendency to scab as shown by the formula.

The casting shown in Fig. 108 was made in a mould using the same sand as that used to obtain the mass of scabs shown in Fig. 108(b), except that in one case 5% coal dust and in a second 1% pelleted pitch was added and the mould was filled in 25 s. The castings obtained are shown in Fig. 109.

Coal dust and pitch are equally suitable for preventing scabbing and the decision as to which to use depends on the kind, and particularly the range, of castings being made and on whether the high dry strength conferred by pitch is beneficial or harmful. In general pitch is more suitable when castings varying widely in size have to be made in the same system sand. It is less satisfactory when conditions are such that the poured moulds are left for a long time before shaking out; in such cases the pitch melts and later re-solidifies, raising dry strength to such an extent that there are difficulties at the shake-out and in the sand preparation plant.

Any substance which melts to form a viscous liquid over the appropriate temperature range can be used in place of coal dust or pitch and several inorganic substances have been so used on an experimental scale. They would not be suitable in the foundry, since on cooling down, the melt re-solidifies, producing sand with a very high dry strength which could not be handled in the sand preparation plant.

Under some very difficult conditions, coal dust and pitch are not alone sufficient to prevent scabbing. These conditions are discussed later.

It was shown in Chapter 9 that wood flour increases the deformation of dry sand at all temperatures and is far more effective than coal dust or pitch. All fibrous materials have this effect, but in practice, pliable fibres are more satisfactory than stiff fibres because the latter tend to segregate

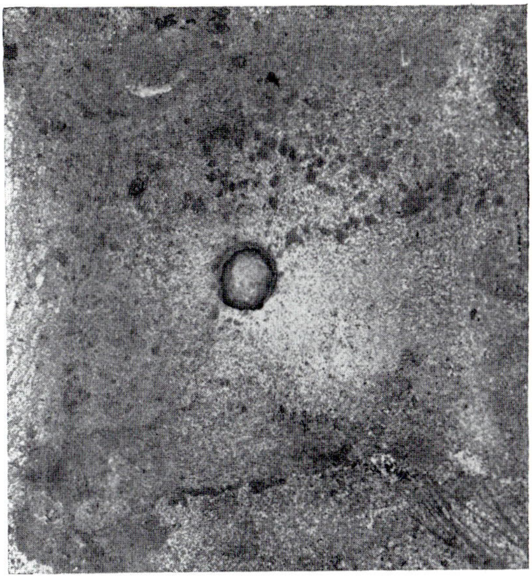

(a)

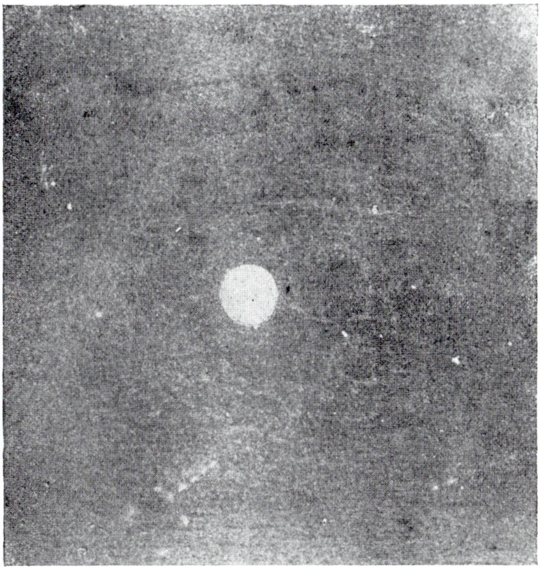

(b)

FIG. 109. Elimination of scabbing by means of coal dust and pitch; (a) 6% coal dust, (b) 1% pitch.

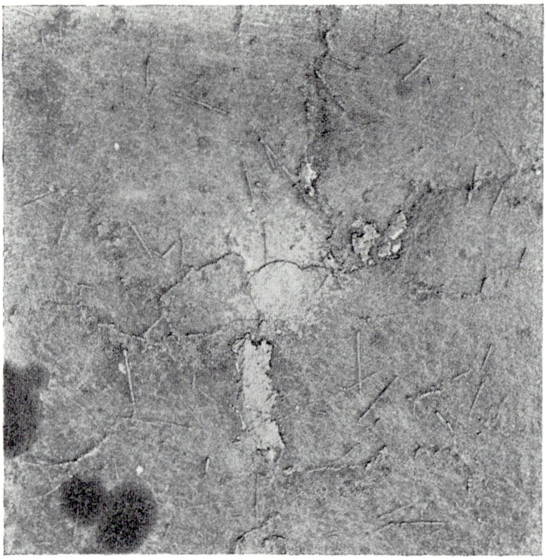

(a)

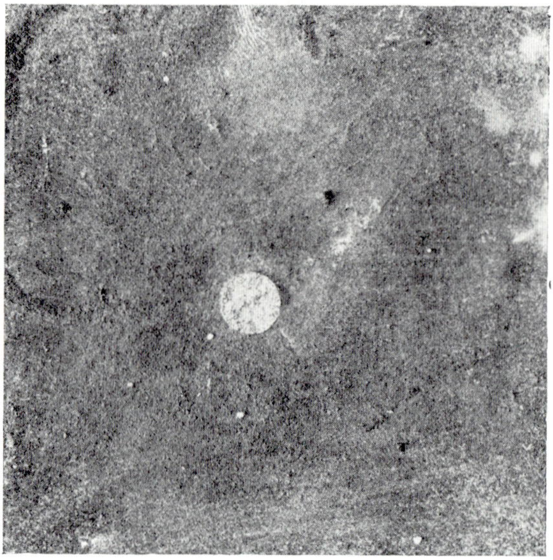

(b)

FIG. 110. Fibrous materials and scabbing; (a) stiff fibres, (b) pliable fibres.

when the mould is being rammed. The sand used to prepare the mould for the casting shown in Fig. 108(b) was mixed with (a) $\frac{1}{2}$% stiff fibres obtained by cutting the stiff stalks of grass bents into $\frac{1}{2}$ in lengths, and (b) the asbestos fibre used by laboratories for filtering, also cut into $\frac{1}{2}$ in lengths. These sands were used to produce the castings shown in Fig. 110(a) and (b). Pouring time was 60 s.

Many of the stiff fibres are spread over the surface of the mould and are ineffective. In consequence, scabbing has not been eliminated entirely although it has been greatly reduced. An examination of the broken mould showed that there was a tendency throughout for the fibres to align themselves at right angles to the direction of ramming. This did not occur when pliable fibres were used and the only limitation here was that if the length exceeded 1 in, there was a tendency for the fibres to collect on the outer plough of the mill.

Materials with a dual effect

The most important example is starch, which is one of the most effective materials for suppressing scabbing.

It was shown in Fig. 48 that starch widens the moisture range of clay-bonded sands and it follows that the fall in strength which occurs when moisture condenses immediately behind the hot face is less when starch is present. At high temperatures, the starch melts and forms a viscous liquid, thus reducing the risk that the dried and heated layer on the mould face will crack. The effect on wet strength is more important.

Other organic binders such as dextrin, molasses and sulphite lye increase hot deformation, but only slightly, since the amount which can be used rarely exceeds 1%. They have no effect on wet strength and are not particularly effective in reducing scabbing.

The addition of even 1% starch raises the shatter index of all clay-bonded sands and that of many sands used in the foundry to 100. This is an advantage when using high pressure moulding machines with some form of contoured head, and in this field starch is particularly useful. When, however, the moulds are rammed on squeeze machines using a flat head or on most jolt–squeeze machines, differences in hardness between different parts of the mould are increased and there is an increased risk of unsoundness in the castings, due to the presence of soft areas in the mould.

Effect of size distribution of sand grains

It was shown in Chapter 3 that sands composed of grains all of one size have a high permeability. At the same time, these sands are brittle when bonded with clay and dried, with a deformation appreciably lower than that of a sand with a wide grain size distribution. They therefore have a greater tendency to scab. Moreover, when a mould face is heated by radiation, uniform grains reach the temperature of 573°C and undergo a sudden expansion together. The shock due to this expansion is then at a maximum and this factor may also increase the tendency to scab.

The 'four screen sand', *i.e.* sand with the major part of the grains spread more or less evenly over four neighbouring sizes in the AFS series, which is widely recommended in America, is a compromise between the uniform sand with its higher tendency to form scabs and the sand with a wide grain size distribution with its low permeability.

Difficult castings

Scabbing is not a serious problem in the majority of foundries and the appearance of the defect is frequently an indication that the amount of coal dust should be raised. There are, however, conditions which are so severe that coal dust alone is not effective and increasing the amount used beyond that normally present merely changes the appearance of the scab. Four kinds of mould give particular difficulty.

1. Very hard moulds.
2. Moulds with a flat top, which are poured slowly.
3. Moulds in which metal flows over a raised cod of large area.
4. Moulds, particularly those of considerable size, in which the heated face is free to expand as in Fig. 106.

1. *Very hard moulds*

As mould hardness increases, expansion also increases, but deformation to fracture of both green and dried sand decrease. The incidence of scabbing therefore increases and the defect occurs on castings on which no defect would be expected when using a clay-bonded sand containing 5% or so of coal dust and ramming on a jolt–squeeze machine.

An example of this has been described by Nicholas and Outhwaite.[16] The casting concerned was one of the usual type of motor end shield, a casting made in large quantities using moulds containing about 5% coal dust and a variety of natural and synthetic sands and rammed on jolt–squeeze machines. An occasional scab may occur if there is a pronounced fall in the amount of coal dust present, but the defect is rare.

Serious scabbing occurred when the mould was made in a synthetic sand containing 7% of a sodium bentonite when the mould was made on a high pressure squeeze machine. Raising the amount of bentonite to 12% reduced the size of the defect somewhat. The addition of wood flour eliminated the defect when the sand contained 7% bentonite, and peat, starch and dextrin, in descending order, were somewhat less effective. Coal dust and pitch were effective when the sand contained 12% bentonite, but no direct comparison with wood flour is possible owing to the difference in clay content.

Usually, coal dust or pitch will be found to be effective with a majority of castings, even when made in hard moulds. When they are not adequate, wood flour should also be added. For the reasons given in Chapter 6, it is safest to aim at 1% wood flour, particularly when the moulds are rammed hard. The rammed sand shows traces of elasticity when the amount of wood flour reaches 2% and this would lead to dimensional inaccuracy.

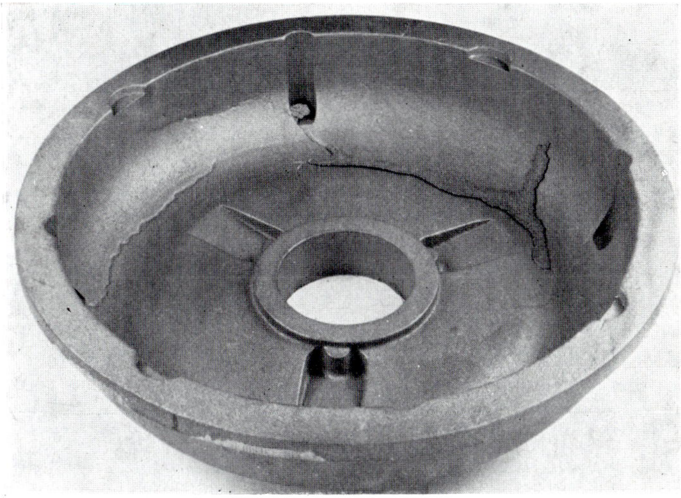

FIG. 111. Scabbed end-shield casting. (After Nicholas.)

The increased tendency to scab which follows hard ramming can be seen from the type of casting on which Nicholas obtained severe scabbing when using a sand containing 7% bentonite and a water content of only 2·7%. This is shown in Fig. 111.

2. *Moulds with a large flat top poured slowly*

Under these conditions, a mould containing pitch or coal dust may scab, but the appearance of the scab changes and a somewhat different mechanism seems to be involved.

When the pouring time of the casting shown in Fig. 109(b) and made in sand containing 1% pitch, was raised to 60 s, a large scab appeared, but its appearance differed from that of the scabs on moulds without pitch, obtained when pouring in half the time. The scab was raised further above the casting, probably because the wet layer had moved further back into the mould, and the depression in the casting was rounded. The appearance suggested that the dried layer, weakened by the presence of a large amount of liquid, had sagged under its own weight. Increasing the amount of pitch would be expected to cause a still greater fall in strength and a still

(a)

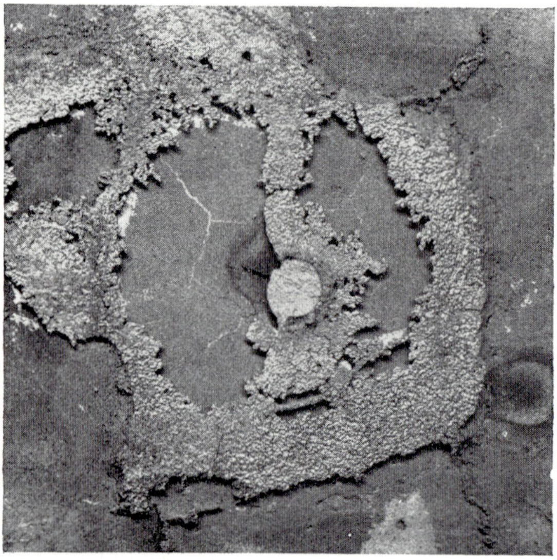

(b)

FIG. 112. Scab obtained with very slow pouring; (a) 1% pitch, (b) 2% pitch.

deeper scab, and this effect was obtained when the pitch content was raised to 2%.

The two castings are shown in Fig. 112.

Under these conditions, the amount of coal dust or pitch should be reduced to the minimum required to give a reasonable surface, and 1% wood flour added.

3. *Metal flowing over a large cod*

Scaled-down test castings are not particularly useful for studying conditions such as this because a number of difficulties arising from mere size cannot be reproduced. Perhaps the most useful model is the ordinary household bathtub, which is produced in very large quantities.

There are two methods of running. The bath is invariably cast upside down and with the one method the metal flows through two downgates into two horizontal runners, one on each side of the bath and extending its whole length so that metal enters at the bottom of the mould along each side. From the point of view of avoiding casting defects, this is the simplest method, since the impact of the falling metal is taken outside the casting, dynamic pressure inside the mould cavity is small and uncovered sand is not strongly heated. The drawbacks of the method are that larger boxes must be used to make room for the runners, more sand must be used and the casting yield is lower.

When metal flows directly on to the top of the cod of sand forming inside of the bath and flows in streams down its side, the risk of erosion is increased, the mould has to be hard enough to resist dynamic as well as static pressure, and uncovered parts of the mould alternate with those strongly heated by actual contact with molten metal. Under these conditions, it might be expected that the expanding sand beneath a stream of metal would extend laterally and 'ride' over the weaker undried sand at the side, forming a scab similar to that obtained under the conditions shown in Fig. 106, but in practice scabs which form are buckles.

Under such conditions, prevention of scabbing is difficult. If a synthetic sand bonded with a sodium bentonite is used without any additions, serious scabbing occurs even when the pouring time is less than 5 s, while slower pouring spreads the defect over a large proportion of the surface. In one particular case, a bath mould was prepared from a sand containing 6% Wyoming bentonite and poured in 6 s, not a particularly slow pour, and the total length of typical buckle scabs on the sides, all of them approximately 1 in width, was about 30 yards. In addition there were shorter and wider scabs near the runner.

The first requirement is to reduce the pouring time to below 5 s. Such rapid pouring increases the dynamic pressure and if mould hardness is 70 or below, the casting will be overweight and there is likely to be a waviness on the vertical sides which shows up when the enamelled casting is viewed at a suitable angle. When this defect can be accepted, it is possible to get by with an addition of coal dust, but otherwise it is necessary to raise mould hardness to 75–80 and coal dust is not successful. The scabbing can be eliminated by adding wood flour, but the amount needed, which depends on the design of the bath, may be as much as 2%.

Whenever metal flows over green sand, and particularly when only part of the affected area is covered, a scab is likely. Conditions are rarely as severe as when a bath, with its large surface area, is being made, but the remedial measures are always the same — pour fast and when coal dust does not suppress the defect, add wood flour or a fibrous material.

4. *Moulds with areas which are free to expand*

Scabs which occur on the kind of casting illustrated in Fig. 106, although rare, are difficult to deal with when they do occur. Boenisch showed that the risk was reduced by using a synthetic sand in which all the clay was a sodium montmorillonite, but this procedure is not completely certain. Moreover, foundries using a natural or semi-synthetic sand cannot make use of this finding, unless they change their practice or use a synthetic facing sand for the mould concerned.

Since the expanding layer of sand is restricted at one end only, and is free to expand at the other, the addition of a material such as coal dust or wood flour, which increases the deformation to rupture of a sand section restricted at both ends, is of no help.

The only effective additives are fibrous materials of sufficient length to reach from the dried out face of the mould, through the wet layer, to the unchanged sand. Pliable fibres are more satisfactory than stiff ones, owing to the tendency of the latter to arrange themselves in a direction parallel with the mould face. The length of the fibres should be several times greater than the distance between mould face and unchanged sand; otherwise only those which are at right-angles to the mould face are effective.

Scabbing on dry sand moulds

The object of drying a mould is to reduce the risk of mould failure due to one of the causes described in this Chapter and, since these risks become more serious as the size or weight of the casting increase, drying is usually reserved for the foundry making large work. Drying increases strength, rigidity and resistance to erosion and greatly reduces the amount of gas evolved.

Drying also greatly reduces the risk of scabbing, mainly because it suppresses, at least partly, the wet layer which forms behind the mould face when a casting is poured. The only water available when the mould is dry is the combined water in the clay, which is much less than the amount present as free moisture in green sand and is given off only slowly, at temperatures exceeding 400°C. When the dry sand mould is poured, a small amount of well-superheated steam travels away from the mould and any wet layer which may form is farther away from the mould face, more diffuse and less intense than when the mould is green.

Nevertheless, owing to the large area of sand which is heating, cracking may occur unless the mould can deform. In view of the very small effect of any moist layer formed, the use of a sodium montmorillonite as a bond offers little advantage, and in any case, many makers of heavy castings prefer to use a natural or semi-synthetic sand on account of the wider spread of grain size obtainable and the resulting lower permeability and greater resistance to metal penetration.

Coal dust and pitch are rarely added to dry sand moulds and one of the fibrous materials is preferable. Wood flour and even coarse sawdust are frequently used, but both have the drawback that they tend to produce a mould with a friable surface. Horse manure was at one time almost universally used as the cheapest source of pliable fibres and the laments of old moulders concerning its virtual disappearance are mainly because the two fibres which have frequently been provided as alternatives — chopped straw and chaff — are stiff and tend to tear the surface of the mould, particularly when a strickle is used. The other fibres described in Chapter 6 are more satisfactory.

Prevention of scabbing

The defect rarely causes serious trouble, even on green sand moulds, except when certain difficult castings are being made. The following are the most important factors in suppressing the defect.

1. The rate of pouring should be as fast as possible. When the defect occurs on only one or two types of casting, it is frequently simpler to alter the gating system than to prepare a special sand or change all the sand in the foundry.
2. The less refractory natural sands have a lower tendency to scab. The decision on which kind of sand to use is controlled by factors other than the risk of scabbing, but should a change be made in the kind of sand used for any purpose, the possibly increased risk of scabbing should be considered.
3. Among synthetic sands, those containing a clay consisting wholly of sodium montmorillonite have the lowest tendency to scab.
4. Prolonged milling in order to produce a mouldable sand with a minimum amount of clay slightly increases the tendency to scab.
5. In most cases, scabbing can be eliminated by the addition of coal dust or pitch, but where pouring is slow and a large flat area of sand is heated by radiation, a scab may form which is much larger than usual. In this case, increasing the amount of coal dust or pitch produces a still deeper scab.

6. In difficult cases, including that described in 5, wood flour or a fibrous material should be added.
7. In large dry sand moulds and in green sand moulds when the scab is of the kind due to expansion of an unrestricted part of the mould face, only the longer fibres are effective.

Rat tails

A rat tail is a narrow depression, usually aligned along the edges of the incoming metal stream, especially on the bottom face of a flat mould. Sometimes a long scab may narrow at one end to form a typical rat tail and the defect, like the scab, is due to expansion.

The defect occurs mainly on very hard moulds with flat surfaces and it can be eliminated by the methods used for getting rid of scabs. The most effective additions to the sand are the fibrous materials and wood flour, followed by starch, coal dust and pitch. When a synthetic sand is used, a sodium bentonite is to be preferred to any other clay, but its use does not guarantee freedom from rat tails unless one of the additions given above is used.

Alternative form of rat tail

A rat tail identical in appearance was well known in the light castings foundries, particularly in the Falkirk area, at the time when small domestic gas or electric cookers were an assembly of iron castings. The defect was confined to the flat plates forming the sides or door, but there were features which indicated that the mechanism of formation might differ from that involved when scabs and rat tails form on hard moulds.

The following are the most important features.

1. The defect occurred only on the thinnest castings and was infrequent when the thickness was as much as $\frac{1}{4}$ in.

2. The moulds concerned were rammed by hand and laid out on the floor for pouring. They were not particularly hard.

3. A facing sand was used, but whether this was milled or hand-mixed seemed to make no difference.

4. The incidence of the defect was sporadic, but the number of castings affected was large. It was unusual to have an occasional defective casting.

5. Changes in the amount of coal dust added or in the method of running had no obvious influence.

When the defect arose, the incidence was so severe that only obvious changes could be made — increase in the amount of coal dust and change in the method of running. Eventually the defect disappeared, but frequently re-appeared some time later, when it was sometimes ‘cured’ by going back to the original amount of coal dust and the original runner system.

The defect finally disappeared, but only because the thin castings were replaced by steel sheet and the true cause was not unearthed. Since the same natural sand and the same amount of coal dust were used, both when the defect was present in almost disastrous proportions and in its absence, and there was no tie-up with moisture content or other sand properties, it is difficult, in retrospect, to find a direct relation between the defect and the sand used. However, in all the foundries where the defect occurred, a separate facing sand was used and there was no record of the proportion of new sand, and there is some not very conclusive evidence that sand in which dead clay is absent, or present in very small amount only, may be more prone to expansion defects. On one occasion, there were numerous rat tails on thin places cast in moulds made from a red sand in which there was more than 50% of new sand. These disappeared when the sand had been re-used a number of times without adding any new sand.

A similar effect has been reported on one or two occasions in connection with scabs on steel castings. Sand recovered by dry cleaning has been found to have a slightly lower tendency to scab than completely clean new sand under the same conditions.

REFERENCES

1. Hoar, T. P. and Atterton, D. V. (1950). *J. Iron Steel Inst.*, **66,** Sept., pp. 1–17.
2. Longden, E. (1932). *Foundry Trade J.*, **46,** Jan., p. 395.
3. Womochel, H. L. and Sigerfoos, C. C. *Trans. Amer. Foundrymen's Soc.*
4. Greenhill, J. M. (1962). *BCIRA J.*, **10,** No. 2, pp. 163–165.
5. Toriello, L. T. and Wallace, J. F. (1963). *Modern Castings*, **44,** Aug., pp. 404–417.
6. Nicholas, K. E. L., Roberts, W. R. and Guthrie, A. (1964). *BCIRA J.*, **12,** No. 6, Nov., pp. 808–819.
7. Morgan, A. D. (1963). *BCIRA J.*, **11,** Jan., pp. 20–57.
8. Tonks, W. G. (1956). *BCIRA J. Res. Devel.*, **6,** June, pp. 214–225.
9. Dawson, J. V. and Smith, L. W. L. (1955). *BCIRA J. Res. Devel.*, **6,** pp. 226–248.
10. Mountford, F. A. (1965). *BCIRA J.*, **13,** pp. 230–248.
11. Boenisch, D. and Patterson, W. (1966). *Modern Castings*, Oct., or (1967). *British Foundryman*, **LX,** Nov., pp. 438–447.
12. Levelink, H. C. (1958). *Giesserei*, **45.**
13. Patterson, W. and Boenisch, D. (1958). *Giesserei*, **45,** p. 465.
14. Boenisch, D. and Patterson, W. (1967). *British Foundryman*, **LX**, Nov., p. 438; also (1966). *Modern Castings*, Oct.
15. Lawrie, W. D. (1955). Discussion on paper by Parkes and Godding. *Foundry Trade J.*, **99,** Oct. 27, p. 473.
16. Nicholas, K. E. L. and Outhwaite, J. G. Unpublished work at BCIRA.

CHAPTER 11

Oil–Sand Cores

The term 'core' was originally applied to the rammed sand form used to shape the interior of a hollow casting. Sometimes the design of the casting is such that the interior can be shaped by means of a projecting section of the mould, which is then said to 'leave its own core', but more frequently, the core is made separately and later inserted in the mould.

Such 'cores' may also be inserted so as to shape part of the exterior of a casting, usually in order to simplify and speed up production. For example, when a vertical face of a mould has a number of projections or recesses, it may be quicker and safer to make an oversize mould with a flat face and insert a core, than to have the appropriate part of the pattern made as a loose piece. Sometimes the whole of the mould may consist of a number of separately made sections fastened together, usually without a box, and the complete mould is known as a core assembly.

The methods of production and many of the essential properties are the same whether the 'core' is for internal or external use and it is convenient to consider all such separately prepared sand masses as cores.

Occasionally, cores are made in green sand, but this practice is limited to those of moderate weight and uniform section. Large, bulky cores may be in dried clay-bonded sand, but for the most part, the properties needed cannot be obtained by the use of clay as a binder.

THE ESSENTIAL PROPERTIES OF CORES

Owing to their often complicated nature and the conditions of use of internal cores, the requirements of the separately made core are far more diverse than those of the mould. Each part of the process from ramming up to final removal from the casting presents its own problem.

Ramming

The first need is for a sand which can be easily rammed and if any form of mechanical ramming is used, the shatter index must not be too high. Even so, the methods which are successful with clay-bonded moulds, such as slinging, jolting and squeezing, can be used only for very simple shapes and elsewhere, the only successful methods are blowing or 'shooting'. When very tough sands are used, it is necessary to ram the cores by hand.

Stripping

A core must be capable of being removed from its box when the only force acting on it is that due to its own weight, a more difficult operation than forcing a mould contained in a box away from the pattern.

Handling

At a later stage, the core must be handled when it is placed in a mould and any sand grains detached during this operation are likely to appear in the casting.

Accuracy

In some cases, a number of cores may have to be assembled and any inaccuracy in the base core may lead to much greater errors elsewhere. Although it is truism to say that all moulds and cores must be as accurate as possible, it is equally true to say that some cores must be still more accurate.

This is one of the greatest difficulties of the core maker. In order to obtain a good strip under the conditions existing, a high deformation is necessary; to make ramming possible, high toughness must be avoided. Strength must therefore be low and a decrease in strength and increase in deformation limit the accuracy attainable.

Permeability, surface finish and metal penetration

Many cores are almost completely surrounded by metal and any gases given off when the core is heated must escape through the comparatively small area of the core prints. The effective permeability must be extremely high.

At the same time, surface finish must be reasonably good and metal penetration must be avoided.

These requirements are directly opposed and it is frequently necessary to increase the effective permeability by venting and reduce the possibility of metal penetration by coating the core.

Erosion

The actual strength of a core is so high that erosion is not a problem so long as the strength is maintained. In practice, the bonding materials used in the oil–sand core are organic and these break down when heated to high temperatures. Premature breakdown may lead to a complete failure of the bond in the surface layers and to serious erosion.

Distortion

Many cores are quite thin and of considerable length and in use, may be supported only at opposite ends. Under these conditions they may sag under their own weight; when the core is immersed in molten metal, the upward thrust is three or four times greater than the weight of the core and the risk of distortion increases on that account. Frequently, conditions are such that a sand core has not the rigidity required and auxiliary means have to be used, such as putting a wire in the core or using chaplets to act as distance pieces, but each of these methods leads to some other risk of failure and should be used as sparingly as possible.

Breakdown

When the casting has cooled down, the core must be removed. Frequently, the shape is such that no direct force can be applied beyond that which can be obtained by vibrating the casting, and for this to be effective the binder must distintegrate almost completely. Under these conditions the presence of even a small proportion of clay in the sand mixture might make core removal impossible.

THE RELATION BETWEEN THE COMPOSITION OF OIL–SAND CORES AND THEIR PROPERTIES

The requirements given above can be met by using a clay-free silica sand with an organic binder. In practice, there is no single binder which can be used to give the very high strength needed in the finished core and also the properties needed to enable the core to be stripped. Two binders have to be used — a drying oil which develops very high strength after baking, and a green binder, usually a carbohydrate, to give moulding properties.

As in the case of clay-bonded sands, each of the constituents affects certain properties of the mixed sand and of the cores, and although a property which is mainly controlled by one may be slightly influenced by another, it is convenient and in no way misleading to consider the three constituents — sand, oil and green binder — separately.

1. The sand grains

The sand grains control permeability and surface finish. They have an effect on strength, both in the green and in the baked condition, although these properties are controlled mainly by the binders used, and under some conditions, can influence the severity of finning, a defect arising from penetration into a crack in a core.

Permeability and surface finish

The relation between grain size and distribution on the one hand, and permeability on the other, is the same whether the sand is coated with clay or an organic binder. This has been discussed in Chapter 3. The rule that surface finish deteriorates as permeability is raised holds good.

Green strength

In the absence of clay, the green strength of oil–sands is very low — so low that the weight of a core more than a few inches tall imposes on the lower layers a load higher than its green strength. However, the mixed sand is so plastic that the core does not immediately collapse, but flows slowly, like a liquid of high viscosity. Eventually, complete collapse would follow, provided that no change in strength occurred, but in practice, the surface layers lose water on standing, as a result of which they become so strong that a dried layer a few grains in thickness prevents any further movement. However, by the time this has occurred, the core has sagged and lost its original shape.

The extent of sagging depends on the rate of movement of the core and of development of the dried layer. The sand grains influence the rate of movement in two ways.

1. When the sand consists of fine grains or a mixture of coarse and fine grains, the number of points of contact is greater than when only coarse grains of uniform size are used. Strength is slightly higher, but the most important factor is that internal friction is higher and the rate of movement slower. As the rate of air-drying on the surface is little changed, sagging is less.

2. As the grain size of the sand increases, the amount of moisture required to develop the plasticity of the green binder decreases slightly. A similar relationship can be seen in clay-bonded sands, but is not of great importance in a sand containing 3–5% water. Many oil–sand mixtures need little more than 0·5% water and under these conditions, an increase in grain size without a reduction in moisture content, may lead to an increase in sagging.

Baked strength

The baked strength of a core is affected by the sand grains in two ways, due respectively to total surface area and to grain shape.

Surface area. The amount of drying oil used is small, usually about 1·0%, and this has to be spread over a very large surface area. Fine sands have a large surface area and a correspondingly large number of points of contact, but a thin coating. These factors tend to cancel each other out unless a very fine grained material is present, when it becomes difficult to maintain a complete coating and strength is reduced. Quite small amounts of silica flour and, even more, clay seriously reduce baked strength.

Differences between sands differing in size are small provided material less than 50–100 μ in diameter is absent, but a mixture of a coarse and a fine sand, each containing the same proportion of oil, has a higher baked strength than either alone.

Grain shape. The importance of grain shape was first pointed out by Robertson and Emodi[1], who also described a method of measuring it. They described the property as 'rugosity', a term which embraces both surface roughness and grain shape, but differences in the roughness of the sands used in core making are usually small and the most important factor is shape. The calculation relies on the fact that a sphere has a smaller surface area than another particle of equal volume, but different shape.

The method was developed further by Davies and Rees[2], who also measured the angularity of a number of silica sands used in core making.

Care is needed in the interpretation of the results obtained by the method used. The coefficient of angularity is defined as the ratio of the total surface area of the sand grains to the total surface area of the same number of grains, if each grain was spherical and had the same volume as the corresponding grain in the sand under examination. The total surface area can be measured by means of an air permeability method, originally devised for measuring the total surface area of powdered cement, and apparatus suitably modified for dealing with sands can be obtained from suppliers of sand testing equipment. The corresponding area of the true spheres is calculated, assuming that the diameter of the grains is halfway between the aperture size of the sieve through which the sand has passed and that on which it is caught. This may be a dangerous assumption.

For example, 80% of Southport sand is caught on a 152 μ (100 mesh) sieve with about 10% on each of the next coarser and next finer sieves in the series. The 'average' grain size of the main fraction is about halfway between 211 and 152 μ. One of the Leighton Buzzard sands contains rather under 10% caught on the 152 μ sieve, with the remainder on the next three coarser sieves. In this case, the finest fraction of the sand only just passes the 211 μ sieve and its diameter is only a little less than 211 μ. If the two fractions from the two sands caught on the same sieve are examined side by side under a low-powered microscope, the difference in size is clear. For the purposes of the above calculation, the two fractions are the same size, but in fact, the fraction from the Southport sand has the larger surface area, not because of greater angularity, but because of smaller grain size. An even greater difference would have been observed if the fraction from the Leighton Buzzard sand had been compared with a similar fraction from a sand containing only a small proportion on the 152 μ sieve with the remainder passing through.

Davies and Rees showed that angular sands are weaker than rounded grains, particularly in tension, the difference being much less in compression, and pointed out that this was the result of 'angle to flat' contacts such as that shown in Fig. 113(a) which predominate over the 'flat to flat' shown in Fig. 113(b) which is obtained with rounded grains.

By packing the angular grains in the appropriate manner, it would be possible to obtain a dense compact, with many 'flat to flat' contacts, which would have a low permeability and high strength in compression and tension, but this form of packing can be obtained only by placing the grains one at a time. With random packing, there are many large voids and the permeability of an angular sand is higher than that of a rounded sand and the strength, even in compression, somewhat lower. The most pronounced difference is in the lower tensile strength.

Surface roughness. Robertson and Emodi referred to the surface area ratio as a 'coefficient of rugosity', implying that surface roughness is a significant factor, but the work of Davies and Rees showed that roughness was not important in any of the wide range of silica sands they examined.

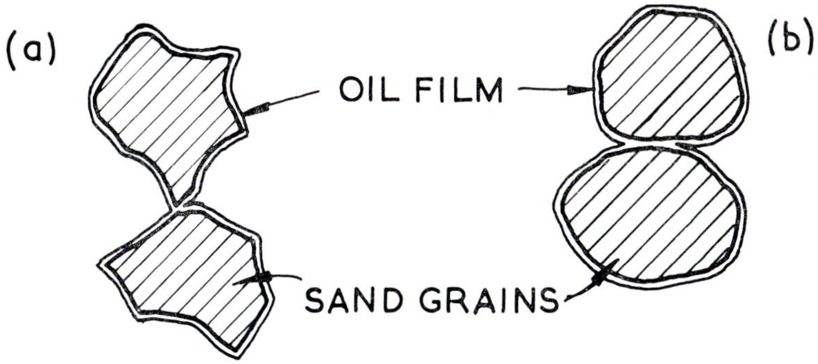

FIG. 113. Contact between round and angular grains.

If, however, the grains are first deeply etched with hydrofluoric acid, there is an appreciable fall in strength, both in tension and compression, although it is difficult to find evidence as to the mechanism involved. It may be that when the grains are smooth, but not when they are rough, capillary action draws part of the binder to the points of contact between grains, or that a significant proportion of the very small amount of oil is lost in the depressions on a rough surface. The green strength of a clay-bonded sand is rather higher, not lower, when the grains are etched, but in this case, there is a much thicker layer of a less mobile binder.

Whatever the cause, we can agree with Robertson and Emodi that rough grains would have a lower strength, and with Davies and Rees that the sands used for core making are too smooth for the difference to be significant. The most important factor is grain shape, and two points are of particular importance.

1. Very angular sands should be avoided whenever possible in the core shop.

2. When comparing the strength of cores with different base sands, the tensile test should be used rather than the compression test, since it shows more clearly any differences due to angularity. Moreover, stresses on cores are usually in transverse, when the stress has a major tensile component.

Sagging

Sagging has a serious effect in the foundry and is one of the major reasons for the replacement of the oil–sand core by cores cured in the core box.

Sagging depends mainly on the rate of air-drying which is controlled by the type of green binder used and by atmospheric temperature and humidity, but the size of the sand grains has an important influence, and Sheehan[3] and Curry[4] have reported extensive sagging in the core shop when a coarser sand replaced a finer one. There are two reasons for the increases reported.

1. The mixed sand can be regarded as a viscous liquid in which are suspended large numbers of solid particles. Whenever two solid particles are in contact, friction between them reduces the rate of flow. The greater the number of solid particles, the greater the number of points of contact, and fine sands flow more slowly than coarse and sag less.

2. Whatever type of binder, clay or an organic substance, is used, an increase in surface area increases the amount of moisture needed to develop green plasticity. Since the total surface area in a given mass of fine sand is greater than that of the same mass of coarse sand, the finer sand requires a higher moisture content. The difference is particularly apparent on sands bonded with an organic binder such as dextrin or molasses, both of which are very sensitive to small changes in moisture content. If, therefore, a fine sand is replaced by a coarser one and the moisture content and bonding material remains constant, the sand will appear to be damper and behave in a manner very similar to that found if the sand and binder had been unchanged and the moisture content raised. There will be an increase in sagging.

The example reported by Curry is particularly interesting. The sand used had a fairly wide range of grain size and after drying was stored in a deep hopper. Segregation occurred and it was found that when coarser sand was withdrawn from the hopper, sagging increased; when segregation was prevented, sagging returned to normal.

Even a small amount of clay raises green strength and reduces sagging, and the preference of some core makers for a Congleton sand with about 0·5% clay — and only part of that true clay — rather than a completely clay-free sand, is due to this factor.

Finning and veining

Under some conditions, a core may crack when it comes into contact with molten metal and a fin may form as a result of the entry of metal into the crack, as illustrated in Fig. 114.

The fin usually has a narrow base and tapers to a sharp edge. On rectangular cores it usually occurs at an edge, but can also occur on round or flat surfaces. Castings of copper base alloys, in particular, may show a network of fins over the whole of a flat surface, when the defect is known as veining.

The defect is due to the expansion of the sand grains and disappears when the silica sand is replaced by zircon sand or other granular material

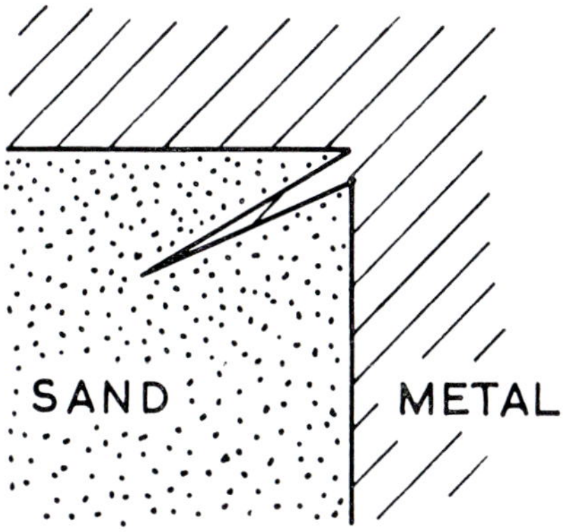

FIG. 114. Formation of fin.

with a low coefficient of expansion. The incidence of the defect is influenced by both the sand grains and the binder, but the exact mechanism by which it occurs is still not clear, and in practice, the defect can be very troublesome should it occur in a sensitive position on the casting. Many experiments have been carried out without resulting in complete agreement among observers, but most agree on some points.

1. A fin is more likely to occur when the sand grains are of uniform size. Figure 115 shows two H section castings, the flats being 4 in apart, in which the channels were formed by cores. The core for the casting shown in (a) was made from Somerford sand, 60% of which remains on one sieve, with most of the remainder on the two adjacent sieves. The core for (b) was made from a mixture of Somerford sand with two Leighton Buzzard sands, one coarser and one finer, the mixed sand being spread more or less evenly over five sieves.

The casting cored out with the use of uniform sand has a fin along its whole length; no fin occurs when the sand was spread over five sieves.

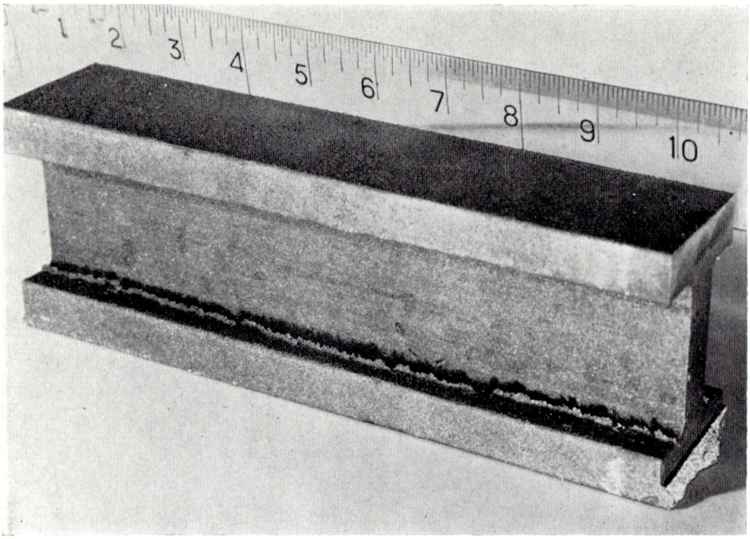

(a)

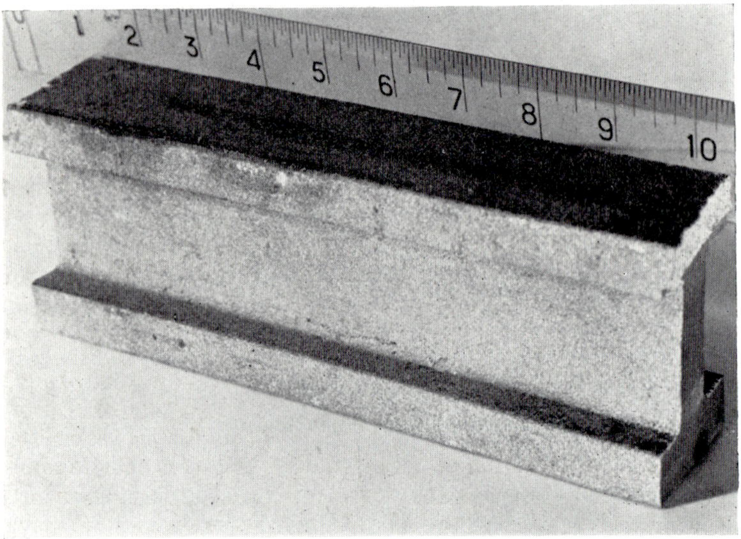

(b)

FIG. 115. Fin on casting. (a) Fin resulting from sand of uniform grain size; (b) absence of fin with sand of wide grain size distribution.

2. Sand grains consisting only of quartz sand are more likely to fin than less pure sands. This is to be expected, since quartz has a higher expansion.

Southport sand has a lower quartz content than any other of the silica sands widely used by core makers. It is also the most uniform in grain size. It appears that the lower expansion is decisive.

On the other hand, serious finning has occurred when using Southport sand.

3. The addition of finely divided iron oxide reduces finning, although it is doubtful whether the chemical composition of the powder is relevant. In one large foundry, additions of iron oxide and of silica flour were used alternately over a period of years without leading to a definite conclusion as to which was the more effective. Either addition reduced the incidence of the defect, but neither was completely successful. In view of the objections to the use of silica[5–7] when alternative materials are available, iron oxide is to be preferred. On the other hand, Sylvestro[8] found that while Fe_2O_3 was successful, Fe_3O_4 was not, which indicates that chemical composition is important.

The defect has, at various times, been the subject of many investigations. It still is.[9,10] The nature of the sand grains is not the only factor, or, except in so far as it controls expansion, the major factor and the nature of the bond is, in many cases, of more importance.

2. The oil

The strength of the finished core depends mainly on the amount and kind of oil and on the baking conditions. The oils used are 'drying' oils and baking is frequently described as 'drying', but although water is removed during the operation, the increase in strength is chiefly due to a change in the structure of the oil.

The oils may be of animal or vegetable origin. There are many such oils, but the mechanism of 'drying' is the same for all. The most widely used is linseed oil, but this is because the amount produced can be increased or decreased in accordance with the demand by foundries and the many other industries which use it. Several fish oils are excellent, but the amount which can be produced is not under control.

Baking of linseed oil. The hardening of linseed oil is due mainly to the presence of linolenic acid, which is present is considerable quantity and has the following composition:

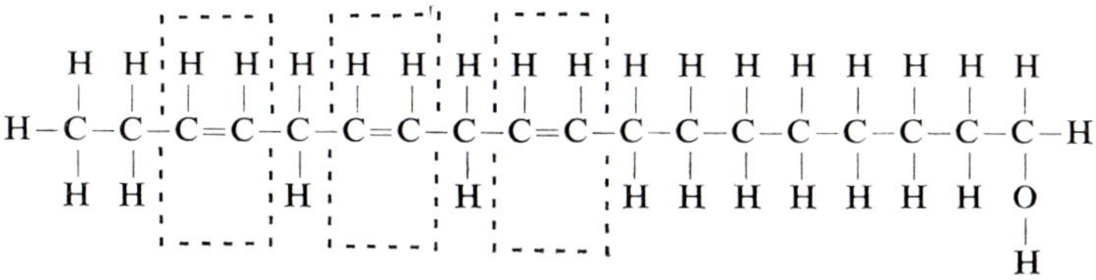

This constituent consists of a chain of 18 carbon atoms, the first of which is also attached to three hydrogen atoms and the last to two hydrogen atoms and a hydroxyl group. Each of the remaining carbon atoms is attached to two hydrogen atoms, except for the three pairs marked, each of these six being joined to one hydrogen atom only. The strength developed when linseed oil is baked is due to the presence of the six unsaturated carbon atoms; if each of these had two hydrogen atoms the oil would be relatively inert except for the hydroxyl group at one end of the chain.

When linseed oil is heated in the presence of oxygen, an oxygen atom is shared between unsaturated carbon atoms in two neighbouring chains, which are then firmly joined together.

The reaction is a complex one and more than one carbon atom in each chain is involved. However, whatever the mechanism, the joining together of two chains still leaves unsaturated carbon atoms which can join up with other chains, and as the reaction proceeds, the single chains combine to form aggregates or polymers of increasing size.

When the polymers reach a certain size, they become unstable and if maintained at a high temperature, decompose, not into the original chains, which cannot be re-formed, but into free carbon and a number of other compounds, many gaseous. The very dark colour of 'over-baked' cores is due to free carbon.

Linolenic acid — and linseed oil — is a liquid of fairly low viscosity. As the polymers increase in size, the oil becomes more viscous and eventually solid and it is at this point that the strength of oil-bonded sand increases rapidly. The size continues to increase and the polymers become harder and more brittle, until decomposition begins when the strength of the core falls.

The change in strength is not the only factor which must be considered: the change in freezing point is equally and, under some conditions, even more important. Linolenic acid is a liquid with a very low freezing point. As polymerisation goes on, the freezing point increases and when it exceeds the temperature of the core, the strength is high. However, if it were possible to stop polymerisation at a time when the melting point of the solidified oil was only a few degrees above stove temperature, the core would have a high strength, but the binder would soften almost immediately when the core was reheated to a slightly higher temperature and a thin core immersed in molten metal would become plastic and weak.

In order to obtain at the same time the highest strength and the highest melting point, it would be necessary for all the polymers to be of the maximum size obtainable without breakdown, but this is impossible. The original carbon chains do not line up in single file, form two deep, then fours and so on. Polymerisation proceeds in a random fashion and there are present at the same time polymers differing widely in size. Maximum strength is obtained before any breakdown occurs, at which time much smaller polymers with low melting point are present; maximum hot rigidity occurs when no small polymers are present, by which time the largest have broken down and strength has fallen.

The condition of the bond has an effect on finning which is as important as that of the base sand, sometimes more so. Cores which soften at high temperatures are free from finning, but those which are rigid fin badly.

As the baking temperature is raised, the rate of polymerisation increases and so does the rate of breakdown. The strength of oil–sand cores after baking follows a similar pattern. With a high baking temperature, strength rises rapidly to a sharp peak and falls equally rapidly. At a somewhat lower temperature, strength rises more slowly, remains fairly constant for a time and falls fairly slowly. At a still lower temperature, a long time is required to reach maximum strength, but once it has been reached, little change takes place for a considerable time, after which there is a very slow fall. The cycles of change are shown in Fig. 116.

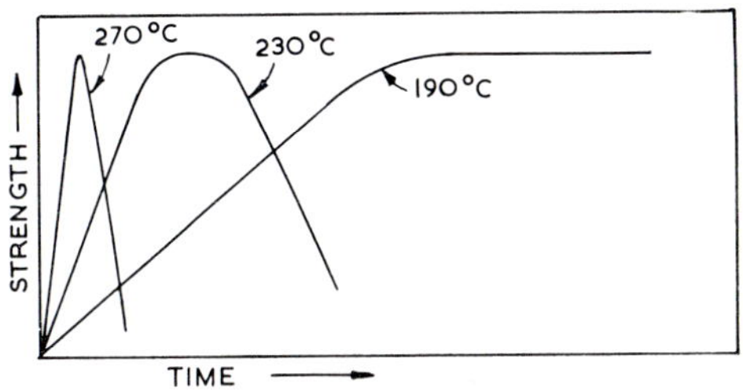

FIG. 116. Effect of baking conditions on core strength.

These changes influence baking conditions in the foundry. Small cores of uniform section can be baked at a high temperature for a short time, always provided temperature is controlled accurately. When core section is not uniform, or cores differing in size have to be baked together, a lower temperature and longer baking time are needed. In the type of large core stove, fortunately disappearing, into which cores of all shapes and sizes were loaded, a coke fire was blown up for a short time and the dampers almost completely closed, the temperature was low and most of the cores ended up along the flat of the right-hand curve in Fig. 116. So long as conditions were not particularly difficult, the cores were usable.

The baked strength of oil–sand cores is far greater than that of any dried clay-bonded sand. Using the clay-free silica sands with a permeability in the range 150–250, the kind most widely used, an addition of 1% linseed oil gives a tensile strength exceeding 300 lb in^{-2} and a compression strength exceeding 1000 lb in^{-2}.

Other drying oils. Many other natural oils, vegetable and animal, can be used in place of linseed oil and occasionally these come on to the market at an attractive price. For example, at a time when linseed oil was available only at £200 a ton and rationed at that, a considerable amount of tung oil became available at £130 a ton. The foundry should avoid these other oils. Usually, there is a slight difference in composition which requires some change in baking procedure — for example, tung oil contains some compounds with four pairs of unsaturated carbon atoms and bakes more rapidly than linseed oil, but gives a rather brittle bond. By the time the appropriate baking conditions have been arrived at by trial and error, the glut may be over. The best way to take advantage of these oils is to replace part of the linseed oil and if the amount so replaced does not exceed 30%, it is usually difficult to detect any change in behaviour. This job is better carried out by the makers of proprietary binders who mix oils in bulk rather than by a foundry measuring out a few pints at a time.

Quick-drying oils. Certain petroleum polymers produced by oil refineries polymerise in much the same way as linseed oil, but much more rapidly. These oils are suitable for use in the core shop and are particularly useful when stove capacity is inadequate, since drying times can be reduced considerably.

They can also help when cores differing widely in size have to be dried together in the same continuous stove. By using linseed oil for the smaller cores and a quick-drying oil for the larger, the different drying times which would otherwise be necessary are eliminated. The use of two such different oils calls for additional supervision, but is less wasteful than having to send the larger cores twice through the stove.

One drawback is that the oils are a little sticky and more frequent cleaning of the core boxes is necessary. The difference in behaviour is not very marked and is less than would follow a change from dextrin to molasses as green binder, but it is there.

Extenders. There are organic substances which have negligible bonding properties themselves, but which can be used to replace part of the linseed oil without any loss of strength in the baked core.

Even a paraffin has some extending action and up to 10% linseed oil can be replaced by a paraffin without weakening the core. Moreover, the paraffin improves stripping. A paraffin should not be added to any of the proprietary binders, since the maximum useful amount is usually present.

Other extenders have had varied success. Wood resin was at one time used on a considerable scale and was quite effective in amounts up to 15%. One drawback was that the solubility of the resin falls rapidly as temperature falls below 10°C and after a winter shut-down, a hard layer of resin crystals several inches deep formed on the bottom of the container.

In the laboratory, the most effective extenders were the petroleum extracts developed in 1944, at a time when the supply of linseed oil was limited and sometimes critical. Half of the oil could be replaced without loss of baked strength and since the extracts were viscous liquids, part of the green binder could be omitted without loss of moulding properties. In the foundry, the extenders were a disaster. Minor difficulties were that it was impossible to obtain the necessary temperature control in a foundry stove and that a viscous oil was evaporated in the hot zone and condensed in the cool zone, from where it dripped on to the cores and left a messy deposit or on to the fire box and set the whole affair alight. The most serious difficulty was that the baked cores softened to such an extent in contact with molten metal that dimensional accuracy was lost. In one case, the metal on the upper face of a casting was so reduced in thickness that many castings solidified white and cracked in service, the true cause coming to light only after much time had been wasted checking core boxes.

On the whole, extenders, with the exception of paraffin, are best left alone.

Synthetic resins. At one time, synthetic resins were used in place of linseed oil on quite a large scale, a procedure which received considerable impetus when the price of linseed oil was very high. Unfortunately, the supply of resins was limited and by the time they became readily available, the price of the oil was back to normal and the resins had to stand on their own feet.

Both urea–formaldehyde and phenol–formaldehyde were used and they had a number of advantages and disadvantages.

1. The resins polymerise rapidly once the appropriate baking temperature has been reached. Absorption of oxygen is not a factor. The baking cycle can therefore be shortened.
2. As the resins are emulsions with water, the cores can be blacked before baking. This is particularly useful, since the cores have friable edges if any air-drying occurs, a difficulty which is avoided if the cores are sprayed just as they enter the stove.
3. The baking temperature of the phenolics is a little below that of linseed oil, and the baking temperature of the ureas is appreciably below that of linseed oil. Less of the green binder is decomposed and the amount of resin can be reduced, but against this there is an increase in the volume of gas evolved and careful venting is necessary.
4. Cores bonded with urea resins are almost completely free from finning. However, the cores give off ammonia at high temperature, the ammonia cracking to form nitrogen and hydrogen. The latter dissolves in molten iron, but is insoluble in the solid metal, with the result that numerous pinholes are formed in some castings. This eventually led to the disappearance of the urea resins, although in the earlier tests they had been the more widely used.

It is difficult to explain an apparent anomaly in the formation of pinholes. Urea resins have been discarded by several foundries because of difficulties with pinholes, but in a paper[11] given at a conference on foundry sands, Sheehan described the use of urea resins in a foundry producing castings mainly for internal combustion engines of various kinds. All the cores made contained urea resin with starch as a green binder, the size of the cores varied from a few ounces up to 3 cwt, the castings from very small to more than a ton, and were machined extensively. Reference was made to the very rapid evolution of gas from the cores during casting and the need for careful venting, but pinholes were not even mentioned. Yet even a few pinholes would have resulted in rejection of castings of the kind being made.

The effects of oil or resin binders

The properties of the baked core depend not only on the amount and kind of bonding material used, but also on the extent of polymerisation. In any reference to the effects of a binder, baking conditions must always be taken into consideration.

Strength at room temperature. Strength depends on the amount and kind of binder and on the baking conditions. Moreover, a second binder is almost always needed in order to improve the moulding properties of the sand and all the available binders, on drying, have a strength which is very much less than that of the baked oil, but is still appreciable. They break down at temperatures lower than those needed to cause decomposition of the polymerised oil and raise the strength of a lightly baked core, but may reduce the strength of an overbaked core by causing discontinuities in the oil film.

The strength of the baked oil is highest when the baking conditions are such that polymerisation has proceeded to the point where no liquid remains, and begins to fall as soon as the largest polymers begin to decompose.

As a result, the maximum tensile strength which can be obtained from a clean silica sand containing 1% linseed oil and 1% dextrin, or similar green binder, is about 400 lb in^{-2}. In the foundry more prolonged baking is needed and strength falls to about 300 lb in^{-2}.

Hot strength and rigidity. All organic binders decompose when maintained at a high temperature for a sufficiently long time, but when a core is heated very rapidly, there is a change in strength which is not due to decomposition. When polymerisation has proceeded to the point where all the oil is just solid at room temperature, cold strength is at a maximum, but even a moderate increase in temperature is sufficient to cause re-melting and there is a rapid fall in strength. The cores become weaker and, since a viscous liquid is formed, plastic.

After more prolonged baking to the point where all polymers have reached their maximum size, cold strength is lower because this stage can only be reached when some decomposition has occurred. On rapid heating there is little re-melting of the bond or loss of strength or rigidity.

Breakdown. The rate of breakdown after casting depends on the amount and kind of binder. Oils break down more slowly than phenolic resins and these in turn more slowly than urea resins. Lightly baked cores break down less readily than hard baked cores.

Gas evolution. Polymerisation does not lead to any great change in the volume of gas evolved or in the rate of evolution and a major part of the gas is, in any case, due to decomposition of the green binder.

However, linseed oil also contains organic materials which decompose without polymerisation at the baking temperature. In addition, polymers which have decomposed leave a residue consisting mainly of carbon which is inert in the reducing atmosphere inside a core. Both the gas evolved and rate of evolution continue to decrease as baking proceeds.

The volume of gas given off from a sand containing 1% linseed oil when the core has been baked so as to give maximum strength is about 6 cm^3g^{-1}; under the baking conditions usually employed in a foundry this falls to 3 cm^3g^{-1}.

Finning. The defect arises when a core is rapidly heated and the sand grains expand. It becomes more pronounced as the cores become more rigid or less plastic at the temperature at which the break occurs.

If it were possible to bake a core so that all the polymers had reached their maximum size without any breaking down, strength would be high and, since there would be no constituents which melt readily, little loss of strength and increase in plasticity would take place on heating, but the risk of finning would be greatly increased. In practice, lightly baked cores show less tendency to fin than hard baked cores, but the need for dimensional accuracy and a low rate of gas evolution usually prohibit underbaking as a means of overcoming finning.

The problem is a complex one. Cores bonded with a urea resin are almost completely free from finning and when designing a test core which can be used to compare the incidence of the defect in different kinds of core making materials, it is difficult to find one which will show any signs of finning, although there is little softening on heating. Various mechanisms have been suggested whereby the more rapid decomposition of the urea resins can be used to explain the absence of finning, but none is completely satisfactory. It is not difficult to postulate a theory to explain why a more rapid rate of breakdown should inhibit finning over a certain range of conditions, but the urea resins are successful for most conditions — core shape and size, pouring temperature and metal thickness.

Core making. The rate at which cores can be produced is to some extent dependent on the ease with which they can be stripped, leaving a clean core box, and the lubricating action of the oil tends to reduce the sticking of individual grains to the box, which might otherwise occur. Even so, a further improvement can be obtained by adding about 0·1% paraffin to the mixture, one of medium viscosity such as a spindle oil giving the best results.

With an oil-free mixture, paraffin is a necessity and the time at which it is added becomes important. The sand and binders should be mixed and the oil added not more than a minute before the mixed sand is discharged.

Dielectric baking. Synthetic resins polymerise rapidly when the appropriate temperature is reached and, by using a dielectric heater, it is possible to bake a core in a few minutes. Heaters of this kind are used by makers of plastics in order to warm up moulding powders which consist of a synthetic resin and a filler.

The earliest attempts at baking cores by this method were not really successful. The equipment used was similar to that employed by the plastics makers and had a frequency of 10 Mc s^{-1}, but although a synthetic resin can be heated quite readily at this frequency, the loss factor of silica is so low that the rate of heating is negligible. In consequence, heat was developed in the small amount of resin and was then transferred to the much larger quantity of sand, which made the whole operation slow.

Later, equipment working at a much higher frequency was produced and was quite successful in a number of foundries, mainly in America and Germany.

The main object of the exercise was to bring core making time into line with mould making time and so reduce the need for carrying a large stock of finished cores and the attendant demands on storage space and transport inside the foundry. The development of the hot box process provided a more convenient means of obtaining this result.

3. The green binder

The essential constituents of the baked core are the sand and cured oil, but a mixture of sand and oil has no green strength and the unbaked core would collapse as soon as the support of the core box was withdrawn.

Even water gives a little cohesion to the green sand, and the green strength of a mixture of sand, oil and water may be higher than that of one which also contains a green binder. On occasion, quite complicated cores are made without a binder, but this requires sectioned top half core boxes and using the bottom half box as a drier shell, and is rarely worth while. To obtain reasonable stripping properties under the unfavourable conditions present when having to remove a small core, using its own weight as the only means of applying a load, a deformation higher than that necessary in mould making is required.

In order to obtain such a high deformation by means of clay, it would be necessary to use more than is present in a moulding sand. Since the addition of as little as 2% clay reduces baked strength by a factor of as much as four or five, such a method is out of the question.

The required increase in plasticity is obtained by the addition of one of a number of organic binders which have two properties in common.

1. They give an increase in deformation which is of a higher order than that which can be obtained by adding clay.
2. They give a very low green strength.

Four kinds of binder are used:

starch;
dextrin;
molasses;
sulphite lye.

Starch. The starch most widely used in foundries is obtained from maize. After grinding, the grain is suspended in water, the husk and germ removed and the suspension dried by feeding in a thin stream on to a steam-heated cylinder. The solid starch is then ground. As a result of this treatment, the starch is 'pre-gelatinised', individual grains being burst open so that the starch readily forms a solution in, or emulsion with water.

The composition of starch is $C_xH_{2y}O_y$, where x and y are very large numbers. The molecule is therefore large. Since a similar empirical composition can be obtained by mixing carbon and water, in the appropriate proportions, starch is known as a 'carbohydrate'.

Dextrin. This is prepared by heating starch alone or in the presence of a mineral acid. The treatment breaks the starch down into smaller — although still quite large — molecules without changing the basic composition. Dextrin is also a carbohydrate.

Molasses. This is a concentrated solution of cane or beet sugar, another carbohydrate, with the composition $C_{12}H_{22}O_{11}$. It is the residue left when all the sugar which can be taken out by repeatedly evaporating the cane extract and allowing to cool for the sugar to crystallise, has been removed. There are other substances, inorganic and organic, present in the extract, and after each evaporation the solution becomes more concentrated. As a result, the boiling point goes up until it approaches the temperature at which sugar decomposes. The residual solution is used for a number of purposes, including core making.

Sulphite lye. This is the concentrated liquor left after making paper from wood pulp. It contains both organic and inorganic substances derived from the wood or added during treatment and is a viscous liquid used in several industries as an adhesive. Sulphite lye can also be obtained as a solid, but since removing the water costs money, and water then has to be added in the foundry, it is more economical to buy a solution.

There are three kinds. One contains an appreciable amount of sodium, a second calcium, and a third ammonium, salts, depending on what was used in the treatment of the pulp. The name of the particular base is used to define the product, *e.g.* soda lye. The properties of the different lyes differ in some respects. Little ammonium lye is used in Britain.

Green binder and sand properties

Green deformation. This property is not measured, but the amount of binder used is adjusted so as to give, as nearly as possible, constant stripping properties. The method of core making also has an influence, hand core makers preferring a tougher sand than any which could be handled by a core blower.

If 1% dextrin were needed, then it would be necessary to use 1½–2% starch to replace it. Molasses and the sulphite lyes contain quite large amounts of water and the amount used has to be raised to above 2%.

Stripping. Some green binders are a little sticky and it may be necessary to clean the core box at intervals. The cleanest strip is given by dextrin and when this is used, a core box need be cleaned only when it is taken from store, particularly when a little paraffin is added to the mixture. Starch and calcium lye are only a little inferior, but molasses and soda lye produce stickiness.

Sugar is not inherently sticky, but molasses contains a little more water than is desirable and all the organic binders stick to the box given the slightest excess of water. If the excess in molasses is taken up by adding a small amount of dextrin without any water, there is no sticking.

Green strength. The green strength of a sand containing linseed oil and dextrin and the appropriate amount of water, as measured on one of the standard testing machines, may be as high as 0·5 lb in^{-2}. This is a false figure. If a load of 4 oz in^{-2} is applied and maintained for a time, under conditions such that there is no loss of water by evaporation, creep followed by collapse takes place.

Exceptionally, starch gives a green strength almost double that obtained when dextrin is used.

Water affects the green strength of sands with organic binders in the same way as it affects that of clay-bonded sand. As the moisture content is reduced, strength increases, but deformation decreases and the sand becomes too brittle for moulding; as the moisture content increases, strength falls and the sand shows an increasing tendency to stick to the core box.

Increasing the amount of binder appears to raise green strength, but this again gives a false impression. The true change is analogous with that of raising the viscosity of a liquid — flow takes place under the same stress, but more slowly. To regain the original rate of flow, it is necessary to raise the stress and with a constant speed of loading, strength seems to be higher.

Moisture range. The difference between moisture content at which the sand becomes sticky and that at which it becomes brittle, is very small with all the binders except starch, and so is the total amount of water needed — usually less than 0·5% with a permissible variation of about 0·1%. Sands containing starch, on the other hand, require between 2·0 and 4·0% water.

The addition of a borate to starch increases both the minimum amount of water needed and the permissible moisture range. In one patent,[12] a core binder is described which consists of starch, pregelatinised by a specified treatment, to which 10–15% by weight of borax has been added, together with a core oil.

By using mixtures of this kind, it is possible to get by with undried silica sand. However, 'get by' is the operative phrase, and more uniform core properties can be obtained by using dry sand and adding a fixed amount of water.

Loss of water on standing. The rate at which water evaporates from the surface of a core on standing depends on two factors:

(1) the vapour pressure of the water;
(2) the temperature and humidity of the atmosphere.

1. The vapour pressure of the water depends on (a), the temperature, and (b), the concentration of any substances in solution. As the temperature rises from 10 to 30°C vapour pressure rises from 9·2 to 31·8 mm of mercury, a more than threefold increase. Dissolving any substance in water reduces its vapour pressure, the reduction depending on the number of molecules. Starch and dextrin are large molecules and contain no soluble impurities and have little effect. Molasses is a solution of sugar, a smaller molecule, and also contains such a large concentration of salts derived from the original cane juice, that it loses no water at any room temperature likely to be found in the foundry. The sulphite lyes have an intermediate rate of loss.

2. As the humidity of the air falls, the rate of evaporation rises, the increase being proportional to the degree of unsaturation, *i.e.* to (100–relative humidity). If humidity falls from 80 to 30% a change which may occur in a single day, the degree of unsaturation rises from 20 to 70%, and the rate of evaporation increases by a factor of 3·5. When there is also a rise in temperature, this factor is raised still further.

As a result of these changes in temperature and humidity, the rate of loss of water by evaporation has been known to increase eightfold in

June between 8.0 a.m. and 3.0 p.m. when a hot day has been followed by a cool night.

Rate of air-hardening. The rate of air-hardening depends on how much water is present and how fast it evaporates.

Cores bonded with only dextrin and oil contain less than 1% water and have a high rate of evaporation and air-harden very rapidly. When starch is used, there is a similar high rate of evaporation, but as the amount of water to be removed is 2–4%, air-hardening is much slower. With molasses, there is only a small amount of water, but as none evaporates there is no air-hardening. The sulphite lyes lose water in the order: calcium (most rapid), sodium.

The rate of air-hardening has an important influence on core shop practice and particularly on the accuracy of cores.

Maintaining a constant rate of air-hardening. By using a mixture of dextrin, which air-dries rapidly, and molasses, which does not air-dry at all, in different proportions, it is possible to obtain any intermediate rate of air drying. This is a method used, with variations, by producers of mixed proprietary binders, who can increase or decrease the rate of air-drying when requested. Unless, however, a foundry is prepared to carry a stock of core binders in which these differing proportions are present, the changes due to variations in temperature and humidity during a single day cannot be dealt with.

When using dextrin or starch it is possible to maintain reasonably constant conditions. The comparative rates of air-drying are first obtained from the temperature and humidity, and then the concentration of some soluble salt, required to reduce this rate to a suitable constant figure, is obtained by experiment. The most satisfactory soluble salt is ammonium nitrate, since it is fairly cheap, has a very high solubility, and decomposes into two gases at stove temperatures so that nothing remains in the core. Sheehan[3] has given the concentrations required to maintain a constant rate of air-hardening of cores containing dextrin.

Absorption of water after baking. When baked cores are allowed to stand in air after reaching room temperature, they absorb water and lose strength. It is on this account that in all methods recommended for the testing of baked cores it is specified that the test cores must be cooled in a desiccator. When the cores are placed in a green sand mould they are kept in a saturated atmosphere which favours absorption; the prints are in actual contact with green sand and the rate of absorption is still higher. So far as dimensional accuracy of the castings is concerned, the rigidity of the baked cores is a major factor and any weakening of the prints is harmful.

All the green binders absorb water on standing, and the amount absorbed under any given conditions depends on how much binder has remained undecomposed after baking. A lightly baked core absorbs water more readily than one baked for a longer period or at a higher temperature.

When the binder contains soluble salts, some of which are hygroscopic, the rate of absorption is very greatly increased. Molasses followed by sodium and calcium lyes, in that order, has the greatest rate of absorption of water on that account.

Gas evolution. All organic binders evolve gas when heated to a sufficiently high temperature, and in cores, the volume given off by the green binder

Table 1. *Properties of green binders*

Green binder	Amount needed (%)	Green strength (oz in $^{-2}$)	Water needed (%)	Rate of loss of water	Rate of air-hardening	Stripping properties	Resistance to water absorption
Dextrin	1	4–6	0·4–0·6	Very rapid	1 (most rapid)	1 (best)	1 (highest)
Starch	1·5–2	8–16	2–4	Very rapid	3	2	1
Molasses	2–2·5 includes water	4–6	Nil	Nil	Nil	4	4
Soda lye	2–2·5 includes water	4–6	Nil	Slow–moderate	4	4	4
Calcium lye	2–2·5 includes water	4–6	Nil	Moderate–fast	2	3	3

depends on the amount of binder used and the extent to which it has been decomposed during baking. In very lightly baked cores, the gas derived from the green binder is usually greater than that from the oil, but the difference decreases with further baking.

The use of excessive amounts of binder and underbaking leads to blown castings (*see* Chapter 12).

The properties of the various green binders are summarised in Table 1.

The above figures are only comparative. For example, green strength depends on moisture content and rate of application of the load, whilst the rate of loss of water by the mixed sand and of water absorption by the baked cores depends on atmospheric conditions.

4. Mixed binders

A considerable proportion of the binder consumption by iron foundries in Britain is in the form of proprietary mixtures.

Some of these are mixtures of natural drying oils or of the 'quick-drying oils' and are used in exactly the same way as linseed oil. Many contain a little paraffin and there is no point in adding more.

The most widely used are complete binders and contain both the drying oil and the green binder. The use of such binders has the disadvantage that if one constituent has to be raised to suit any particular conditions, the other must be raised in proportion. Should it be necessary to increase the green binder in order to make a particularly difficult core, the oil is also raised; should more oil be needed to produce an exceptionally strong core, the amount of green binder must be raised. This practice is not only wasteful, but increases the risk of obtaining scrap castings because of the high rate of gas evolution.

The difficulty is partly met by having two kinds of mixture, semi-solid and creams, in which the proportions of oil and green binder differ.

One advantage of the mixed binders is that the producers can vary the composition in order to change the rate of air-hardening, a job not easily carried out by small foundries.

There are three kinds of such mixtures:

(1) semi-solids;

(2) creams;

(3) impregnated powders.

Semi-solids are emulsions of oil, dextrin and water in which the ratio oil/dextrin is between 1:3 and 1:4. The amount of water is between one-half and one-third that of the dextrin and no further water is needed unless the sand mixture also contains clay or other finely divided material.

Part of the dextrin is replaced by molasses and the proportion can be modified to suit individual foundries. A little paraffin is added to improve stripping.

Owing to the large proportion of green binder, the mixed sand is very plastic and is often preferred by hand core makers on this account. The amount used should not exceed 3%. If the baked cores are not sufficiently strong, the semi-solid should be replaced by a cream.

Semi-solids should always be weighed out, first placing a little dry silica sand on the scale to prevent sticking. Attempts to measure out the semi-solid in a pint pot always lead to inaccuracy, and inaccuracies in measuring are always on the high and never on the low side. If 3% is really needed and the amount used varies between 2 and 4%, it then becomes impossible to make cores when only 2% is used, and the minimum goes up to 3%. This does not improve the accuracy and the maximum goes up to 5 and the average to 4%, an expensive way of producing blown castings.

Creams differ from semi-solids in that the proportion of oil is higher and the proportion of green binder is lower. The sand is less plastic, but baked strength is higher.

In general, although there is no firm dividing line, creams are used for core blowing and semi-solids for hand ramming. Many of the mixtures prepared for hand ramming and containing semi-solids are so tough that they cannot be handled by a core blower.

Impregnated powders consist of starch to which has been added as much oil as possible while still retaining a powder which can be measured out easily. It may be necessary to add more oil to obtain the baked strength required without adding so much starch that it is difficult to ram the mixed sand.

Glucose may have been added to reduce the rate of air-drying, or a borate to increase the moisture range.

5. The effect of clay

Whatever the rate of air-drying, it is not, as a rule, possible to produce a heavy core with a depth exceeding 6 in or so in a mixture containing organic binders only, without serious risk of sagging or even of collapse. It is necessary to raise the green strength before the core can even be removed from the box.

The addition of a small proportion of clay increases the green strength and can reduce sagging and distortion during handling. Many core makers prefer Congleton sand to Southport sand because of the presence of a small amount of clay. The 'clay grade' is between 0·5 and 1%, only a part of which is true clay, but even this small amount makes for easier handling of delicate cores.

Larger amounts of clay are needed for bulky cores, but since clay has such a large surface area, there is a rapid fall in baked strength as clay content rises. The effects on baked strength of adding bentonite in amounts up to 2% to a clay-free silica sand with 1% linseed oil and 1% dextrin are shown in Fig. 117.

The addition of 2% of bentonite has reduced baked strength by five-sixths and the core has become friable.

It is, however, neither necessary nor desirable to increase the oil to the extent required to bring the strength back to the original value. It may occasionally be necessary to maintain a strength of 300 lb in^{-2} when the cores are long, thin and complicated, or when two comparatively heavy sections are joined by a thin neck, but in most cases, an appreciably lower strength is adequate. Even so, it is rarely possible to reduce the amount of oil to below 1% because cores are not uniformly baked, thin sections, projections and edges heating up more rapidly than the remainder. By the time the body of the core is baked, these sections are overbaked and an excess of oil is used in order that their strength shall be adequate even in the overbaked condition.

When the amount of oil is raised, the amount destroyed by overbaking is little changed, but the proportion destroyed is smaller, with the result that the fall in strength is less steep. In consequence, when clay is added, it is usually sufficient to increase the oil to the extent needed to bring the baked strength back to 150 lb in^{-2}.

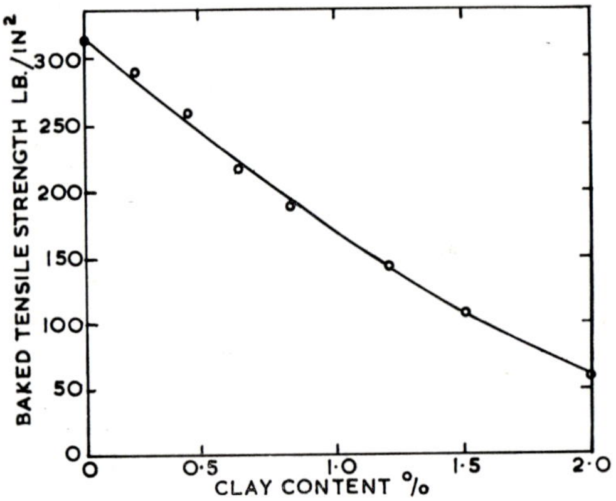

FIG. 117. Effect of bentonite on baked strength of an oil–sand core.

Clay may be added as such, or as a clay-bonded sand. Since the clay-bonded sands contain fine grains they reduce permeability sharply; the addition of 15% of a red sand with a permeability number of 25 reduces the permeability of Southport sand from 150 to 60. When a very high permeability is not necessary, for example, in a thick core with large prints which can be easily vented, the use of a natural sand may be beneficial, since metal penetration is reduced and it may be possible to get by without using a core coating.

6. The effect of heat on cores

When any core, bonded with organic binders only, is heated for a sufficiently long time at a high enough temperature, all the binder decomposes. With access of air, only the sand grains are left, together with any ash from the binders, but in a reducing atmosphere free carbon is also present.

However, the core and the final product are not the only important factors. It has been shown earlier in this chapter that when a core is rapidly heated, part of the solid binder may re-melt and that gas is continually being given off by the heated core. These changes begin while the metal is still liquid and must therefore have an effect on the quality of the casting. Moreover, when solid the castings contract on cooling, the strength of the core may be such that the contraction is hindered, and the casting may distort or crack. Three interim changes must therefore be considered:

(1) hot softening;
(2) gas evolution;
(3) rate of breakdown.

Hot softening
As a drying oil polymerises during baking, its melting point rises continuously. If baking ceases as soon as the melting point of all the polymers exceeds the stove temperature, all the binder is solid and strength is quite high. Should the temperature be raised very rapidly by, for example, 200°C, many polymers will re-melt. Further polymerisation will occur, but the rate of polymerisation is so slow that its effect is small. Not all the

Table 2

Time (h) at 232°C	Cold tensile (lb in^{-2})	Cold compression (lb in^{-2})	Compression at 400°C (lb in^{-2})	Deformation on 2 in at 400°C (in × 10^{-3})	Deformation for load of 100 lb (in × 10^{-3})
$\frac{1}{2}$	370	1100	100	27	27
1	340	1000	250	21	8·4
$1\frac{1}{2}$	150	475	400	17	4·3

polymers melt, because of the wide range in size and melting point, but there will be considerable loss in strength and, since the remelted polymers are viscous liquids, an increase in deformation.

In order to illustrate the kind of change which occurs, a silica sand with the addition of 1% dextrin and 1% linseed oil was baked at a temperature of 232°C for periods of $\frac{1}{2}$ hr, 1 h and $1\frac{1}{2}$ h, producing underbaked, normally baked and overbaked cores. The tensile strength and compression strength at room temperature were measured.

Standard cylindrical test pieces were then heated to 400°C in a dielectric heater in about $\frac{1}{2}$ min and the compression strength and deformation measured. The results obtained are shown in Table 2.

The highest cold strength was obtained after baking for $\frac{1}{2}$ h only, but on reheating to 400°C, compression strength fell from 1100 to 100 lb in^{-2}. Increasing the baking time to 1 h raised the melting point of the polymers, but some decomposed and cold strength fell. In this case, rapid heating to 400°C caused a fall in compression strength to 250 lb in^{-2}. Increasing the baking time to $1\frac{1}{2}$ h reduced the cold strength by more than half and the cores were very friable. However, on reheating to 400°C, there was only a slight further fall, showing that few polymers had a melting point below 400°C.

The most important figures in Table 2 are the deformations in the right hand column, which show that a lightly baked cylinder 2 in diameter and 2 in long deforms 0·027 in under a load of 100 lb in^{-2}. A long, thin core surrounded by molten iron and supported only at the two ends would distort badly. So far as distortion is concerned, the most satisfactory core is that baked for $1\frac{1}{2}$ h, but the edges are so friable that the core cannot be handled without removing grains from the surface. In practice, some rigidity must be sacrificed in order to avoid friability. Moreover, an increase in rigidity favours finning and on both counts, the most useful core is that baked for 1 h at 232°C.

Gas evolution

The amount of gas evolved when a core is heated depends on the amount of oil and of green binder and on the time and temperature of baking. The atmosphere inside a mould whilst a casting is being poured is strongly reducing and when carrying out a test to determine the volume of gas evolved it is important that oxygen be excluded. This is ensured by using an atmosphere of pure nitrogen.

Tests were made on two different core mixtures, each containing 1% linseed oil, with 1% dextrin in one mixture and 2% starch in the other.

One set of tensile test pieces of each mixture was baked for 30 min at 232°C, to show the effect of underbaking and a second set for 1 h at the same temperature, to show the effect of normal baking.

Samples for testing were obtained by rubbing together the broken faces of the tensile test pieces.

The volume of gas evolved and the rate of evolution are shown in Fig. 118.

The following are the most important points shown by Fig. 118:

1. Underbaking leads to a large increase in the volume of gas.

2. More gas is evolved by the green binder than the oil, especially in the underbaked cores.

If the results shown in Fig. 118 for the sand containing 1% linseed oil and 1% dextrin are combined with those shown in Table 2, three important points emerge.

1. The cores baked for $\frac{1}{2}$ h at 232°C have the highest cold strength.

2. The same cores soften and become plastic on heating to 400°C.

3. The same cores evolve large volumes of gas.

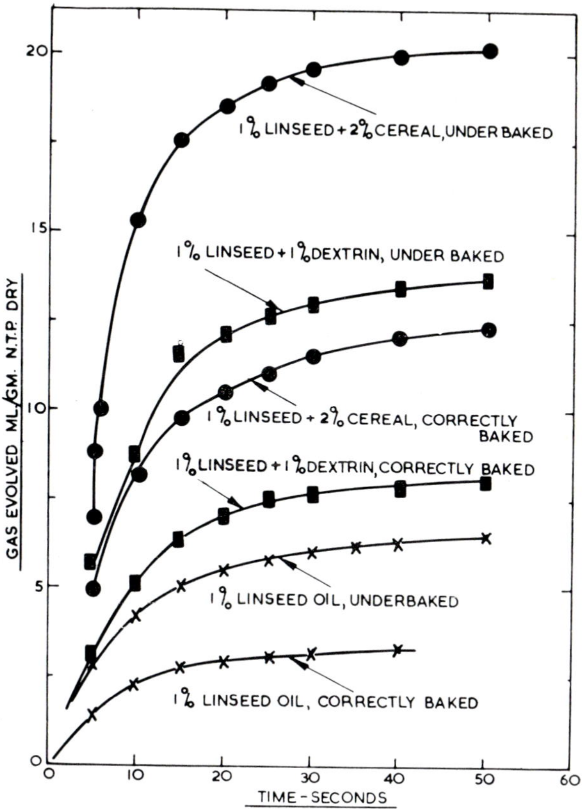

FIG. 118. Volume of gas evolved by cores heated to 850°C.

It follows that *in the foundry, it is necessary to have a baking time considerably longer than that needed to give maximum cold strength.* It also follows that cold strength cannot be used to determine the appropriate baking cycle.

Quick-drying oils and resins and gas evolution

When a quick-drying oil or a synthetic resin is used in place of linseed oil the baking times must be reduced. The change does not in any way hasten the reduction in the volume of gas evolved by the green binder, which remains as high as that from the same binder in an underbaked core containing linseed oil.

Whenever a quick-drying binder is used, gas evolution is high and means must be found for its removal.

Breakdown after casting — hot tearing

Provided that a core is maintained at a high temperature for a sufficiently long time, all the organic binder decomposes and the sand loses all cohesion. If the core is not large and is almost completely surrounded by metal, breakdown is usually complete by the time the casting is cold and the unbonded sand can easily be removed.

Breakdown is not, however, instantaneous even at the temperature of molten iron and, moreover, since sand is a poor conductor of heat, the rise in the temperature of the sub-surface layers is slow. In the period immediately after casting, the strength of a thick core is only slightly reduced and, should the metal be contracting at this stage, there is considerable resistance and the casting may crack.

There is little difficulty with grey iron on this score. The iron expands on solidification and by the time contraction begins, sufficient time has elapsed for core breakdown to proceed. Usually breakdown can keep ahead of contraction.

White iron and steel, on the other hand, contract during solidification and soon after casting there is a very weak mixture of solid and liquid, the contraction of which is prevented by the resistance of a core only slightly weakened by heat. The result may be a hot tear, which is simply a crack occurring while some liquid metal is present. This is discussed in Chapter 12.

REFERENCES

1. Robertson, R. H. S. and Emodi, B. (1943). *Nature*, **152,** p. 539.
2. Davies, W. and Rees, W. J. (1944). *J. Iron Steel Inst.*, **150,** No. 2, pp. 19–47.
3. Sheehan, J. J. (1938–1939). *Proc. Inst. British Foundrymen*, **32,** pp. 43–58.
4. Curry, T. (1952). *The Iron Worker*, **16,** No. 2, pp. 18–27.
5. Morey, R. E. and Catto, J. R. (1946). *Amer. Foundryman*, **9,** June, pp. 53–63.
6. Report of AFS Committee 8J (1952). *Trans. Amer. Foundrymen's Soc.*, **60,** pp. 218–222; also **61**, pp. 150–155.
7. Korpi, R. and Passman, G. (1955). *Amer. Foundryman*, **7,** March, p. 505.
8. Sylvestro (1957). *Trans. Amer. Foundrymen's Soc.*, **65,** pp. 323–339.
9. Horiuchi, Mayaxaki and Natsui (1960). *Imono*, **34,** pp. 469–475.
10. Institute of British Foundrymen Sub-Committee T.S.61 (1964). *British Foundryman*, **57,** Feb., p. 75.
11. Sheehan, J. J. (1949–1951). *BCIRA J.*, **3,** pp. 221–227.
12. British patent No. 543 202.

CHAPTER 12

Coremaking — Defects due to Cores

1. SELECTION OF MATERIALS

Sand

Grain size and distribution

Since many cores in service are almost completely surrounded by metal, a high permeability is necessary, but it is not sufficient to put a lower limit on permeability and leave it at that. Even with the sands most widely used, it is frequently necessary to add a coating to the cores in order to prevent metal penetration and give the kind of surface finish required.

As the permeability of the core increases, the thickness of coating needed to prevent metal penetration increases. In order to hold this coating in position, an organic binder is added as well as clay and this binder gives off a large volume of gas when heated; the volume from a given weight of coating is usually several times greater than that from an equal weight of core. An undue increase in permeability may therefore favour the formation of the blowholes which the change should have prevented.

In consequence, most oil–sand cores are made using a sand with a permeability in the range 150–200. There are several deposits of such sand — Erith, Reigate, Aylesford, King's Lynn, Congleton, Southport and Irvine. They are not outcrops from one and the same geological deposit and may differ from each other in grading and mineralogical composition.

Sometimes a coarser sand, such as that from Leighton Buzzard, may be added, but this has little effect on permeability. It does, however, reduce the surface area and, on this account, may increase strength slightly. It may also reduce the incidence of finning.

Grain shape

Very angular sands should be avoided. There are no commercial deposits of spherical or near spherical grains in this country and all the sands are described as 'sub-angular' or 'rounded', terms for which there is no precise distinguishing definition. Sands with obvious edges and corners, even if these are rounded off a little, and sands containing compound grains with re-entrants, should not be used.

Chemical composition

Some of the sands contain 98% or more of quartz, others 96%, and the shore sands as little as 90%. Some foundries prefer shore sands because expansion on heating is lower and there is said to be less finning. The difference is marginal.

All shore sands contain $CaCO_3$ in the form of shell fragments. The $2\frac{1}{2}$–3% present in Southport sand has little effect. Some gas is given off when the casting is poured, but as decomposition does not begin until the sand temperature reaches 800°C and is then slow, its influence on the rate of evolution of gas in the short critical period during, and immediately after pouring, is small.

Sands from the west coast round the Severn estuary may contain as much as 10 or even 12% $CaCO_3$. Again, the overall rate of evolution of gas is not increased to an extent which causes trouble, except where the factor of safety is small, but on the surface of an uncoated core, local pock-marking may occur. In addition, the presence of such a quantity of mainly flat fragments tends to reduce strength.

Shore sands may also be contaminated with soluble salts derived from sea water. Formerly, the sand from Southport was obtained from old dunes, well washed by rain, but at the present time sand is scraped from the shore itself. This is washed by the supplier. The salts are hygroscopic and if any appreciable quantity remains, the baked cores have an increased tendency to pick up moisture on standing. The presence of undesirable proportions of such salts can be detected by drying a sample, allowing it to stand under a bell jar, or in a similar enclosure, by the side of a dish of water and finding the increase in weight. A blank test should be carried out on a second sample which has been well washed with hot water, since all sands pick up a little water in a moist atmosphere.

The presence or absence of clay

Some sands are quite clay-free, but in others, a small amount is present. Much of the sand from the Congleton area contains 0·5–1·0% and occasionally up to 1·5% of clay grade, only part of which is true clay, the remainder being an inert dust.

This small amount of clay slightly reduces the strength of the baked cores, but eases the problem of handling delicate cores in the green state.

Any one of the above sands is suitable, but it is advisable to keep to one and to see that there are no appreciable changes in composition, particularly very fine material. Some shore sands are occasionally contaminated with mud if they are near an estuary and occasionally there has been an increase in fines in other sands, usually when a particular quarry was nearly exhausted.

Oil

The most widely used oil is linseed oil and most proprietary binders contain it. For a large consumer, specifying pure linseed has the advantage that the iodine value, which is a measure of the total unsaturated carbon atoms and can easily be determined in the laboratory, is always between 175 and 195 and a single test is enough to check the quality of a tank load.

The alternative mixed oils are usually reliable and equally satisfactory, but a control test is more difficult. Most contain a paraffin, which has no unsaturated carbon atoms and reduces the iodine value accordingly. However, there are 'semi-drying' oils such as cotton seed which contain two instead of three pairs, and 30% of one of these would have the same effect on the iodine value as 10% paraffin. The oil containing paraffin would behave very much like pure linseed oil, but it would be difficult to harden that which included cotton seed oil.

When testing mixed oils, the most suitable procedure is to use them to prepare sand mixtures and make cores. It is difficult to draw up a suitable specification owing to the pronounced influence of even small changes in baking conditions.

Green binders

The decision as to which green binder to use is more difficult. Dextrin gives the most rapid air-drying and the most accurate cores, and also the best stripping properties, but air-hardening commences as soon as the sand is mixed and the sand may become unusable in a short time. Indeed, in hot, dry weather there is likely to be a pronounced 'crust' on the sand by the time it reaches the core makers.

At the other extreme, molasses does not air-harden and there is no difficulty in using up all the mixed sand, but, except when all the cores are either shallow or contain clay, many have to be baked in drier shells. Apart from the cost of the shells, cleaning is a nuisance and unless this task is performed regularly, baked cores may stick and core scrap increase.

The sulphite lyes and starch air-harden at intermediate rates and are widely used on that account. The lyes are cheaper, but tend to stick in the core box. Starch has to be used in rather larger quantities, which increases gas evolution, but not to a serious degree if the cores are properly baked. One limitation of sand containing starch is that properties change slowly on standing with loss of plasticity. The change is just noticeable if the mixed sand is allowed to stand overnight and is pronounced when it stands over the week-end, even when conditions are such that there is no loss of water. Similarly, if sand containing dextrin or sulphite lye air-dries and is then re-mixed with the appropriate amount of water, the original properties are regained, but sand containing starch treated in the same way loses some of its plasticity.

2. SAND PREPARATION

Mixing equipment

The function of the mixer is to spread the bonding material uniformly over the surface of the sand grains, whether that material is a combination of organic materials or a clay. There are, however, four factors which have an influence on the kind of mill most suitable for sands with organic binders which differ from those which are important when mixing clay-bonded sands.

1. There is no returned sand which may be moist or dry and this factor does not have to be taken into account when deciding on the kind of mill or the duration of milling.

2. An intensive rubbing action is not needed. By the time the binders are spread uniformly throughout the sand, milling is complete.

3. When a mill with rollers is used for mixing oil–sand, the bottom of the mill becomes highly polished, and before mixing is complete, these rollers slide instead of rolling, pushing a heap of sand along. Mixing then ceases.

Sliding can be prevented by placing metal strips across the mill bottom or using a renewable checker plate, but a further difficulty arises when using a semi-solid or very viscous binder. This binder tends to form balls up to 1 in or so in diameter which become coated with sand grains. The ploughs pile up the sand in an inverted V in front of the rollers and the balls roll down the V, in much the same way as coarse grains roll down the side of the cone formed when feeding dry silica sand into a hopper. In this way, they may escape the rollers for quite a time and mixing is not complete until they are all dispersed.

4. The permissible moisture range is very narrow and there is a risk that with prolonged milling, the loss of water may be sufficient to make the sand useless.

In consequence, a paddle mixer is usually preferred. The action is rapid and, unless excessive binder is used, the power consumption is moderate.

Mixing and the order of additions

Mixing time should be as short as possible, owing to the risk of loss of water. All the energy put into the operation eventually appears as heat and, even with a paddle mixer, there may eventually be a rise in temperature which increases still further the amount of water lost by evaporation. Indeed, when using dextrin, which air-dries rapidly and has a very narrow moisture range, it may be necessary, particularly in hot weather, to add rather more water than is needed and continue mixing until the excess water has evaporated. There is a continuous loss of water from the mixer which depends on atmospheric temperature and humidity, which can vary widely even during a single shift, and it is not easy to foresee the exact amount of water which will be lost. However, there is a pronounced change in the appearance of the sands when the excess water has gone, and an experienced operator can decide with safety when the mill should be discharged.

The order of additions is important only when the green binder is added as a powder. In this case the powder should not be added first since mixing of two granular materials differing in size as much as sand on the one hand and powdered dextrin or starch on the other, is not effective. A liquid should be added first and the powder next, but apart from this, the order has only a small effect. Some foundries add the oil first, followed by the powder and, after a short interval, the water; others add water first, followed by the powder, with the oil last. When the green binder is readily soluble in water, *e.g.* dextrin, the order has no effect on the final strength, but some observers claim that when starch is used the baked core is stronger when the oil is added after the starch; others have been unable to detect any difference. If it does exist it can only be minimal.

The problem does not arise when using a complete binder, since all are added together. In spite of this, mixing time is not reduced and, indeed, when using a semi-solid, it may well be longer.

3. COREMAKING

The ramming of cores is in one way simpler and in another more complex than the ramming of moulds.

It is simpler in that the wide range of moulding machines of different kinds used for moulds is not also used for cores. Large cores of simple shape may be made by slinger or by jolting, but for the most part, the choice is between ramming by hand or by compressed air, either rapidly in a shooter or a little less rapidly in a core blower.

It is more complex owing to the wide range of cores to be made. At one end of the range are the simple cores which are essentially blocks of sand of regular shape, and at the other end, cores which are of such shape that a straightforward strip is not possible, or that it is not easy to see how the core can be made with certainty. An example of the latter is shown in Fig. 119.

The long twisted section is less than $\frac{3}{8}$ in thick and the shape is such that the core must be supported until it is baked. When supported at the two ends, it sags under its own weight, and when immersed in molten metal, bends upwards. Gas has to travel a long way along a narrow core if it is to escape through the prints. Indeed, the three core prints along the thin section are necessary to hold the core in position and enable gas to escape.

This is a particularly difficult core and illustrates many of the defects which must be guarded against. In spite of the high strength of baked oil–sand, a wire is needed to avoid breakage during handling and distortion after pouring. The wire has only a thin layer of sand for cover and should there be a fine crack, which might easily occur during handling in the

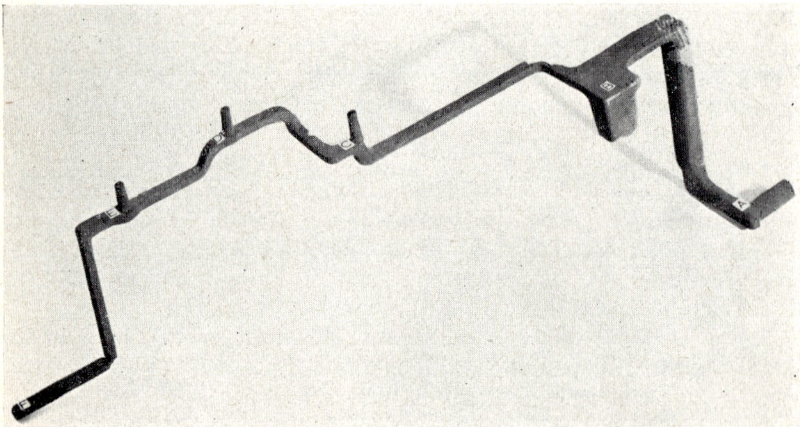

FIG. 119. Difficult core.

green state or during baking as a result of the different expansions of metal and sand, molten metal must reach the wire which cannot then be extracted and the casting is defective. The very high core permeability needed cannot be obtained by using a very coarse sand, since in this case, a very thick layer of blacking would be needed to prevent metal penetration, and such a coating gives off a large volume of gas and may itself crack, thus producing the very defects it should prevent. It is therefore necessary to raise permeability by venting and to accommodate a wire and a vent along a core only some $\frac{3}{8}$ in square needs very accurate placing if metal is not to reach the wire or block the vent, either of which would inevitably cause the casting to be rejected. It is doubtful whether any of the newer core making processes which have already replaced oil–sand on a large scale could be used successfully here.

The importance of venting

The total volume of gas given off from a heated green sand mould is much greater than that given off by a heated core, but most of the gas is water vapour which condenses a short distance behind the mould face — and condensing a gas eliminates gas pressure as effectively as removing the gas. Owing to the low conductivity of sand, the rate at which the moist layer is subsequently converted to vapour is slow. Moreover, the area of the mould through which the gas can escape is large and, provided that precautions are taken to see that the first rush of gas due to the increase in volume of air in the mould cavity and the evaporation of water from the surface is taken care of, and that there is a means of escape through the bottom of the mould, defects due to mould gas can be eliminated without much difficulty.

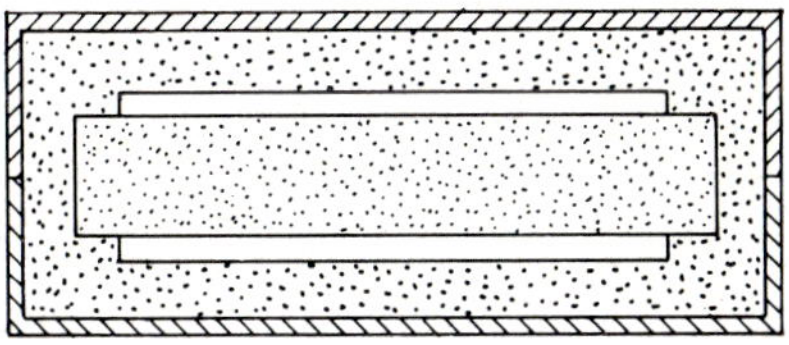

FIG. 120. Core with 4 in diameter prints.

The total volume of gas from a core is less, but the volume of permanent gases is larger. In addition, with thin cores the centre soon reaches a temperature exceeding 100°C and water vapour is then as important as the permanent gases. Moreover, at the high temperature, in the centre of a thin core, the gas occupies a larger volume and has a higher viscosity than in the cooler conditions which exist in a mould some distance behind the face. This large volume of gas may have to escape through narrow core prints. Core permeability must be higher than that adequate in a mould.

Suppose a plain cylindrical casting 5 in diameter has a core 4 in diameter, as shown in Fig. 120. Any gas given off by the core has to escape through the two ends, which have a combined area much less than that of the mould surface, and a higher permeability is needed. Even so, such a core is often made in the same green sand as is used for the mould, but nevertheless some moulders would take the mainly unnecessary precaution of placing a vent through the centre of the core.

If, now, the casting is modified as shown in Fig. 121, retaining the same overall dimensions, but reducing the size of the prints to 1 in diameter, the area through which the gas must escape is reduced to one-sixteenth of the original area, and a corresponding increase in permeability is needed. If a permeability of 25 was just sufficient for the first core, this must be raised to 400 for the second.

Suppose, further, that the overall dimensions of the mould are retained, but the plain core is replaced by a complicated one. The surface area of the core is increased and the rate of evolution increases in consequence. If the surface area is doubled, permeability must be raised to 800; in addition, if

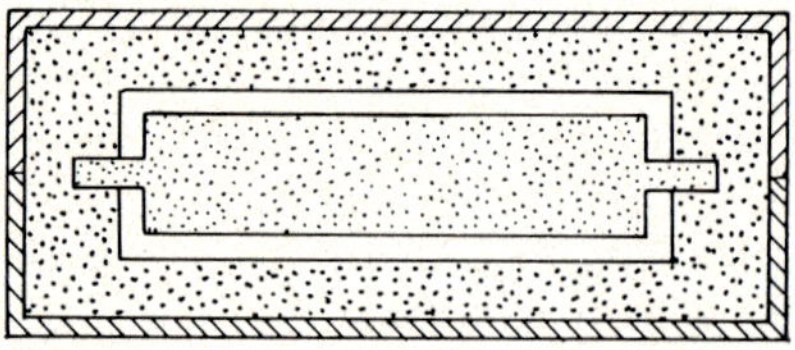

FIG. 121. Core with 1 in diameter prints.

the green sand is replaced by oil–sand, there is a further increase in the volume of gas to be removed.

It would be possible to obtain a sand which would enable core permeability to be raised to 2000 or so. The core would then be permeable to molten metal or to blacking, either of which would then reduce permeability to 0.

The effect of venting

The effect of venting on permeability and on the pressure of gas inside a sand core can be seen in Fig. 122.

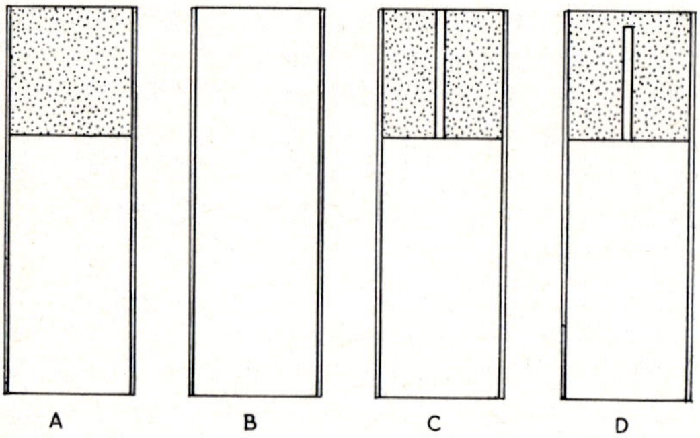

FIG, 122. Effect of venting on permeability and gas pressure.

Figure 122A represents a normal red moulding sand rammed up to form the standard test piece. If the sand has a permeability number of 25, the time required to pass 2 l of air is 120 s, and the pressure recorded by the water manometer is 10 cm.

If, now, the sand is removed from the tube as in Fig. 122B, and the test repeated, the time required to pass 2 l of air falls to 2 s and the pressure to almost 0.

Placing a hole $\frac{1}{4}$ in diameter right through the test piece, as in Fig. 122C, has the same result. This does not show that air can pass as readily through the hole as through the empty container, but that it can flow through either more readily than along the circuit from the bell of the permeability apparatus to the sealing device.

If, however, the hole is bored only part of the way through the test piece, leaving a layer of sand $\frac{1}{4}$ in thick at the end, as in Fig. 122D (this operation must be done with care or there may be a crack through the $\frac{1}{4}$ in layer), the time required to pass 2 l of air is 50 s, the manometer reading is 10 cm and the permeability number 60.

The vent passing part of the way through the test piece has increased permeability from 25 to 60, but a more significant figure is the pressure, since it is the pressure which decides whether the gas will force its way through the molten metal. This has not been reduced sufficiently to show on the manometer.

It is therefore important to see that there is a clear path from the end of the vent to the outside of the mould. If a vent is placed through the core shown in Fig. 121, and the moulding sand between the end of the core and the box is left undisturbed, the effectiveness of the vent is greatly reduced.

Other forms of venting

The design of a core may be such that it is not possible to place a hole through the section where a vent is needed. It is, for example, impossible to place any but a straight hole through the centre of a core made in one piece. A method used under these conditions is to place a thick ribbon of wax, leading to the core prints in the partly rammed core, and then fill up the box and complete ramming. When the core is baked, the wax melts and is absorbed by the surrounding sand, leaving a passage through the core. It is claimed that the wax decomposes, but this is doubtful. Indeed, occasionally the wax is so near the surface of the core that it decomposes during pouring with such a copious evolution of gas that the risk of obtaining a blown casting is increased rather than decreased.

The risk can be reduced by replacing the wax by the woven nylon tubing now available.

4. CORE BAKING

In order to bake a core, oxygen must be available and the temperature of the core must be raised.

There is no serious difficulty so far as the supply of oxygen is concerned. The amount of oxygen needed is small and short of hermetically sealing all the fortuitous openings into the stove and, where the products of combustion enter the stove, maintaining a reducing flame, the amount available is ample. It is only in the interior of a core that the supply of oxygen is deficient. The oxygen in the pores is soon used up, leaving nitrogen, and replacing this by air simply by diffusion is a very slow process. In consequence, the interior of a thick core loses only water and no further change takes place. Such a core may not only go through the stove without its centre undergoing any change other than loss of water, but may also go through the casting process unchanged and the sand in the centre needs the addition only of water to regain the properties of the original sand mixture. Recovery is not worth while economically, but during a period of severe shortage of core binders, large cores were removed from castings, ground up and re-used with a quite small addition of new sand and binder.

The presence of this unbaked interior does not affect the quality of the casting. The rate of heat penetration into a sand core is slow and the casting is solid before rapid gas evolution begins and, in addition, there is rarely any difficulty in putting very large vents into cores of this kind.

The problem which faces the core shop is that of getting the surface of all the cores to the same temperature at the same time and keeping them so. It is unduly optimistic to hope that a single recorder in a stove records the surface temperature of each and every core; it records only the temperature of the air at one point in the stove and searches made by placing several pyrometers in a batch stove have shown temperature variations as great as 50°C. Uniform baking cannot be expected under these conditions.

There are more uniform conditions in a continuous stove in which all the cores pass through the heated zone at the same speed, but even here, with the recorder showing a uniform temperature, conditions may vary. When recorders were first introduced, core stove attendants were suspicious, but soon came to regard them as giving useful evidence of their success in maintaining a constant temperature. However, an experienced man on a continuous stove soon learns to bring back a wandering temperature by manipulating the feed of both fuel and air and, in so doing may alter the depth of the hot zone and, therefore, the time that the cores are maintained at a high temperature. The 'clock' record is good, but the cores are not uniformly baked.

Core temperature and stove temperature

Even were it possible to maintain a constant temperature in the stove, it does not follow that all core surfaces will be at the same temperature as each other.

The surface of the core receives heat in three ways:

(1) by radiation from hot surfaces;

(2) by conduction from hot air;

(3) by conduction from other bodies with which it is in contact, these in turn being heated by methods 1 and 2.

When cores are placed in a hot zone, not all are so placed as to receive the same amount of heat by radiation. One side of a core may be near the hot wall and the other completely shielded, while cores in the centre of a row are shielded to some extent by those outside.

The transfer of heat from air to cores is by no means uniform. Hot air in contact with a vertical core surface loses heat and passes in a downward direction and is replaced by more hot air, but on a horizontal surface particularly if that surface is dished or contains hollows, the cooled air remains in contact for a longer time and heating of the core is less rapid. When a fan is placed in a batch stove in order to improve temperature distribution, a surface on which the air impinges directly warms up more quickly than a surface in such a position that the air flow is parallel to the face.

Heat is conducted to cores *via* the core plates, and here the composition of the plates is an important factor. The heat is applied underneath since the top is shielded by the core, and the lower the heat capacity and the higher the conductivity of the plate, the more quickly does heat reach the cores. Aluminium has a lower heat capacity and a higher conductivity than iron and heating is more rapid; non-metallic plates have a low conductivity and heat transfer is slow.

Owing to these variations, there is no such thing as a uniformly baked core, but precautions should be taken to reduce the variations as far as possible. Not much can be done to even out heat transfer by radiation, but air distribution can be improved by leaving small spaces between core plates and, particularly when several small cores stand on one plate, drilling holes in the plate.

The end point in core baking

It is not possible to devise a simple quantitative test which can show when a core is baked to the best advantage. As baking proceeds, strength rises to a maximum, but at this point, gas evolution is high and thin cores soften when heated. Further baking results in a continuous fall in baked strength, gas evolved on heating and hot plasticity, and at no point is there a break in any of the changes which would give a definite end-point.

The usual method of deciding whether a particular core is properly baked is by means of the colour. A light yellow core is definitely underbaked and a black core overbaked. A properly baked core lies in between and is dark brown, but how dark is dark? Subjective judgments are very good for picking out obvious failures, but not so satisfactory for sorting out smaller differences which are still important so far as the quality of the castings is concerned. Nevertheless, in spite of its imperfections, colour is still the most useful method of sorting out under- or over-baked cores.

It does not follow that the controls on the stove are set solely in accordance with the colour of the cores. If there is any prolonged failure because of hot distortion or high gas evolution, either the temperature or the time of baking must be raised. The colour of the cores can then be used to check quickly any failure of the newly established controls.

5. CORE STORAGE

By far the best way of storing cores is not to store them at all! Deterioration of a core as a result of moisture absorption commences as soon as a core is cold and continues throughout its storage life. Keeping a core in a warm dry store at a few degrees above foundry temperature postpones serious deterioration for a long time, but such a store is rarely available.

Unfortunately, it is not always possible to keep cores on the move from the stove, through any finishing operations and into the moulds. The time which elapses between when the core is rammed and when it is ready to go into a mould — time needed for baking, blacking, re-drying and inspection — is several hours and a buffer stock is desirable.

The problem should not be serious in a foundry producing long runs of identical castings, since an almost continuous flow can be maintained. It is less simple when small batches are required at intervals, in which case a few cores in excess of the number needed are made, in order to ensure that all the moulds can be poured together, and such cores as are left over are put on one side until the next order is received. These are not always used first when the next batch is required and it is not unknown for the same cores to be held in reserve for long periods, and a core which has been on the back of a bench for some months is not really suitable for use. When cores must be stored they should be used in the order in which they are made and any which have been standing for a few days, or even a shorter period if the atmosphere is cold and damp, should be warmed before use. The temperature used for drying after blacking is suitable; a second baking would lead to friable edges.

6. CORE BREAKDOWN

The wide use of cores with organic binders has been due to the ease with which the binders break down at the temperatures reached after casting, so that the cores can later be removed. For the core to function satisfactorily, it must not break down too quickly, or there may be erosion by the metal stream, but on the other hand, it must not retain so much strength that it restricts contraction of the cooling casting.

The rate of breakdown is therefore important. There are no standard tests for this property and the results obtained by heating a loaded core at a fixed rate until it fails can be misleading. The rate of breakdown in a reducing atmosphere is very much slower than when oxygen is present, and a core may be used successfully in a mould for a casting weighing several tons when a test piece prepared from the same mixture and heated for a short time at 500°C in an oxidising atmosphere disintegrates.

Two extremes must be guarded against: too rapid breakdown, followed by erosion, and too slow breakdown preventing casting contraction and followed by cracking. Too slow a rate of breakdown is due to the use of an excessive amount of binder and underbaking.

Core erosion

When a core is heated in a reducing atmosphere until the binder has decomposed, there is some residual strength, probably due to the presence of very finely divided carbon, since with ashless binders there is no other residue, and so long as there is no direct impingement of metal, there is little risk of failure. The residual strength is much greater when clay is present in the cores and this is helpful when making very large castings, since the cores concerned are usually large and clay is included to reduce sagging. The amount of clay which can be used is limited because of the extent to which its presence reduces baked strength, and the increase in residual strength is not sufficient to enable the core to resist direct impingement. When erosion does occur, overbaked cores must not be used, since these are well on the way to breaking down before pouring starts, but, in addition, it is usually necessary to modify the gating system.

Nothing can be done to reduce impingement on strainer cores and these should be avoided where possible. Should they have to be used, the amount of oil in the core should be raised to 2% and if erosion persists, the core should be replaced by a ceramic.

Cracking and hot tearing

These are really one and the same defect, occurring under different conditions. Some metals, including white iron and steel, contract on solidification, and at the time when both solid and liquid are present, strength is very low. Any appreciable resistance to contraction in the period during which some liquid is present causes the casting to crack. In this case the crack is known as a hot tear. It can easily be identified by examination under a microscope.

A hot tear must develop soon after casting and there is little time for the core to break down unless the casting is really thick. Since an oil–sand core has a very high strength before breakdown, it is a frequent cause of hot tearing, particularly in thin castings.

Grey iron expands during solidification and is not prone to hot tearing.

When solid, the casting contracts, but by this time the surface layer of the core has weakened and, as contraction increases, so does the thickness of the weak layer. The risk of cracking in grey iron is, therefore, much less than that of hot tearing in white iron and rarely occurs unless the casting is thin, the cores large and the rate of breakdown slow.

Core removal

In some castings, the core is almost completely surrounded by metal and must be removed through small prints. It is not possible to attack the core directly and it must disintegrate almost to single grains when the casting is vibrated so that the sand flows like a liquid. The presence of an appreciable amount of clay may be harmful in such a case, since any sand inside the casting may be raised to the temperature at which sintering of the clay occurs and the residual strength will then be several times greater than the dry strength of a sand with the same clay content.

Usually, cores which are almost completely surrounded by metal are fairly thin and by the time the casting has cooled all the binder has decomposed. It would be easier to break a casting than to remove a solid core a foot or more in diameter when the core prints were small and the metal thickness a fraction of an inch.

7. METAL PENETRATION AND CORE REMOVAL

Severe metal penetration may make it almost impossible to remove cores, and this difficulty occasionally arises when making very large castings in grey iron. Here, penetration may occur when there is a thick, solid layer on the casting surface, and the interior consists of part solid and part liquid, and, since the solidifying metal is expanding, pressure on the cores is very high. It is not reasonable to expect the whole of the blacking covering a complicated core to be able to withstand this high pressure without giving way even over a minute area and, since by this time the temperature of the whole of the core is equal to that of the metal, even a very small pinhole allows metal to flow into the core until all the pores are filled. Under these conditions the phosphorus content of the penetrating metal is higher than that of the remainder. This kind of defect is likely to lead to rejection of the casting.

8. DEFECTS DUE TO CORES

An oil–sand core, from the very nature of the materials of which it is composed and the methods by which it is produced, has certain inherent defects which cannot be entirely eliminated, but only reduced to a minimum which enables an acceptable casting to be produced. Indeed, the success of the newer methods, which for some applications have almost completely replaced the oil–sand core, is due to the way in which they avoid these defects. With all of them, hardening takes place in the core box, which gives dimensional accuracy; none requires the use of a green binder, the source of many defects in oil–sand cores. Yet each of the new methods has its own limitations. Heated metal core boxes are needed for some; the rammed core may have to stand in the core box for a long time, or it may be difficult to remove the core from the casting. Venting may be difficult. The consumption of the older types of binder, which is a fairly accurate measure of the weight of cores produced, has fallen by about half during the last 20 years, although there has been a large increase in the production of cores by all methods combined.

The arrival of methods of core making by which a core can be cured in a few seconds in an unheated box may change this, but none of these has been in use on a scale which enables an assessment of its value to be made. In the meantime, many foundries are, and will continue to be, forced to use the oil–sand core and must use all possible methods of minimising its drawbacks.

The most important of these are:

inaccuracy due to sagging;
core cracking;
gas holes;
friability and erosion;
hot distortion;
fins;
cracking and hot tearing of castings;
metal penetration.

Sagging

The strength of unbaked oil–sand is low, but even so the strength test does not give a true indication of the rammed sand to flow under its own weight. When using one of the standard testing machines — a spring balance machine with a low capacity, or a Fischer machine with a low strength attachment, or loading the test piece by means of a shot loader — the green strength of a clay-free silica sand with 1% linseed oil and 1% dextrin is between 5 and 7 oz in^{-2}. Green strength increases as moisture content falls; when it exceeds 7 oz in^{-2} the sand is too dry to strip well, and when it is below 5 oz in^{-2} the sand is too wet and sticks to the core box. The pressure due to its own weight of a column of rammed sand 5 in high is about 5 oz in^{-2}, but nevertheless, a core of this height sags. It gradually subsides and eventually collapses. All oil–sand mixtures, even those containing small amounts of clay, creep in this way under a load appreciably less than their measured green strength. With all such sands, the green strength test is more useful for indicating whether the sand is too wet or too dry than for showing the tendency to sag.

When a core air-dries, its strength increases to such an extent that a dry layer a few grains in depth is sufficient to inhibit further creep, but by this time, dimensional accuracy has been lost. With any particular core, the extent of sagging depends on the rate of air-drying. The least inaccurate cores are obtained when the sand used air-dries as fast as is possible without becoming unusable by the time a batch has reached the core maker and been rammed up.

The weather poses a problem here. The rate of air-drying has been known to increase eightfold on a single day; admittedly this is exceptional, but even wider variations occur over a year. The method described by Sheehan (see Chapter 11) is quite satisfactory for eliminating this variable, but is not particularly easy to apply in small foundries.

The effect of an overhang is to make sagging more obvious, since a section of the core first bends and then falls off. Provided that the sand air-dries, the best method of preventing this is to ram a suitable support in the core box and remove the prop when the core has air-hardened sufficiently. Leaving the support in position until the core has been baked is a nuisance, since frequent cleaning is necessary to prevent sticking and core breakage.

Sagging can be prevented by a liberal use of drier shells, but this practice ncreases the cost of coremaking considerably. Core scrap is high, particularly so if the shells are not cleaned frequently, and wherever possible, shells should be used only for cores of such shape that they cannot stand unsupported.

Core cracking

Cracks are sometimes found in cores when they come from the stove. Alternatively, the cores may crack on contact with molten metal, producing at best a fin. Whenever a large fin occurs on a flat face which has been in contact with a heavy core, the possibility of a crack being present in the core before casting should be examined.

Most of such cracks are formed before the core reaches the stove. The green core sags and continues to move slightly when the layer of air-dried sand is not quite thick enough to prevent all further movement. As a result, a fine crack starts in the dry layer. This may be very thin and very difficult to find without going over the surface with a moderately powerful hand glass.

When the core reaches the hot zone in the stove, the crack usually opens up and becomes easily visible. Alternatively, it may remain invisible to any but a very careful inspection and open up when in contact with molten metal.

Cracks of this kind are usually confined to fairly heavy cores, since the weight of sand which is moving when sagging occurs is sufficient to rupture the thin air-hardened surface. They occur most frequently when the moisture content is at the high end of the useful range. Where it can be applied, the most effective remedy is to raise the actual green strength by adding a small amount of clay.

Gas holes

Whenever the rate of evolution of gas exceeds the rate at which it can escape through the core prints, pressure builds up until gas bubbles through the metal and may be trapped beneath the skin if solidification has commenced, or beneath a core.

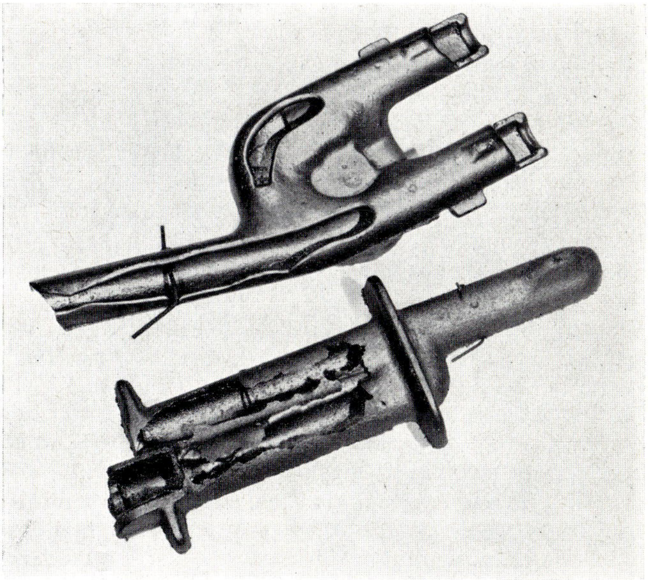

FIG 123. Defective burner castings.

Gas holes from cores are, as a rule, readily distinguished from other holes such as those due to shrinkage, to high gas content of the metal, or to mould–metal reactions. They are usually large and well-rounded, but occasionally the defect may appear in a completely different form. Figure 123 shows two burner castings which are not completely filled.

In the upper casting, the edges of the metal are smooth, rounded and bright and the defect has the appearance of a misrun which might be due to cold metal. In the lower casting, the edge is rough and irregular and for some distance is in the form of a hollow shell, the kind of defect which would result from a run-out. Both defects are due to gas from the core which could not escape through the core print because molten metal entered the vent. In one case, this happened almost immediately after pouring commenced and gas pressure prevented complete filling of the mould; in the second, it occurred a little later, gas pressure was higher and some metal which had entered the mould was blown back. It is important to recognise the cause of defects of this kind, since it does not help to use hotter metal to cure one, or to adjust the moulding machine or the moulding boxes for the other. Whenever castings which are heavily cored are not completely filled, the cores should be examined. The two burner castings shown above are useless, but there were excellent castings of the vents.

Causes of blocked vents

There are three main causes of blocked vents.

1. *Metal in the vents.* Whenever there is only a thin layer of sand over a vent, there is a possibility that the binder will weaken sufficiently for metal to force its way in. The two castings shown in Fig. 123 were obtained by filing the core until the layer of sand over the vent was only a few grains thick. When making a core such as that shown in Fig. 119 in which both a wire and a vent have to be placed on a core $\frac{3}{8}$ in thick, great care is necessary.

Occasionally, the precautions taken to see that the way out for gas is made easy are so effective that there is also a way in for metal. Usually, in this case, metal finds its way to the back of the core print.

2. *Core paste in vent.* When a core is made in separate parts which have to be fastened together subsequently, a core paste or similar adhesive is used for jointing. A ribbon of paste is laid on the one core and the second pressed on. With too liberal use of paste or laying the paste too near a vent, some may be squeezed into the vents. In order to minimise the risk, a slotted template can be used to locate the layer of paste.

3. *Mismatching of vents.* In some complicated cores, it may be necessary to have subsidiary vents leading to the main vent. Any misplacement of these, should it leave even a thin shell of sand as an obstacle, effectively blocks the vent.

Vents may also be blocked by moulding sand. When the vent terminates on the mould joint, it is a simple matter to provide a clear exit to the exterior. When the core vent ends at a high point in the mould cavity, accurate placing of the mould vent is easier if the vent wire is pushed through the mould from the inside face.

Friability and erosion

Sand grains may be removed from the surface of the core during handling, or by the metal stream during pouring. The former usually occurs only when the surface of the core is friable, but the latter depends mainly on the method of running.

Surface friability is almost always due to overbaking. Table 2 of Chapter 11 shows that the core which had been heated for $1\frac{1}{2}$ h at 230°C was not only weak, but brittle, and it is this combination which produces a friable surface. A green core is so weak before air-drying that it may collapse, but as it is plastic it is not particularly friable.

It is only on rare occasions that the whole of a core is overbaked, and the condition is usually confined to edges, especially the edges of thin projections on heavy cores. These edges warm up far more rapidly than the body of the core and are overbaked before the remainder is properly baked. This is quite a problem with some cores. It could be overcome by baking for a very long time at a temperature not exceeding 200°C, but as a rule, the stove capacity does not permit this; frequently, the difficulty is made more serious by a lack of stove capacity, which requires the temperature to be unduly high in order to reduce baking time and increase throughput, a procedure which makes the difference between thick and thin sections more pronounced. The usual way of reducing the friability of edges is to brush a little oil on to the appropriate part of the green core, a method which is successful when carried out by an experienced operator, but is otherwise risky.

Erosion by the stream of molten metal occurs more readily when the cores are already friable because much of the binder has decomposed, but the most important factor is direct impingement of metal throughout the whole of the time required to fill a mould. Oil–sand cores are assembled to make moulds for castings weighing several tons, and so long as the time during which the stream of metal strikes directly on an uncovered core is short, the method is successful; otherwise serious erosion may occur in a casting weighing a few hundred pounds. When the defect does occur, an increase in pouring speed helps by reducing the time available for breakdown of the bond, but it may be necessary to change the gating system.

Sand in a casting is a nuisance, since it is not easy to determine the source of the defect. Nearly all sand inclusions contain some slaggy material, and most other inclusions contain some sand, and in consequence there is doubt as to which kind of defect is to be eliminated. Even when it is clear that sand is the main cause, the source is not easily determined. Loose sand has been spread on the bottom of a mould without being picked up by the metal, but this is not to be relied on and there is always a possibility that the inclusion is derived from sand displaced when the cores were set. When this possibility has been excluded, the whole of the mould, the cores and the running system have to be explored and unless the erosion is deep, this is no easy task. Wherever possible, the gating system should be arranged so that metal does not impinge on an uncovered core for more than a few seconds, in order to eliminate one source of sand inclusions which may be difficult to trace, as it may not be visible on any external surface of the casting.

Hot plasticity and core distortion

When sand is in contact with metal, it absorbs heat far more rapidly than when surrounded by air. In the stove, a core warms up slowly; in the shell moulding process, sand is heated to a depth of $\frac{1}{8}$ in or so in 20 s. The rate of heating in contact with molten metal is still more rapid and the centre of a thin core is heated rapidly enough for there to be an appreciable temperature rise before solidification of the casting commences.

It was shown earlier that when a core has been baked so as to give maximum cold strength, rapid reheating to a temperature of 400°C causes much of the polymerised bond to re-melt. In the particular core tested, re-melting of the bond was accompanied by a fall in strength to a tenth of its original value, and the hot core had a high deformation.

Effects of hot plasticity

When a thick core is in contact with molten metal, only a comparatively thin layer is heated quickly, but the interior remains cool. The surface

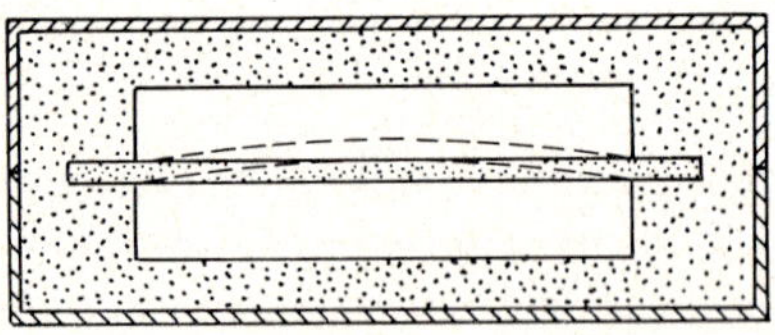

FIG. 124. Deformation of a straight core.

softens, but there is no appreciable distortion because the interior remains rigid. The higher rate of gas evolution may produce blowholes and softening of the surface may reduce the incidence of finning, but dimensional accuracy is not affected.

In the case of a core $\frac{1}{4}$ in thick and with metal on both sides, the centre is nowhere more than $\frac{1}{8}$ in distant from the hot metal. There is a far more rapid rise in temperature when the core is in contact with molten metal at 1100°C than in the shell moulding process and a temperature of 400°C is reached in a few seconds.

When sand is immersed in molten iron, it is subjected to an upward thrust equal to about four times its own weight. If the core consists of a long thin rod supported at the two ends, it bends upwards under this load, as shown in Fig. 124.

There is no difficulty in preventing distortion by placing a wire along the centre and many core makers would do this automatically.

If, now, the core is in the form of a grid, the pattern of deformation is more complicated, but is of the general form shown in Fig. 125.

In practice, a part of the core near the point of entry of metal is heated while the remainder is still rigid, and deformation takes a slightly different form. If the rods are replaced by plates, it becomes difficult to foresee the pattern of deformation; when the cores have both thin and thick sections it becomes impossible. Wires can be used to a limited extent only.

Should the casting be sectioned in one place, it may appear that the core dimensions have increased; if sectioned in another, they seem to have decreased, and there have been claims that cores have 'grown' or 'shrunk'.

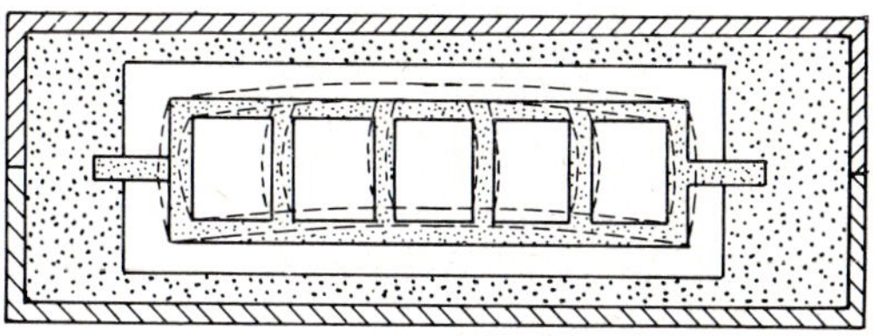

FIG. 125. Deformation of a grid.

In practice, they do neither, and the apparent change in dimensions is due to distortion. Sand expands on heating and contracts on cooling, and it is possible that the kind of bonding material has an influence on these volume changes: binders used in, for example, oil–sand, shell and hot box cores, are very different and may influence the expansion and contraction, but the differences are small compared with those due to the distortion of oil–sand cores which soften on heating. Moreover, a change from one process to another is not made from day to day, whereas the baking conditions in a core stove can change from hour to hour and from place to place.

Dimensional inaccuracies due to core distortion are particularly serious, since they cannot, as a rule, be detected by inspection of the casting, as the external dimensions are not affected. They may lead to a reduction in wall thickness such that white iron is present and the casting may be hard to machine or, what is worse, fail in service.

Occasionally the inaccuracy is due to cores shifting because they are not securely anchored, but there is rarely any difficulty in deciding when this has happened since one end of the core moves bodily.

Core distortion is invariably due to underbaking.

Fins

Finning is the most difficult of all defects to eliminate completely. It is one of the effects of sand expansion and can be entirely eliminated by replacing the quartz by zircon or some other material which expands less on heating. This is rarely a practicable solution.

The many 'cause and cure' investigations referred to in Chapter 11 have failed to bring agreement, but at least have thrown up suggestions for reducing the severity of the defect. The sands composed only of quartz grains have the highest expansion, but among the sands suitable for core making, only the shore sands have an appreciable proportion of grains other than quartz, and some observers have found that finning has been less when using Southport sand. Others have found that sands with grains of uniform size are more prone to the defect than less uniform sands. In one series of tests, it was found that finning was severe with Leighton Buzzard sand, less severe with Southport sand, and less still with a mixture of the two. The addition of Leighton Buzzard sand increased the quartz content of the sand, but any harmful effect was more than offset by the beneficial effect of the wider grain size distribution.

The effect of the various additions made by investigators is also confusing. Some have found that finely divided Fe_2O_3 inhibits finning but that Fe_3O_4 does not, which indicates that chemical composition is decisive. On the other hand, silica flour and iron oxide have been used, turn and turn about, on a production run without any great difference showing up, which indicates that fineness is the important factor.

Underbaked cores are less likely to fin than others, but to secure a worthwhile improvement, gross underbaking is needed and blowing is almost certain. This practice is useful only for showing that a plastic material is less likely to crack than a brittle one, if any evidence is needed.

Frequently, the kind of base sand to be used is controlled by considerations other than the risk of finning, and in these circumstances, the most promising solution is the addition of a fine powder. There are objections on health grounds to the use of fine silica, since, although it can be obtained in the moist condition and no dust arises during handling, it may become airborne when the cores are removed. The addition of 2% ferric oxide is the most promising of the methods which have been tried.

Cracking and hot tearing

As a casting cools, it contracts and anything which restricts the contraction causes the casting to stretch if it is ductile, and crack if it is brittle.

Suppose the casting is in the form of a bar with arms at each end as illustrated in Fig. 126.

When the bar is solid, it contracts and a compressive stress is imposed on the core and a tensile stress on the bar. Under the compressive stress the core deforms. If the decrease in the length A–B of the core is less than the contraction of the bar, the bar must crack or stretch. A brittle bar must crack if, at any time after pouring, the load required to produce a deformation of the core equal to the contraction of the bar exceeds the tensile strength of the bar.

At the temperature of molten metal, the core binder decomposes and the core becomes weak and easily deformable, whilst oil–sand itself is exceedingly strong and resistance to deformation is high. When the casting is thin, little of the core breaks down and an oil–sand core is the most likely material to cause cracking; when the casting section is increased and much of the core loses cohesion before contraction is appreciable, it is least likely.

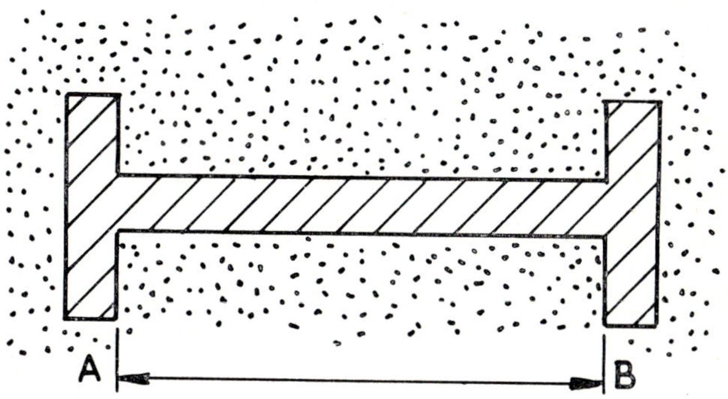

FIG. 126. Contraction hindered by core.

In practice, cracking of grey iron, solely as a result of the restraining effect of the core, is not frequent. The metal expands on solidification and before it is solid and begins to contract, the binder has burned out in the surface layers of the core. Very severe cracking has, however, occurred in castings of considerable size and rather less than 1 in thick. The internal core was large and clay was added to prevent sagging. The amount of oil used was increased in order to bring the strength of the core up to 300 lb in^{-2}. Since this involved more than doubling the amount of oil, breakdown was slow, the resistance high and several castings ended up in two halves.

The risk is much greater when the metal concerned contracts on solidification, since the core must be deformed at a time when the metal is part liquid and part solid and is very weak. Hot tearing of this kind is a frequent source of trouble when the castings are of white iron or steel.

Underbaking the core in order to take advantage of the temporary softening which occurs when the core is first heated is occasionally successful in eliminating hot tears, but the method is not to be recommended for general use. The margin of safety is small and a slight increase in core size or reduction in metal thickness may bring back the defect. It is safer to overbake the core as far as is possible without making it too friable to handle. The rate of breakdown is thus raised and, since the core as a whole is weakened, the resistance to deformation decreased.

In spite of all that can be done, hot tearing of white iron castings remains a problem when oil–sand cores are used. The popularity of shell cores with makers of malleable castings is due to their lower resistance to the contracting metal and the resultant decrease in hot tearing.

Metal penetration

Owing to the high permeability of oil–sand cores, it is necessary to coat with blacking or some other fine-grained material when a smooth surface finish is required. The provision of such a coating also reduces the risk that erosion will begin as a result of the removal of a few projecting grains. In itself, the loss of a few grains may not be a serious matter, but behind the surface layer of sand grains, the core is less tightly packed and is less resistant to erosion and metal penetration.

The surface coating, therefore, fulfils a useful purpose in addition to the improvement in appearance, but it cannot stand up against very high pressure. In one case described by Wright *et al.*,[1] penetration into the cores in an anvil block weighing 50 tons was so severe that all the pores were filled with metal and the casting had to be scrapped although the permeability of the clay-bonded sand used was below 30 and the coating was carefully inspected.

This was not a case of penetration under the pressure due to the weight of metal. The phosphorus content of the iron used was less than 0·3% and that of the penetrating metal above 0·6%, and solidification had proceeded for quite a way before penetration took place. In the next casting, the cores were made with crushed burnt fireclay so graded that dry permeability was less than 2 and there were few open pores. The cores were finally fired for some hours at 900°C, producing what was essentially a firebrick with a permeability of 0 and no open pores.

This illustrates an important distinction between cores used externally and cores which reach to a point well inside the casting. A sand with an organic binder and a permeability of 100 or more is used successfully for cores outside castings weighing 20 tons or so, and with a reasonable core coating, an excellent surface is obtained. There is little prospect of using an oil–sand core which projects well inside a grey iron casting of this kind. With a normal core permeability, the risk of penetration under the pressure developed after partial solidification is so high that it should not be taken; reducing the permeability to the very low figure needed to prevent penetration would require such a large increase in the amount of oil required to maintain adequate strength that the core would disintegrate, probably with considerable violence, in contact with molten metal. Although, therefore, oil–sand cores can be used to shape the outside of a quite heavy casting, it does not follow that they are suitable for the internal cores when the metal used expands on solidification.

REFERENCE

1. Wright, C. B., Godding, R. G. and Parkes, W. B. (1962). *BCIRA J.*, **10,** pp. 309–314.

Index